THERMODYNAMIC PROPERTIES OF PROPANE

NATIONAL STANDARD REFERENCE DATA SERVICE OF THE USSR: A Series of Property Tables

1. Thermodynamic Properties of Helium
2. Thermodynamic Properties of Nitrogen
3. Thermodynamic Properties of Methane
4. Thermodynamic Properties of Ethane
5. Thermodynamic Properties of Oxygen
6. Thermodynamic Properties of Air
7. Thermodynamic Properties of Ethylene
8. Thermophysical Properties of Freons, Part 1
9. Thermophysical Properties of Freons, Part 2
10. Thermophysical Properties of Neon, Argon, Krypton, and Xenon
11. Thermodynamic Properties of Propane

In Preparation

Thermodynamic Properties of n-Hexane
Thermodynamic Properties of Natural Gas
Transport Properties of Nitrogen, Oxygen, and Air

THERMODYNAMIC PROPERTIES OF PROPANE

V. V. Sychev
A. A. Vasserman
A. D. Kozlov
V. A. Tsymarny

Theodore B. Selover, Jr.
English-Language Edition Editor

⬤HEMISPHERE PUBLISHING CORPORATION
A member of the Taylor & Francis Group
New York   Washington   Philadelphia   London

THERMODYNAMIC PROPERTIES OF PROPANE

Originally published as Termodinamischeskie svoistva propana, Izdatel'stvo standartov, Moscow, 1989.

1 2 3 4 5 6 7 8 9 0   B R B R   9 8 7 6 5 4 3 2 1

This book was set in Times Roman by Hemisphere Publishing Corporation. The editor was Michael Folker; the production supervisor was Peggy M. Rote; and the typesetter was Wayne Hutchins.
Printing and binding by Braun-Brumfield, Inc.

A CIP catalog record for this book is available from the British Library.

Library of Congress Cataloging-in-Publication Data

Termodinamicheskie svoistva propana. English.
    Thermodynamic properties of propane / V. V. Sychev . . . [et al.].
       p. cm. — (National standard reference data service of the
USSR)
    Translation of: Termodinamicheskie svoistva propana.
    Includes bibliographical references and index.

    1. Propane—Thermal properties—Tables. I. Sychev,
V. V. (Viacheslav Vladimirovich). 1933-   . II. title.
III. Series.
TP761.P94T4713 1990
665.7'72—dc20
                                         90-46509
                                           CIP

ISBN 0-89116-932-6

CONTENTS

PART II

PART III

PREFACE TO THE SERIES

This treatise is part of a continuing series on thermodynamic properties of technologically important fluids. These very important contributions by scientists and engineers working through the aegis of the Soviet National Service for Standard Reference Data have been released to Hemisphere Publishing Corporation for translation to make them available to the English reading technical community. The authors are Soviet experts in the field.

While a team of translators was involved in producing the English versions, the overall series is being published under the editorship of T. B. Selover, Jr.

Each volume presents a comprehensive survey of the world's literature up to its publication date. A special effort has been made to give a thorough presentation of Russian as well as other work. Many studies not previously known to Western counterparts are included. The results have been to broaden the range of applicability of data and to improve upon equations of state to provide accurate computation methods for generating smoothed tables of properties.

For some volumes there are no equivalent comprehensive surveys available in English. Thus, a valuable service has been fulfilled to workers in the fields of process design, equipment development, custody transfer, and safety.

Each volume is set up in the same way with Part I dealing with a study of all necessary aspects of experimental data interpretation and analysis. Then in Part II the fundamental constants, symbols with units, and data tables can be found. The use of SI units is consistent throughout.

The section on experimental data is particularly important because it covers, in detail, the key studies. Possible errors in measurement or data analysis that could have led to inaccurate tabular results in publications are pointed out.

Methods of constructing the equation of state and procedures for computing the data tables are covered thoroughly. There is a detailed error analysis of data generated from the equation of state relative to literature values.

Properties of each fluid in phase equilibria are treated at the freezing curve and the saturation curve with tables given for both temperature and pressure dependence. Properties cover the range of triple point to critical point in the condensed phase. Single phase properties in each volume cover a range of temperature and pressure wide enough to include most practical applications. Ideal gas property equations and calculation methods are included. The scope of properties generated and covered in the tables is more comprehensive than is typically found in any one English-language treatise. This gives the advantage of internal consistency.

An extensive bibliographic listing is included with each volume. All Russian citations have been translated. In some cases the availability in English of translated Russian sources is given.

Frequent reference to key English-language references has been made to help in providing correct translations. In some cases the original symbols have been changed to avoid confusion within the text or to avoid misuse of terms commonly found in English. Where mistakes in the original Russian text have been found, they have been corrected and so noted as editor's changes. Added descriptive clarification has been used in some places for tables, figures, and text to clarify meaning.

Although we recognize that key review papers have been published in 1986 for some of the fluids in this set, the series stands on its own merit. It represents a vast accumulation of knowledge never before available to Western countries. Through careful study of these volumes, workers in the field will develop an appreciation for both the scope of studies and where differences exist. Hopefully, this will lead to more dialogue between the authors and their Western counterparts.

Theodore B. Selover, Jr.

PREFACE

Propane (C_3H_8) is a saturated hydrocarbon; under normal conditions it is a combustible, colorless, odorless gas. Propane can be found as a component in natural gas and associated crude oil gas; it can be obtained from petroleum and its products by a cracking process and by the Fischer-Tropsch process.

Propane is widely used as a solvent and is the raw material for the production of ethylene, propylene (by pyrolysis), nitromethane, and carbon black. Pure propane and a mixture of propane and butane are high energy fuels, commonly used in public transportation. Propane and its chlorine and fluorine products are refrigerants. Very pure propane can be used for bubble chambers in physical experiments.

A very low temperature of the triple point (about 85 K) and a very steep slope of the melting curve make propane a convenient medium for transmitting pressure at low temperatures.

Propane mixtures with air are explosive in concentrations from 2.1 to 9.5 vol.%; at 739 K propane is subject to spontaneous ignition. Therefore, propane belongs to flammable (explosive) hazardous substances; its storage and usage require special precautions.

The thermodynamic properties of propane, first of all p, v, T relations, were investigated in a relatively small numbers of papers published over the past 50 years. The scope of the investigations was limited in temperatures and pressures. Low-temperature data are especially scarce. This applied particularly to measurements of the vapor pressure. Data on the vapor density for temperatures below 290 K are nonexistent. Investigations of the thermal and acoustic properties of propane are very limited; each of the properties was usually studied in 1–2 papers. The almost complete absence of Soviet authors in the list of experimental investigations of the thermodynamic properties of propane should be pointed out.

In these circumstances, compiled values recommended in monographs can be very important. However, such publications are almost absent in the Soviet and world practice, with the exception of possibly the only extensive publication, by the U.S. National Bureau of Standards [44].

The derivation of the equation of state and tables of the thermodynamic properties and control over their validity are difficult when the amount of experimental information is limited. In addition, a very broad temperature range of the liquid phase makes very difficult the task of approximating its thermodynamic functions by a single equation of state. The properties of the liquid can be determined from continuous integration of the equation over the isotherm from the ideal gaseous state to the liquid state including the transition through the two-phase region. Lack of reliable thermal data for the liquid phase forced the present authors to exclude from the tables the low temperature values of the specific heat and the velocity of sound, despite the fact that the method of equivalent equations, which was so successful in the previous monographs of the series, was used.

This monograph includes the experimental investigations carried out at the Odessa Institute of Naval Engineering. Thermodynamics researchers of the Moscow Power Institute, the All-Union Science Research Center of the U.S.S.R. State Committee for Standards, and the researchers from the Odessa Institute of Naval Engineering derived a series of the equivalent equation of state, obtained the average equations, and calculated the tables of the thermodynamic properties of propane on the boiling and condensation curves and in the single phase region. For convenience they are presented in the form of isobars for the pressure range from 0.01 to 100 MPa and for temperatures from 100 to 700 K.

The present authors are grateful to the staff of the above mentioned institutions, especially to A. Ya. Kreizerova, Yu. V. Mamonov, and O. V. Svetlichnaya for their great help in the preparation of this material.

THE EXPERIMENTAL DATA
ON THE THERMODYNAMIC PROPERTIES
OF PROPANE

Practically all of the published measurements of the thermal, thermodynamic and acoustic properties of propane are mentioned and analyzed in this chapter. The data on properties of propane at low temperatures were complemented by a contribution of one of the present authors. Nevertheless, the amount of information for the lowest temperature is insufficient, especially for thermal data.

1.1 INVESTIGATIONS OF THE DENSITY OF PROPANE IN THE SINGLE-PHASE REGION

The results from measurements of the density of gaseous and liquid propane are listed in Table 1.1. In the majority of these studies, a method of compression is used with a sample of a known mass in a vessel of a known volume. A measured amount of mercury (mass or volume) is introduced into the vessel under pressure. A simple chronological order is used for the description of studies in this review.

Sage et al. [89] measured the specific volume of liquid and gaseous propane by using a rather crude apparatus; the accuracy of their thermostat did not

Table 1.1 Experimental Investigations of the Density of Propane for the Single-Phase Region

Year	Author	Parameter range		Number of points
		ΔT, K	Δp, MPa	
1934	Sage et al. [89]	294–378	0.2–21	smoothed
1937	Beattie et al. [14]	370–548	2.4–31	111
1940	Deshner and Brown [35]	303–609	0.1–15	275
1949	Reamer et al. [84]	311–511	0–69	306
1949	Cherney et al. [26]	323–398	1.1–5.0	25
1960	Dowson and McKetta [33]	243–348	0.05–0.18	18
1962	Dittmar et al. [36]	273–413	1.0–103	337
1964	Kahre and Livingston [58]	233–353	1.0–9.6	78
1966	Huang et al. [53]	273–373	p_s–34.5	36
1970	Babb and Robertson [13]	308–473	64–1073	63
1978	Ely and Kobayashi [38]	166–323	0.3–43	222
1978	Golovskii et al. [4]	88–273	1.5–61	154
1978	Warowny et al. [106]	373–423	0.3–6.2	51
1982	Thomas and Harrison [101]	258–623	0.5–40	798
1983	Haynes [47]	90–300	0.6–37.5	196
1984	Kratzke and Müller [67]	247–491	2.2–61	60
1984	Straty and Palavra [97]	363–598	0.2–34	144

exceed 0.1 °C. The compression of a propane sample in the cell was carried out through the introduction of mercury. The volume of mercury was measured by the movable electrical contact with the accuracy of about 0.02 ml. The measured pressure includes the pressure due to the difference between the mercury level in the cell and in the mercury-oil separator. The accuracy of the pressure measurements amounts to $\sim 7 \cdot 10^{-2}$ and $7 \cdot 10^{-3}$ bar for the pressures from 207 to 20.7 bar, respectively. Unfortunately, the authors of [89] did not estimate the overall accuracy of the experimental data; however, some indirect information allows one to make an approximate estimation. The uncertainty in the determination of the volume of the sample was estimated taking into account its different constituents. It could not be less than 0.05%, and in many cases it probably reached 0.1%. The accuracy in the determination of the mass of the sample was about 0.2 mg. The uncertainty in the compressibility values can be estimated as 0.03–0.2% as the temperature and pressure changed from their minimal to their maximal values.

Beattie et al. [14] measured compressibility with the apparatus similar to that employed in [89]. The main differences are the use of a calibrated compressor for supplying mercury and the correction for the compressibility of mercury. A piston manometer was calibrated at 0 °C by using the equilibrium vapor pressure of carbon dioxide. The accuracy of the manometer amounts to 0.004%, and the overall accuracy of pressure measurements varied from 0.01% at the minimum pressure and room temperature to 0.03% at the maximum pressure and temperature. The thermostat maintained the stable temperature within 2 mK. The accuracy in the temperature measurements over the range of 100–325 °C reached 0.01–0.02 °C.

The authors of [14] pointed out the danger of the thermal decomposition of propane. According to their data, no perceptible decomposition took place at 250 °C during a period of six hours. However, it became noticeable during heating of the sample over the same period of time at 275 °C.

Two series of measurements were made in [14]. The pressure values in the second series were higher than those in the first one by 0.02% at 100 °C, by 0.2% at 125 °C, by 0.3% at 150 °C, and by 0.4% at 175 °C. This temperature-dependent deviation is rather high and cannot be explained only by the errors in the method. Without explaining the reasons for this discrepancy and for the choice made, the authors of [14] gave preference to the results of the second series, which agree with the calculated values of pressures, obtained by the Beattie-Bridgeman equation at $\rho < \rho_{crit}$.

Deschner and Brown [35] paid much attention to the purification of the investigated samples. They performed four-step fractional distillation of the initially chemically pure propane. The middle fraction was recharged and redistilled at each step. Propane obtained by this method was then frozen with liquid air, and all noncondensed gas was removed by a vacuum pump. This purification procedure was designed to obtain good results. The purity of propane was checked by its density at standard temperature and pressure; the results did not verify the expected purity. If ethane was the main impurity, then the purified propane was only 98.3% pure.

Deschner and Brown measured the density of the gaseous and liquid propane at temperatures of up to 336 and 93 °C, respectively; they gave corrected values of compressibility. Comparison of data in this work with those of Sage et al. [89] and Beattie et al. [14] showed that the data of [89] at $T = 104.44$ °C and $p < p_{crit}$ appear to be too low. However, the detailed comparison is complicated because of the necessity to interpolate the data of [89]. The discrepancies between the data of [35] and [14] average to 0.2%, except near the critical temperature, where the differences reach 2%. The density values of liquid propane given in [35] are about 2% smaller than the corresponding data of [89].

On the whole, the results of Deschner and Brown [35] are not sufficiently reliable. This is probably due to the considerable amount of impurities present in the samples. The lack of details about the experimental procedure makes the evaluation of the obtained results quite difficult. The statement of [35]—that the thermal decomposition of propane at temperature 336 °C is absent—is questionable, because already at 275 °C Beattie et al. [14] observed such a decomposition.

The experimental procedure used by Reamer et al. [84] is similar to that used by Sage [89]. The mass of the introduced sample was determined by weighing-bomb technique, and a platinum resistance thermometer was used. The initial propane sample contained less than 0.1% impurities and was subject to fractionation. The purity of the final product was checked by using the values of boiling and condensation pressure at $\sim 37.8\,°C$. The difference between these values did not exceed $13 \cdot 10^{-3}$ bar, which corresponds to the impurity concentration of not greater than 0.03%. The authors of [84] gave the corrected values of the specific volume. The uncertainty in these values did not exceed 0.2% according to the estimation made in [84].

Cherney et al. [26] used an experimental apparatus which was similar to that used by Beattie et al. [14] and propane of 99.99 moles% purity. The thermostat and the system for introducing mercury to the bomb by means of a compressor were improved. The motion of the compressor piston was determined by a micrometer screw with an accuracy corresponding to the volume change of $3.6 \cdot 10^{-3}$ cm^3. The pressure was measured by a manometer with two pistons of different sizes.

The inaccuracies in the measurements of the temperature and pressure amounted to 0.02 K and 0.02%, respectively. However, an additional uncertainty $\delta p = 0.05\%$ should possibly be taken into account. It was caused by the vapor pressure of mercury and by the different levels of liquids in different parts of the apparatus; as a result, the overall uncertainty was $\delta p_{sum} = 0.07\%$. The volume of the bomb and the mass of the sample were measured with the accuracy of 0.04 cm^3 and 0.001 g, respectively. Thus, the corresponding relative accuracy of measurements was 0.15% and 0.03%, respectively. The overall uncertainty in the coefficient of compressibility was estimated in [26] as 0.25%.

The authors of [26] compared the obtained data with the results of Beattie et al. [14], after the data of [14] were corrected by using a more accurate molecular mass for propane. The discrepancies reached 0.097% at 125°C and 0.33% at 100°C. They always lay within the sum of accuracies of the data being compared.

Dowson and McKetta [33] determined the compressibility of propane at pressures of up to 2 bar for the temperature range from -30 to $+75\,°C$ by a method based on the principle of hydrostatic weighing. In each experiment the equilibrium of the balances was achieved twice by filling the apparatus with the tested and the calibration gases. At these conditions the densities ρ_x and ρ_c are equal at constant temperature, this yields:

$$z_x = z_c \cdot \frac{p_x M_x}{p_c M_c}$$

where z, p, M, are the compressibility, pressure and molar mass, respectively; subscripts x and c refer to tested and calibration reference gases.

According to the data from the chromotographic analysis of the gas phase, the propane investigated in [33] contained not more than 0.04% air.

The experimental data were used to determine the second virial coefficient B. The data could be described by a polynomial $z = 1 + Bp$ for each isotherm.

The value of B was evaluated by the least squares method, but with a low accuracy because the number of the experimental points for each isotherm was small (from 2 to 5). The isotherms can be considered linear within the error limits of [33]. According to the estimation of [33], the total error in the experimental values of z was 0.1%.

Dittmar et al. [36] investigated the dependence of $p(\rho, T)$ and used an unloaded piezometer with a diaphragm and a needle valve. The authors determined a correction, related to the thermal deformation, for the volume of the piezometer. The pressure in the exterior chamber was maintained by a mechanical oil pump and was measured by one of the five spring manometers that had different scales. The constant temperature was maintained by an oil thermostat with a controlling contact thermometer. The temperature was measured by a mercury thermometer with a scale division of 0.1 °C and by an optical device which increased the accuracy of readings to 0.01 °C. The mass of the substance in the piezometer was found by weighing a detachable container after the substance was frozen from the piezometer into the container at -80 °C. The piezometer volume was obtained with the accuracy of 0.01 cm^3 through calibration with water and mercury. Repeated fractional distillation was employed for the purification of propane until its melting temperature was constant. The accuracies of measurements were estimated in [36] as 0.1% for the volume, 0.1 °C for the temperature, and 0.6% for the pressure. The authors of [36] did not give the overall uncertainty of the density data. They only noted that the uncertainty in the mass measurements was the major contributor to the total uncertainty because the cleaning process of the connecting pipes was not stable during freezing. The latter information combined with the estimated $\delta p = 0.6\%$ leads to the conclusion that the data of [36] had a low accuracy. It should be mentioned that study [36] gave only a table of the smoothed values corresponding to the rounded off values of t (°C) and p (kg/m^2).

Kahre and Livingston [58] refined the compressibility of liquid propane near the saturated vapor pressure to provide a more accurate estimation of the gas quantity flow, since liquified gases are often transported or pumped through pipes under the pressures which are above the ambient vapor pressure. They worked with a commercial propane without additional purification and used a thermostat with a liquid, a copper-constant thermocouple, and a potentiometer. Propane in the cylinder was compressed by a calibrated pump which permitted to measure volume changes with the accuracy of 0.01 cm^3.

Study [58] tabulated the experimental and smoothed values of the compressibility coefficient

$$\beta = \frac{1}{V} \cdot \frac{dV}{dp} \approx \frac{1}{V} \frac{\Delta V}{\Delta p}$$

Huang et al. [53] measured the viscosity of propane, and the data on the density of the liquid were obtained as a by-product. This explains the low precision of the density values which was estimated in [53] as $\pm 0.5\%$.

Babb and Robertson [13] measured the density of liquid propane at extremely high pressures. The measurements were made with filling up the piezometer twice; however, the second series of the experiments was interrupted because of an accident in the apparatus. The results obtained in the second filling were criticized because of a possible contamination of the content of the piezometer. The method used in [13] could be considered as indirect, because the mass of a substance in the piezometer was calculated from the measured temperature and pressure, and the published values of density [36, 89]. The authors of [13] estimated the accuracy of the density values as 0.5% at 100 and 200 °C and as 0.3% at 35 °C.

Ely and Kobayashi [38] were probably the first to measure the density of propane at cryogenic temperatures. They obtained extensive information on the density of the gaseous and liquid states for a sample of 99.98 vol.% purity. According to the estimation of the authors of [38], the uncertainty in the density values does not exceed 0.1%.

Golovskii et al. of the Odessa Institute of Naval Engineering [4] measured the density of the liquid propane for a wide range of temperatures and pressures. Their measurements extended to even lower temperatures than those of Ely and Kobayashi [38]. The measurements were made by an unloaded piezometer of constant volume [5] which was placed inside a thermostat made of a copper block with a three-section heater. A photo-thyratron controller maintained the temperature within ±0.01 K. The uncertainty in the temperature measurements by a platinum thermometer and a potentiometer did not exceed ±0.01 K. The accuracy of pressure measurements by a piston manometer and a membrane separator changed from 0.02 to 0.4 bar when the measured pressure increased from 2 to 600 bar. The overall uncertainty in the determination of density of the liquid propane does not exceed 0.05%. Some changes in the density values on the pseudo-isochores were caused only by the thermal expansion of the piezometer. The overlapping data of [4] and [38] agree within 0.15%. This value exceeds the sum of the uncertainties claimed by the authors of these studies; the data of [38] are systematically lower.

Table 1.2 lists the experimental data of [4]. The original paper was deposited and is difficult to obtain.

Warowny et al. [106] determined the compressibility of propane by Burnett method for the temperature interval 373–423 K at moderate pressures, using the traditional apparatus and method of measurements. Quartz and mercury thermometers were used for temperature regulation, and a platinum thermometer for its measurements. The uncertainty in the temperature measurements in terms of the IPTS-68 does not exceed 0.01 °C. The pressure was measured by a piston manometer, which was calibrated through a direct comparison with a reference manometer, with the accuracy of better than $(2-3) \cdot 10^{-4}$ bar. The investigated propane was 99.9% pure.

Thomas and Harrison [101] made a substantial contribution to the experimental study of the density of gaseous and liquid propane. Study [101] was performed very carefully. The authors of [101] purified their sample by concentrated sulphuric acid with a catalyst, then they removed the traces of the sulphuric acid and moisture and thus purified the sample from traces of propylene. The chromotographic analysis at this stage showed traces of nitrogen which was removed by repeated freezing of the samples with removal of the gaseous part by a vacuum pump. The propane resulting from this procedure was 99.998 vol.% pure. It should be noted that the initial propane sample was rather pure (99.993 vol.%). Thomas and Harrison worked with probably the purest propane of any other investigators.

The temperature regulation ($\pm 5 \cdot 10^{-4}$ K) and measurements were carried out at the modern level. The temperature readings of the platinum thermometer were corrected to IPTS-68, because the thermometer was calibrated in terms of IPTS—48. Different piston manometers were used for the pressures less than 100 bar and those in the range from 100 to 400 bars. The sensitivity of the manometers was $0.5 \cdot 10^{-5}$ and $0.2 \cdot 10^{-4}$, respectively. the construction of the manometers was improved to reduce friction. The correction of the effec-

Table 1.2 Experimental Data for the Density of Liquid Propane [4]

p, bar	T, K	ρ, kg/m^3	p, bar	T, K	ρ, kg/m^3
15.45	219.07	597.1	35.50	203.71	615.6
99.83	226.82	596.9	121.60	210.64	615.4
195.54	235.91	596.6	209.08	217.84	615.3
281.45	243.93	596.4	292.24	224.75	615.1
372.85	252.60	596.2	409.92	234.86	614.8
469.93	262.22	595.9	503.67	242.66	614.6
569.57	271.88	595.6	593.11	250.58	614.3
16.28	217.49	598.6	20.99	197.80	621.0
117.29	226.90	598.3	100.62	203.69	620.8
202.21	235.01	598.1	205.74	211.85	620.6
303.03	244.44	597.8	314.79	220.65	620.4
405.41	254.13	597.6	416.59	228.82	620.2
500.14	263.35	597.3	504.26	236.00	620.0
588.20	271.99	597.1	595.66	243.64	619.7
33.73	218.19	599.4	42.56	189.31	631.0
125.23	226.81	599.2	126.90	195.30	630.8
204.96	234.28	599.0	221.63	202.16	630.7
296.36	242.65	598.8	329.90	210.22	630.4
385.01	251.11	598.5	403.05	215.90	630.3
492.29	261.52	598.3	489.94	222.47	630.1
590.16	271.20	598.0	597.81	230.66	629.9
41.09	210.15	609.5	27.65	181.75	637.7
113.46	216.34	609.3	123.86	188.67	637.6
212.90	224.83	609.1	209.27	194.63	637.4
315.48	233.82	608.8	291.65	200.36	637.2
419.33	242.88	608.6	411.88	208.91	657.0
488.18	248.82	608.4	518.58	216.55	636.8
595.07	258.56	608.1	591.93	221.93	636.6

Table 1.2 (*Continued*)

p, bar	T, K	ρ, kg/m^3	p, bar	T, K	ρ, kg/m^3
50.99	174.51	646.6	37.27	122.80	697.8
132.29	180.00	646.5	117.68	126.23	697.7
209.47	184.97	646.3	199.66	129.73	697.7
293.41	190.53	646.2	297.44	133.96	697.6
392.17	197.02	646.0	399.52	138.41	697.6
497.10	204.11	645.8	517.79	143.42	697.3
605.27	211.37	645.6	602.72	147.49	697.2
32.75	166.75	653.4	32.95	114.80	706.1
117.09	171.90	653.2	110.82	117.71	706.1
210.65	177.73	653.1	202.41	121.29	706.0
318.03	184.39	652.9	305.38	125.18	705.9
401.39	189.69	652.7	404.03	129.11	705.8
496.80	195.77	652.6	496.41	133.02	705.7
582.32	201.35	652.4	594.68	137.20	705.6
20.99	159.44	659.5	26.48	106.99	714.0
124.74	165.73	659.4	113.36	110.01	714.0
194.76	170.11	659.3	216.53	113.83	713.9
318.32	177.43	659.1	316.56	117.45	713.8
410.70	182.98	658.9	388.15	120.19	713.7
501.12	188.55	658.8	504.65	124.77	713.6
598.99	194.47	658.6	599.97	128.35	713.6
40.99	152.09	668.2	38.64	101.48	719.7
129.06	157.08	668.1	118.07	104.22	719.7
223.40	162.39	668.0	197.31	106.99	719.6
314.40	167.66	667.8	299.89	110.71	719.5
399.52	172.38	667.7	396.78	114.27	719.4
480.72	177.06	667.6	501.90	118.07	719.4
598.01	183.83	667.4	597.81	121.64	719.3
30.20	144.64	675.7	34.52	94.84	726.2
114.54	148.90	675.6	116.31	97.59	726.1
201.43	153.28	675.5	212.61	100.79	726.1
301.46	158.59	675.3	299.89	103.54	726.0
404.33	164.03	675.2	403.25	107.01	725.9
487.98	168.54	675.1	495.24	110.07	725.9
489.94	168.55	675.1	600.95	113.77	725.8
592.13	174.09	674.9	41.78	89.47	732.0
42.66	137.73	683.1	115.33	91.82	731.9
123.76	141.54	683.0	213.20	95.02	731.9
192.41	144.82	682.9	327.93	98.63	731.8
310.48	150.60	682.8	415.61	101.42	731.7
407.07	155.36	682.7	513.67	104.39	731.7
519.56	161.16	682.5	597.03	106.81	731.6
520.56	161.18	682.5	157.69	88.24	736.6
596.44	164.97	682.4	252.42	91.28	736.6
34.52	130.25	690.0	306.36	92.96	736.5
113.36	133.79	690.0	406.78	95.98	736.5
203.59	137.98	689.8	500.92	98.94	736.4
300.87	142.33	689.8	598.40	101.68	736.4
387.56	146.40	689.7			
492.88	151.31	689.5			
609.97	156.84	689.4			

tive piston area for pressure was made, though the pressure values were moderate.

The method used in [101] consisted of compressing the sample in a thin-walled pycnometer with mercury introduced from a calibrated compressor. The influence of the mercury vapor pressure was taken into account.

In order to find a temperature of thermal decomposition of propane, the experiments on isotherms were performed in two ways. First, the pressure was increased to about 400 bar, then it was decreased, and the control measurements were made at the density of 0.8 mol/dm^3. The difference in pressures was observed at this density only for the isotherm 623 K, and it was 0.03%; this value does not exceed the uncertainty of the measurements. This fact can be interpreted as evidence of the absence of decomposition in propane. The decomposition at 550 K, mentioned in [14], was probably caused by the presence of impurities.

Thomas and Harrison [101] made two series of measurements. In the second series, contrary to the first one, the piezometer was not degassed by heating and evacuating. As a result, propane was contaminated by air in the amount of 0.004%. This value is comparable with the results obtained during purification. Such amount of the impurity caused the uncertainty in pressure of less than 0.017%. Nevertheless, a corresponding correction was made to the final calculations.

In [101] the critical region was investigated in detail, including the critical isochore (ρ = 49.55 mol/dm^3) and also the region adjacent to the equilibrium curve for vapor-liquid. Some of the data were obtained for the supercritical region (212 points at temperatures 370–623 K and pressures 20–40 MPa).

The authors of [101] estimated the total uncertainty in the experimental values of the compressibility of propane to be at most 0.03% at the lowest pressures and densities and vary to at most 0.3% at the highest densities, pressures and temperatures.

The results of Haynes [47] are an excellent supplement to the data of Thomas and Harrison [101] for the low temperature range. Study [47] had the objective of finding a more accurate calculating procedure for the mass of the liquified gases during transportation. One of the most widespread methods of measuring the mass flow rate involves measuring the volume flow rate and density. The simultaneous precise measurements of the density and the dielectric constant are necessary because a one-to-one relation exists between the density and the Clausius-Mossotti function; in the industrial environment, measurements of the dielectric constant are relatively simple. Such measurements were made in [47]. A densimeter with a magnetic suspension and a concentric cylindrical capacitor were installed in the experimental cell to measure the dielectric constants. Temperatures were measured with a platinum resistance thermometer, calibrated according to the IPTS-68 scale; pressures were measured by a piston manometer. A null detector containing a diaphragm was used to separate tested samples from the oil in the gauge. The magnetic suspension of the densimeter is the most efficient for the isothermal conditions, since the magnetic moment of the buoy has a strong temperature dependence. The isothermic method allows determination of the current necessary to maintain the buoy at a desired height immediately before or after each isotherm measurement.

In this experiment commercially available research-grade propane was used. Its purity was determined by the chromotographic analysis to be 99.99 moles %; additional purification was not performed. The temperature was maintained by the thermostat to better than 0.001 K; however, the uncertainty in temperature measurements varied from 0.01 K at 90 K to 0.03 K at 300 K. The uncertainty in the pressure measurements was ~0.01%, and at the maximum pressure, it was even greater. The overall uncertainty in the experimental values of density was estimated in [47] as ±0.1%.

The data of [47] agree with the data of Ely and Kobayashi [38] and Thomas and Harrison [101] within the overall uncertainty limits of the experiments.

Kratzke and Muller [67] measured recently the density of the liquid propane on a pseudo-isochore. The accuracy of the temperature measurements by a platinum thermometer was ±10 mK. The pressure was measured by a piston manometer, but its precision was not included. However, from the results in other papers of the same authors, the relative uncertainty in pressure was evaluated to be better than 0.05%. The purity of the investigated propane (99.954 moles %) is sufficient, but it is not record high. The overall uncertainty in the density values was estimated in [67] to be equal to 0.05%.

The work of Straty and Palavra [97] is very interesting; they used the Burnett method and a method involving a piezometer with a constant volume. On the basis of careful analysis, the authors of [97] decided that the Burnett method is more reliable for the region of moderate temperatures, and used it for determining the specific volume at the 250 °C isotherm. At higher temperatures up to 600 °C the p, v, T data were measured for pseudo-isochores by a piezometer with a constant volume. Although well-known methods were used for collecting the p, v, T data, the study [97] is original in its use of a computer for the experimental control and data acquisition. All measurements with a length of up to 60 h were performed without an operator; this increased considerably the productivity of the method. For the duration of the experiment it was possible to control and monitor it from a remote site by using a personal computer with a modem. However, the results of Straty and Palavra [97] are only used for comparing them with the calculated values, because they are not quite reliable enough to be used for deriving the equation of state.

Table 1.1 did not include the studies of Burrell and Jones [20], of Heuse [50], and of Massie and Whytlaw-Gray [73]. Study [20] has little value because the results cover a narrow pressure range (951–5723 mm Hg) at 150 °C, and were obtained by an obsolete experimental method. In [50] and [73] the mass of one liter of gaseous propane was measured at the normal conditions. The authors of [73] used a non-traditional, very complex method of a buoyancy microbalance. They measured pressures at which propane and oxygen have equal densities, and extrapolated the results to zero pressure. Thus relative molecular masses of the two gases were obtained and the mass of one liter of gaesous propane at the standard temperature and pressure was calculated. It should be noted that the values obtained in [50] and [73] differ by about 0.5%.

The virial coefficients also characterize the *p, v, T* relation. The virial coefficients, and first of all, the second virial coefficient, can be useful for deriving the equation of state. Table 1.3 lists studies in which the second virial coefficient was determined; the third virial coefficient is given only in two studies, marked by the letter "C" in Table 1.3.

The monograph of Dymond and Smith [37] reviewed all of the data published before 1967 for the second virial coefficient of propane. In addition to studies listed in Table 1.3, Dymond and Smith used the *p, v, T* data of Sage et al. [89], Cherney et al. [26] and Reamer et al. [84] for the calculation of the virial coefficient. Deshner and Brown [35] obtained rather low values which have not been considered in [37] for the final calculation of the value of *B* in the temperature interval of 260–550 K. After the monograph [37] came out, a number of new studies, providing the values of *B*, were published. The analysis of these studies was made by Barkan [2] who criticized Dymond and Smith for ignoring the distinct features of separate studies and for not applying a weighting factor to individual studies when averaging the values of *B*. The author of [2] mentioned that considerable discrepancies between various data for the second virial coefficient of hydrocarbons are typical (except methane, ethane and

Table 1.3 Investigations of the Virial Coefficients

Year	Author	Temperature range ΔT, K	Number of points	Comments
1938	Jensen and Lightfoot [56]	273; 323	2	
1942	Hirschfelder et al. [52]	303–570	17	Data of [14, 35]
1950	Bottomley et al. [17]	295	1	
1951	Kretschmer and Wiebe [68]	273–323	3	
1960	Dowson and McKetta [33]	243–348	6	
1962	McGlashan and Potter [75]	295–413	12	
1963	Kapallo et al. [59]	244–321	4	
1963	Huff and Reed [54]	311–511	5	Data of [45]
1967	Brever [18]	248–298		
1969	Dymond and Smith [37]	260–550		
1970	Strein et al. [98]	296–493	11	
1974	Hahn et al. [46]	211–274	4	C
1974	Schäfer et al. [91]	296–512	6	
1978	Warowny et al. [106]	373–423	6	
1982	Thomas and Harrison [101]	323–623	23	C
1983	Barkan [2]	220–560	18	

ethylene). He explained this fact by the systematic errors in the experimental values of z, w, etc., which were used in the calculation of the virial coefficient, or by the errors in calculating procedures. The second virial coefficient was recalculated in [2] from the data of Beattie et al. [14], Reamer et al. [84], Cherney et al. [26]. The obtained results agree well with the new data of Warownye and Willopolski [106].

The values of B at $T < 273$ K are the least justified, because for this temperature range only the data of Hahn et al. [46] are available. However, the data of [46], combined with those for the region of $T > 273$ K, form a smooth continuous curve. Barkan [2] summarized the values of B for the 220–560 K temperature range, which is wider than the range given in [37]. Figure 1 shows that the values of B found by Barkan [2] and Dymond and Smith [37] agree well despite the critical comments made in [2] about the method used in [37].

1.2 THERMAL PROPERTIES OF PROPANE IN THE SINGLE-PHASE REGION

The number of experimental investigations of the thermal properties of propane is limited. Several studies (Table 1.4) give the values of the specific heat, c_p^0, for the ideal gas state. These values were obtained by correcting the values of c_{p1}, measured at the atmospheric pressure, or by the extrapolation of the isotherm data to a zero-pressure value.

Beeck [16] measured the accommodation coefficient of the molecular beam and calculated the values of c_{p1}. The calculation was based on a number of assumptions, the correctness of which had not been proven. The author of [16] estimated the accuracy of c_{p1} values as $\pm 1\%$.

Sage et al. [90] measured the specific heat c_{p1} through an adiabatic expansion of gas to the atmospheric pressure by letting the gas flow out through a fast acting valve. The authors of [90] mentioned that in the earlier study by Sage and Lacey [88], the results were underestimated by about 2.5%, because the quality of thermocouples used for temperature measurements was poor, and because the uncertainty in the p, v, T data, used for the calculation of the values of c_{p1}, was high. In addition, the data of [88] are given only in the form of small-scale graphs.

Kistiakowsky et al. [63] found the specific heat c_p^0 from the measurements of the heat conductivity at rather low pressures. The accommodation coefficient of the gas at the surface of the heated wire, used for the heat conductivity measurements, must be taken into account. At small pressures, molecules of gas do not achieve thermal equilibrium with the solid during the first impact with surface of the wire. Neglecting the accommodation coefficient would result in errors in the heat conductivity values and, consequently, in the values of the specific heat. This is the weak point of the experimental method because accurate values of the accommodation coefficient are usually unknown. Kis-

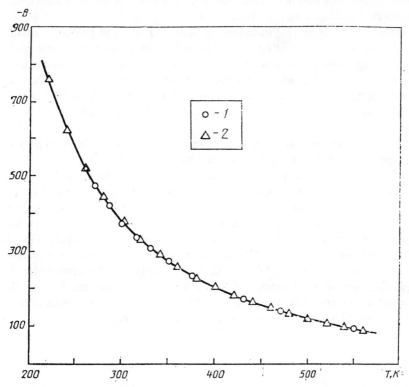

Figure 1 Smoothed values of the second virial coefficient according to the data of: *1*) Dymond and Smith [37]; *2*) Barkan [2].

Table 1.4 Experimental Investigations of the Isobaric Specific Heat of Propane at Atmospheric Pressure for the Ideal Gas State

Year	Author	Temperature range ΔT, K	Number of points
935	Sage and Lacey [88]	311–341	Graph
1936	Beeck [16]	273–573	4
1937	Sage et al. [90]	294–444	10
1940	Kistiakowsky et al. [63]	148–259	14
1940	Kistiakowsky and Rice [64]	272–369	4
1943	Dailey and Felsing [30]	344–693	8
1970	Ernst and Büsser [40]	293–353	4
1973	Chao et al. [25]	0–1500	*

*The calculation is based on the spectral data.

tiakowsky and Rice [64] determined the specific heat c_{p1} by the adiabatic expansion method and calculated the values of c_p^0.

Dailey and Felsing [30] measured the specific heat c_{p1} of propane containing less than 0.1 moles% of ethane and butane impurities at atmospheric pressure by using an advanced flow calorimeter. The Berthelot equation was used to correct the value of c_{p1} when the value of c_p^0 was calculated. This correction, which is greater at lower temperatures, does not exceed 0.1% and the corresponding additional uncertainty is much smaller than the uncertainty of measurements.

Ernst and Busser [40] obtained the values of c_p^0 by extrapolating to $p = 0$ the experimental values of c_p obtained at high pressures.

Chao et al. [25] calculated c_p^0 by using the methods of statistical thermodynamics and compared the results with the experimental data mentioned above (Fig. 2). When necessary the authors of [25] introduced corrections for the transition between the experimental values of c_{p1} to c_p^0. Figure 2 shows the deviation of the experimental values of c_p^0 from those calculated in [25]. Figure 2 shows that the data of Beeck [16] are systematically lower, and the deviation from the data of Sage et al. [90] increases considerably with the temperature. It is possible that the shortcomings of the earlier study by Sage and Lacey [88] have not been completely corrected in [90]. The remaining experimental data deviate within 1% which is quite acceptable for such measurements.

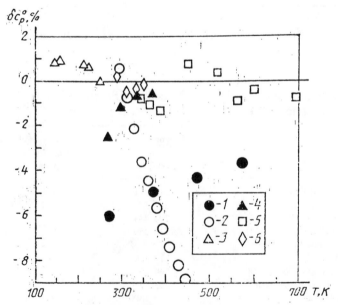

Figure 2 Deviations of the specific heat values c_p^0, calculated by Chao et al. [25], from the experimental data of: *1*) [16]; *2*) [90]; *3*) [63]; *4*) [64]; *5*) [30]; *6*) [40].

Table 1.5 Investigations of the Isobaric Specific Heat of Propane in the Single-Phase Region

Year	Author	Parameter range		Number of points
		ΔT, K	Δp, MPa	
1969	Yesavage et al. [107]	311–422	1.7–10.5	86
1969	Ernst [39]	293–353	0.05–0.4	28
1970	Ernst and Büsser [40]	293–353	0.05–1.4	36

The isobaric specific heat of propane at pressures above the atmospheric pressure was obtained only in three studies (Table 1.5). The measurements of Yesavage et al. [107] were made for the parameters close to the saturation curve, and in the region of the maxima of c_p. In this study the enthalpy values were determined from the c_p data and the throttling effect.

Ernst [39] measured the c_p values at pressures of up to 4 bar, and later Ernst and Busser [40] increased the range of measurements up to 14 bar. Studies [39, 40] can be regarded as one investigation, because they cover the same temperature interval, and the second study added only a few points at $p = 0.4$–1.4 MPa. A direct comparison of data of [107] and [39, 40] is impossible because their parameters do not overlap.

The isochoric heat capacity of liquid propane was measured at high pressures by Goodwin [43] (Table 1.6). Studies of Goodwin are well known; earlier he measured the specific heats of hydrogen, methane, fluorine, and oxygen. The high methodological level of Goodwin's studies makes his data quite reliable. Fifty-eight values of c_v for nine isochores were obtained in [43]. The uncertainty of the results depends on the measured parameters and the accuracy of measurements. For an isochore, the uncertainty increases with decreasing temperature, but in most cases it does not exceed 2%.

Anisimov et al. [1] measured the isochoric heat capacity of propane in the two-phase region and partially in the single-phase region. The authors of [1] used an adiabatic calorimeter and took into account the influence of temperature and pressure on the volume. A correction to the measured results imperfect for the nonisochoric conditions was performed; this correction is especially notice-

Table 1.6 Investigations of the Isochoric Specific Heat of Propane in the Single-Phase Region

Year	Author	Parameter range		Number of points
		ΔT, K	$\Delta \rho$, kg/m^3	
1978	Goodwin [43]	100–337	496–719	58
1982	Anisimov et al. [1]	271–374	159–531	52

able for the single-phase region when $\rho > \rho_{crit}$. For five isochores, the authors of [1] estimated the uncertainty in the c_v values as being 1.5% for the points close to the equilibrium curve and 0.48% for points at a distance from the curve. For the isochore that corresponded to the smallest density (0.157 g/cm³), these uncertainties amounted to 2% and 0.68%, respectively.

It should be pointed out that a general characteristic of [1], given in Table 1.6, does not fully reflect the completeness and thoroughness of this study. Studies [1] and [43] have an almost equal number of experimental points; however, the former is less informative because for each of the six isochores studied in [1], data for only a small number of points in a very narrow temperature range (0.3–7 K) was obtained. The temperature ranges for five of the isochores given in [1] are close the critical temperature, and the densities of two of the isochores are close to the value of ρ_{crit}. Thus most of the data of [1] were obtained for the region of parameters where the accuracy of the description of the c_v values by the virial equation is small.

The standard reference data for the isochoric specific heat of propane at $\rho = 3 \cdot 10^{-3}$ m³/kg are given in Tables of GSSSD 38-82 [8]. Smoothed values of c_v were obtained from the experimental values, measured in a standard vacuum adiabatic calorimeter calibrated from the isobaric specific heat of a solid standard reference sample. The values of specific heat c_x of a sample, determined from the experiment, were related to c_v through an expression containing the values of dp/dT taken from [32]. The values of dv/dp for the calorimeter were determined experimentally, and the values of dv/dT were calculated from the linear expansion coefficient for the material that the calorimeter was made of. The overall uncertainty was estimated as $\delta c_v = 2\%$.

The velocity of sound in liquid propane was measured in two studies (Table 1.7). Lacam [70] used a commercial research-grade propane without performing an additional purification. According to [70], the relative uncertainty in the values of w, obtained from diffraction of light through standing ultrasonic waves, amounts to 0.2–0.4%, increasing with pressure.

Younglove [108] used a superposition method and measured the velocity of sound with accuracy of 0.05%, using a sample containing not more than 0.03 moles% of impurities. The combined data of [70, 108] cover a rather wide region of parameters, supplementing each other.

The coefficient of the adiabatic compressibility $\beta = 1/(w^2\rho)$ was calculated

Table 1.7 Investigations of the Velocity of Sound in Propane

| Year | Author | Parameter range | | Number of points |
		ΔT, K	Δp, MPa	
1956	Lacam [70]	298–498	1–101	200
1981	Younglove [109]	90–300	1.9–35	162

in [70, 108] from the measured values of w; in study [108] the specific heat ratio c_p/c_v was also calculated.

1.3 THERMODYNAMIC FUNCTIONS ON THE CURVES OF MELTING AND EVAPORATION

1.3.1 The Curves of Melting and Evaporation

The information on the curve of melting for propane was given only in the study by Reeves et al. [85]. The objective of [85] was to select an appropriate pressure-transmitting fluid in the 10 kbar work range below room temperature. In 1964 the authors of [85] pointed out that such information was very limited; however, the situation did not really change over the last twenty years.

Propane has a very low melting temperature; for example, it is $-105\,°C$ at 10 kbar, while for heavier saturated hydrocarbons (C_4, C_5, . . .) this temperature lies between -30 and $-50\,°C$.

Reeves et al. [85] used a rather wide-spread method of a plugged capillary. The U-shaped stainless capillary was soldered into the manganin cell chambers at each end. The capillary was placed into a cylindrical copper thermostat. The manganin coils were installed for pressure registration in the two legs of the capillary. The absolute pressure values were measured by a manganin mano-metric coil calibrated by the pressure of frozen mercury at three temperatures. The temperature of the copper thermostat was measured by a nickel resistance thermometer calibrated by a platinum thermometer. The uncertainty in the tem-perature measurements, which was less than 0.1 K at 273 K and 0.2–0.3 K at 77 K, made a small contribution to the overall uncertainty of the results. The procedure of taking data points was the following: temperature equilibrium was reached, the pressure was raised until the sample was frozen, and then the pressure was lowered until it melted. Technical difficulties caused a relatively high uncertainty in the pressure values during melting at low absolute pressures (1 % at $p \leq 1$ kbar and 0.1 % at $p \simeq 10$ kbar).

The reproducibility of pressure readings at freezing was worse than that at melting, because a supercooling of the sample was possible. Therefore Reeves et al. [85] measured only the pressure during melting.

The authors of [85] did not guarantee the high accuracy of the results. Little attention was paid to the sample purity. Contaminants can decrease the freezing temperature by about 1 K and 1.5 K at 273 and 77 K, respectively. Nevertheless, the propane used in [85] was quite pure, and the reproducibility of measurements during repeated freezing was within 1.5 bar.

The results of [85] are given only in a graphical form. The graphs show five experimental values located within the range of 1.5–10 kbar. Some of the investigated substances were described by the Simon equation; for propane this can be expressed as

$$\frac{p}{7180} = \left(\frac{T}{85.3} \right)^{1.283} - 1$$

The first measurements of the equilibrium vapor pressure of propane, p_s, were made about 70 years ago. Since then more than twenty studies have been published. They cover almost the complete temperature range from the triple to critical points (Table 1.8). Many studies duplicate each other, above 300 K the measurements were performed many times (Fig. 3). It should be mentioned that most of the investigations are more than 30 years old.

The study of Burrell and Robertson [21], mentioned in Table 1.8, is only of a historical value because the experimental technique does not meet modern standards. For example, two pentane thermometers were used for temperature measurements, but their readings at the melting points of chloroform and mercury differed by 2–3 K.

Maass and Wright [72] gave little information on the accuracy of their results. The preparation and purification of the samples were described in detail. However, it is impossible to judge the amount and type of residual impurities, because study [72] only mentioned that the purity was controlled through density measurements. The thermostat control was also very inaccurate because

Table 1.8 Investigations of the Equilibrium Vapor Pressure

Year	Author	Parameter range		Number of points
		ΔT, K	Δp, MPa	
1916	Burrell and Robertson [21]	149–229	0.002–0.1	16
1921	Maass and Wright [72]	230–250	0.11–0.27	6
1926	Dana et al. [31]	210–323	0.04–1.7	21
1933	Francis and Robbins [41]	301–336	1.0–2.4	34
1934	Sage et al. [89]	294–371	0.9–4.3	21
1935	Beattie et al. [15]	323; 348	1.7; 2.8	2
1938	Kemp and Egan [62]	166–231	0.002–0.1	12
1940	Deshner and Brown [35]	302–370	1.1–4.2	36
1940	Gilliland and Scheeline [42]	315–372	1.4–4.4	5
1949	Cherney et al. [26]	303; 323	1.1; 1.7	2
1949	Reamer et al. [84]	313–370	1.4–4.3	10
1951	Tickner and Lossing [102]	108–165	10^{-7}–10^{-3}	13
1953	Kay and Kambosek [61]	278–370	0.6–4.3	10
1955	Clegg and Rowlinson [28]	323–370	1.7–4.3	9
1967	Helgeson and Sage [49]			
1970	Kay [60]	323–368	2.1–4.1	6
1973	Carruth and Kobayashi [24]	95–179	10^{-9}–10^{-3}	12
1973	Kahre [57]	278–328	0.6–1.9	5
1977	Moussa [77]	335–370	2.2–4.3	11
1980	Krazke [66]	312–368	1.3–4.1	
1982	Thomas and Harrison [101]	258–370	0.3–4.3	25
1984	Kratzke and Muller [67]	300–357	1.0–3.4	5

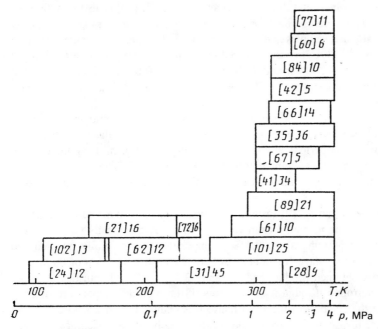

Figure 3 Ranges of parameters of state covered by the measurements of the equilibrium vapor pressure. The corresponding references are shown inside the square brackets; the number of experimental points, outside the brackets.

the constant temperature was maintained in a primitive way, by adding "small pieces of solid carbon dioxide" to different liquids, used for the thermostatic control. A platinum resistance thermometer did not guarantee the necessary accuracy, and the uncertainty in the temperature values reached 0.2 degrees. The study did not give the uncertainty in the pressure measured with a mercury manometer.

Dana et al. [31] prepared a propane sample from natural gas by repeated fractional distillation, withdrawing and saving the middle fraction. The purity criterion was a constant value of p_s for different liquid to vapor ratios. However, the authors of [31] did not identify the quality of the final product. The platinum resistance thermometer was calibrated from the freezing point of mercury, sublimation point of carbon dioxide and boiling point of oxygen. Study [31] gave the adopted values of these reference points which permits adjustment of these points to modern values and the making of necessary corrections. The pressure at $p < 6$ bar was measured by a mercury manometer and a cathetometer (at $p < p_{atm}$ one of the legs of the mercury manometer was soldered and evacuated), and at $p > 6$ bar—by a piston manometer which had the accuracy of 0.3%.

The information on the methods and techniques of measurements used in the studies of Francis and Robbins [41] and Sage et al. [89] is also very limited. In [89] the data were obtained for a very narrow temperature range. The purity of the investigated samples and the uncertainty in the measured values were not provided. Study [89] gave the uncertainty in the pressure measurements (0.07 bar for $20 \leq P \leq 200$ bar and 0.007 bar for $p \leq 20$ bar), and the accuracy of the temperature control ($\sim 0.1\,°C$), but it provided no information on the sample purity.

Beattie et al. [15] investigated the density of the sample in the critical region and measured p_s in order to control the purity of the substance. The sample purity was periodically tested and controlled by measuring the equilibrium vapor pressure at temperatures of 50 and 75 °C. The uncertainties in the pressure and temperature measurements were 0.03% and 5 mK, respectively.

Kemp and Egan [62] measured a number of thermodynamic properties of propane, including its equilibrium vapor pressure. A chemically pure commercial propane containing about 0.1% of impurities was carefully purified and dried during a two stage fractional distillation with the selection of the middle fraction; the obtained product had less than $2 \cdot 10^{-3}\%$ of impurities. The purity was controlled through measurements of the melting temperature of crystals, while the fraction of the solid phase was changed from 5% to 20%. The temperature measurements were made with a relatively high accuracy of a few thousandths of a degree. However, in [62] 0 °C corresponds to 273.10 K, which requires corrections to the measured temperatures and pressure values of p_s.

Deshner and Brown [35] purified commercial propane; however, the final product had about 1.7% of impurities, which is not satisfactory. The authors of [35] did not identify the impurities and only mentioned that the purified propane "did not have condensate impurities."

A number of studies can be combined through a common objective, the investigation of the properties of mixtures. In these studies the data on the individual substances were obtained as a by-product, and as a result they do not always have a high accuracy. These studies include works of Gilliland and Scheeline [42], Kay and Rambosek [61], Clegg and Rowlinson [28], Kay [60], and Kahre [57]. Study [42] is very brief, and it is difficult to judge properly the quality of the propane samples and the obtained results. Other studies gave the data on propane mixtures containing 0.01–0.1% of impurities. In the most recent study of Kahre [57] the mixture contained 0.2% of impurities.

Cherney et al. [26] measured the values of p_s for 99.99% pure propane at two temperatures. The accuracy of temperature and pressure measurements was estimated in [26] as 0.02 K and 0.07%, respectively.

Reamer et al. [84] worked with propane containing less than 0.1% of impurities. A piston manometer, calibrated from the equilibrium vapor pressure of carbon dioxide at 0 °C, had an accuracy of better than 0.05%. A platinum resistance thermometer was also often checked at 0 °C; however, the method of achieving the reference temperature and the overall accuracy of measurements was not specified.

The studies of Tickner and Lossing [102] and Carruth and Kobayashi [24] form a group of their own because of the range of pressures over which measurements were made, and because of their experimental technique. In [102] a mass-spectrometer was used for measuring very small absolute pressures, and in [24] an effusion method was used for the same purpose. The information on the gas-liquid curve at pressures of less than 10 mm Hg is very limited for practically all substances, including even well studied ones. Such information is important for answering fundamental questions regarding the structures of substances, and the high quality purification by a vacuum fractional distillation. The authors of [102] and [24] used in their investigations samples containing 99.7% and 99.99% of propane, respectively. The results of these two studies agree well, despite such a difference in the purity of their samples.

Four studies were published during the last ten years: Mousa [77], Kratzke [66], Thomas and Harrison [101], Kratzke and Muller [67]. The results of the graduate thesis of Teachman completed in 1978 was not available to the present authors.

The modern achievements in the experimental technique have not been used in the recent work of Mousa [77]. The initial sample of propane containing about 0.1% of impurities was further purified to remove moisture and volatile impurities. The purity of the final product was checked by the difference between the bubble and dew point pressures, which was found to be "negligible." Unfortunately such qualitative description of purity is not specific. The simplest means: a thermocouple and a spring manometer were used for measuring the temperature and pressure. The accuracy of the thermocouple reading, 0.01 K, and the division on the manometer scale, ~ 1.4 kPa, permitted to obtain acceptable results; however, the actual uncertainties in the temperature and pressure measurements were quite high: ± 0.2 K and 3.5 kPa, respectively.

The extensive investigation by Thomas and Harrison [101] has been mentioned in connection with the measurements of p, v, T relations. The authors of [101] measured the equilibrium vapor pressure while changing the fraction of the condensed substance over a wide range. They averaged the results obtained for the condensation of 25, 50 and 75 wt.% of the sample in the piezometer. In some cases, for purity control, the measurements were made for the condensation of 5 wt.% and 95 wt.% of the sample. Deviations in the values of p_s during such measurements reached 0.01% and 0.03% for two different propane samples at the lowest temperatures. At higher temperatures this difference did not exceed several thousandths of a percent.

Kratzke [66] and a few years later also Kratzke and Muller [67], performed highly accurate measurements of p_s for a similar, relatively narrow, temperature range. These studies used 99.9954 moles% pure propane. The thermostat system was improved and maintained a stable temperature within ± 5 mK. The uncertainty in temperature measurements by a platinum resistance thermometer, calibrated in terms of the IPTS-68, did not exceed 10 mK. The pressure was measured by a piston manometer with the accuracy better than 0.03%.

The authors of a number of experimental studies and reviews compared the results of different investigators. Most of the old publications disagree; the measured points are too scattered, even in individual studies. Deviations and scatter usually increase with temperature.

Data of Dana et al. [31], Deshner and Brown [35], Gilliland and Scheeline [42], and Moles [77] disagree among themselves, and differ from the majority of other data. This should be expected, considering the imperfect experimental technique and insufficient purity of samples in these studies. The data of Cherney et al. [26], Reamer et al. [84], and Clegg and Rowlinson [28] have a satisfactory accuracy and agreement with the results of other investigators. The data of Beattie et al. [15], Kemp and Egan [62], Carruth and Kobayashi [24], Kratzke [66], Kratzke and Muller [67], and Thomas and Harrison [101] are the most mutually consistent; together these studies cover almost the entire temperature range.

The investigations of the liquid-vapor curve include measurements of parameters at the triple point, normal boiling temperature, and measurements of critical parameters. Methods of temperature measurements for phase equilibria at atmospheric pressure are well developed and easy to implement. Nevertheless, the measurements of the melting and boiling temperatures of propane are rare (Table 1.9); they were performed long ago and results vary considerably.

Propane has a very low pressure at the triple point. The experimental data on p_{tr} are absent, and the temperature T_{tr} has been measured only a few times. Seshadri and Viswanath [92] extrapolated the data of Delaplace [34] to the equilibrium vapor pressure at low temperatures and obtained the value of $T_{tr} = 85.5$ K, which the authors of [92] consider to be approximate. Kratzke [66] recently obtained a more accurate value of $T_{tr} = 85.47$ K. Using his value of T_{tr} and the equation for $p_s(T)$, Kratzke determined that $p_{tr} = 1.33 \cdot 10^{-4}$ Pa. This

Table 1.9 Investigations of the Melting and Boiling Temperatures at Atmospheric Pressure

Year	Author	T_{melt}, K	T_{boil}, K
1920	Anderson [12]	85.35	—
1921	Maass and Wright [72]	83.25	228.65
1921	Timmermans [103]	85.35	—
1933	Francis and Robbins [41]	—	230.85
1936	Hicks-Bruun and Bruun [51]	86.05 ± 0.1	—
1938	Kemp and Egan [62]	85.45 ± 0.05	—

value differs from that adopted in [44] by almost 20%, although the absolute deviation is very small.

Kemp and Egan [62] performed detailed investigations of the specific heats of crystalline and liquid propane. They determined the temperatures of phase transformations including the melting point which happens to be very close to the triple point. The authors of [62] obtained the temperature of the phase transformation (85.45 ± 0.05 K) using an outdated temperature scale, according to which the melting point of ice corresponds to 273.10 K. Unfortunately study [62] does not carry sufficient information to allow corrections for the type and quantity of impurities and for the modern temperature scale.

After Pavese [81] discovered several modifications of solid ethane, Pavese and Besley [82] made a similar discovery for propane and found a stable (β) and a metastable (α) crystalline phase. The authors of [82] used a very pure propane sample and a carefully controlled cryostat. The platinum thermometer for temperature measurements were checked frequently against a low-temperature reference point, the melting temperature of argon equal to 83.798 K, which is located near the triple point of propane. A comparator bridge recorded thermograms. Power was supplied at the rate of about 32 mW per step. About one hour was allowed after each step to permit the recovery of thermal equilibrium to better than 0.3 K because it is known that propane can be easily supercooled. The average value of T_{tr} = 85.515 ± 0.001 K was obtained from numerous measurements. This value differs from that obtained in [72] by 0.07 K. This deviation exceeds the sum of uncertainties estimated in [62] and [82], but is acceptable for the reasons mentioned above.

Pavese and Besley [82] also determined the enthalpy for the melting of β-phase Δh_β = 3.50 ± 0.025 kJ/mol. For the α-phase, the temperature of the triple point was $T_{tr,\alpha}$ = 81.222 ± 0.001 K, and the enthalpy of the α-phase melting was Δh_α = 2.4 ± 0.3 kJ/mol.

Table 1.10 lists the investigations of the critical parameters. The critical temperature T_{crit} was most often measured by a method of disappearing meniscus. For determining p_{crit}, the data on the liquid-gas curve were used. For example, the dependence log (p_s) vs. T was extrapolated to T_{crit}. The procedure of determining ρ_{crit} is usually the least reliable. The simple recti-linear diameter method is very popular, but it has been criticized. According to this method, the temperature dependence of the value of $1/2(\rho' + \rho'')$ is extrapolated to T_{crit}.

Table 1.10 does not include outdated studies of low accuracy made by Al'shevskii (1889, 1894) and Heinlen (1895) in the last century. The values of T_{crit} obtained by Lebeau [71], Sage et al. [89], Deshner and Brown [35], and Gilliland and Scheeline [42] are too high in comparison with the results of most of the other authors. It is natural that the above studies gave overestimated values of p_{crit} and ρ_{crit}. Such inconsistent data were ignored during the final selection of the critical parameters.

Table 1.10 Investigations of the Critical Parameters of Propane

Year	Author	T_{crit}, K	p_{crit}, MPa	ρ_{crit}, kg/m^3
1905	Lebeau [71]	370.7	4.56	—
1921	Maass and Wright [72]	368.8	—	—
1934	Sage et al. [89]	373.25	4.435	232.3
1935	Beattie et al. [15]	369.96	4.257	226.0
1937	Beattie et al. [14]	369.96	4.257	225.7
1940	Deshner and Brown [35]	370.00	4.266	224
1940	Gilliland and Scheeline [42]	372.0	4.38	—
1942	Meyers [76]	—	—	219.4
1949	Reamer et al. [84]	369.98	4.257	220
1953	Kay and Rambasek [61]	369.82	4.249	—
1955	Clegg and Rowlinson [28]	369.81	4.249	217
1968	Kudchadker et al. [69]	369.81	4.249	217
1969	Sliwinski [95]	369.84	—	220
1971	Tomlinson [104]	369.82	—	217
1977	Moussa [77]	369.74	4.254	214
1980	Kratzke [66]	369.800	4.239	—
1982	Thomas and Harrison [101]	369.85	4.247	218.5

1.3.2 Density of Solid and Gaseous Propane in the State of Liquid-Vapor Equilibrium

Table 1.11 lists the density measurements for coexisting liquid and gaseous propane.

Most of the investigations listed in Table 1.11 are not limited to densities of coexisting liquid and gas, and also provide densities in a single-phase region or equilibrium vapor pressures. These publications include studies of Maass and Wright [72], Dana et al. [31], Sage et al. [89], Deshner and Brown [35], Reamer et al. [84], Ely and Kobayashi [38], and Kratzke and Muller [67] which have already been discussed above.

Studies of Clegg and Rowlinson [28], Shana'a and Canfield [93], Jensen and Kurata [55], and Kahre [57] provide the properties of hydrocarbon mixtures as their main result. The densities for pure components were determined only occasionally for the calibration of the apparatus or for the later use in calculations. The studies usually give a few values of density, sometimes a single value [93]. The accuracy of the results is largely determined by their intended use. For example, during the apparatus calibration, Shana'a and Canfield [93] determined a single density value for liquid propane, using a very pure sample (99.97%), and achieved a very good accuracy ($\pm 3 \cdot 10^{-5}$ g/cm^3), while Jensen and Kurata [55] could not achieve an accuracy of better than 0.5%.

Sliwinski [94] used a pycnometric method and measured simultaneously the refraction coefficients and the densities of the coexisting gas and liquid. He calculated from these data the Lorentz-Lorenz function. The shape of this function vs. density is in agreement with the statistical data derived from the theory of dense gases.

The method of hydrostatic weighing was frequently used for measuring densities of boiling liquids. This method was used by McClune [74], Haynes and Hiza [48], and Orrit and Laupretre [79].

McClune [74] worked with a propane sample containing 0.01% of nitrogen and *n*-butane. According to [74], the uncertainty in the values of ρ' did not exceed 0.1%. This estimation was confirmed through a comparison with the results of Rodosevich and Miller [86], Kossek and McKinley [65], Shana'a and Canfield [93], Chui and Canfield [27], and Orrit and Olives [80]; the deviations lie within ±0.05%.

Haynes and Hiza [48] used a magnetic suspension buoy which made their apparatus rather complex. According to [48], an advanced electronic suspension system allowed them to decrease the uncertainty in the values of ρ' to about 0.1% at low temperatures and to 0.06% at 300 K.

Orrit and Laupretre [79] used a very simple method of balancing: a buoy, submerged into the tested fluid, was suspended from one arm of the balances. This procedure was performed by hand. A signal from a differential transformer announced the achievement of the equilibrium. The buoy in [79] was

Table 1.11 Measurements of the Density of Liquid (ρ') and Gaseous (ρ'') Propane in the State of Liquid-Gas Equilibrium

Year	Author	ρ' ΔT, K	Number of points	ρ'' ΔT, K	Number of points
1921	Maass and Wright [72]	195–249	13	—	—
1926	Dana et al. [31]	278–329	12	290–323	5
1934	Sage et al. [89]	294–371	21	294–371	21
1940	Deshner and Brown [35]	303–368	14	303–368	14
1949	Reamer et al. [84]	313–368	9	313–368	9
1955	Clegg and Rowlinson [28]	323–368	8	323–368	8
1968	Shanala and Canfield [93]	108	1	—	—
1969	Jensen and Kurata [55]	93–133	5	—	—
1969	Sliwinski [94]	283–369	14	283–369	14
1973	Kahre [57]	278–328	5	—	—
1973	Rodosevich and Miller [86]	91–115	4	—	—
1976	McClune [74]	98–173	17	—	—
1977	Haynes and Hiza [48]	100–289	16	—	—
1978	Ely and Kobayashi [38]	166–288	18	—	—
1978	Orrit and Laupretre [79]	87–244	31	—	—
1982	Thomas and Harrison [101]	258–369	22	323–369	11
1984	Krtzke and Muller [67]	245–325	5	—	—

made of a material with well known properties for different temperatures and pressures. The authors could make the corresponding corrections and decrease the uncertainty of the experimental values of ρ' to $4 \cdot 10^{-4}$ g/cm^3.

Thomas and Harrison [101] measured the densities of coexisting vapor and liquid using the same apparatus, which was employed for investigating the p, v, T relations of propane for a single-phase region. To increase the accuracy of pressure measurements, the authors of [101] used a manometer with a large diameter piston. Other information on the apparatus, the method being used and the sample being tested was already described in the section 1.1.

1.3.3 Thermal Properties of Propane for the Curve of Saturation

The enthalpy of evaporation of propane was measured in a number of studies (Table 1.12). Dana et al. [31] used a copper calorimeter with an adiabatic screen. Propane, evaporated from the calorimeter, was condensed in a specially cooled container for a fixed time, and then weighed. In the experiments, the rate of the evaporation and the ratio of the evaporated substance to the substance remaining the calorimeter was changed over a wide range. As these parameters varied, several values of the enthalpy of evaporation were obtained at constant temperatures. The coincidence of the results was taken as evidence of the reliability of the data. However, the authors of [31] could not explain the considerable spread in the values of r with the difference of up to 4% for the same temperature.

Sage et al. [87], Yesavage [107] and Carruth [23] extended the temperature range of measured parameters to higher and lower temperatures. Several points determined in studies of Staveley et al. [96] and Helgeson [49] and a single value of r at the normal boiling temperature [62] increased the amount of available information, but cannot be regarded as evidence for or against the validity of previously published data.

Table 1.12 Investigations of the Enthalpy of Evaporation of Propane

Year	Author	Temperature range ΔT, K	Number of points
1926	Dana et al. [31]	234–293	14
1938	Kemp and Egan [62]	231	1
1939	Sage et al. [87]	313–348	16
1950	Staveley and Tupman [96]	185; 213	2
1967	Helgeson and Sage [49]	311–330	4
1969	Yesavage [107]	231–367	13
1970	Carruth [23]	111–237	18
1978	Guigo et al. [6]	186–363	12

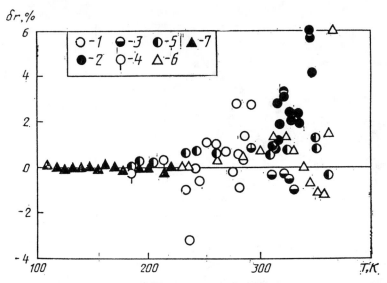

Figure 4 Deviations of the calculated values of the heats of evaporation, r, from the experimental data of: 1) [31]; 2) [87]; 3) [49]; 4) [96]; 5) [6]; 6) [107]; 7) [24].

Guigo et al. [6] measured the enthalpy of evaporation over a broad temperature range. The "true" enthalpy of evaporation r was calculated from the experimental "apparent" values of r* by using the following relation r = r*(1 − v'/v"), where the specific volumes v' and v" were taken from [32].

The equation of Goodwin and Haynes [44], approximating for r(T), was used to compare the results of different authors:

$$r = 24.840848\, x + (x^{0.38} - x)\ [24.166535 + 6.252384\, x - 12.156857\, x^2]$$

where $x = (T_{crit} - T)/(T_{crit} - T_{tr})$.

Figure 4 shows the deviation δr of the experimental values of r from the calculated ones. The equation describes with a high precision the data of Car-

Table 1.13 Investigations of the Specific Heat of Propane on the Saturation Curve

Year	Author	Temperature range ΔT, K	Number of points
1926	Dana et al. [31]	242–292	12
1938	Kemp and Egan [62]	90–231	22
1965	Cutler and Morrison	91–105	7
1978	Goodwin [43]	82–289	78
1978	Guigo et al. [6]	163–363	22

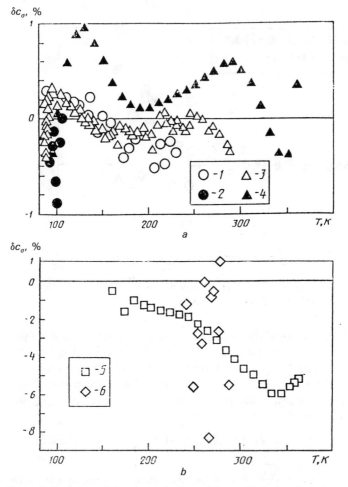

Figure 5 Deviations of the calculated values of the specific heat c_s from the experimental data of: *1*) [62]; *2*) [29]; *3*) [43]; *4*) [107]; *5*) [6]; *6*) [31].

Table 1.14 Investigations of the Velocity of Sound on the Saturation Curve

Year	Author	Temperature range ΔT, K	Number of points
1971	Rao [83]	143–228	10
1981	Younglove [108]	90–290	18

ruth [23], which are the only available data for the low temperatures and which agree well with other authors for 185 K $\leq T \leq$ 235 K. In the range from 235 K to 366 K, the majority of data agree within $\pm 1\%$. While approaching the critical temperature, the spread in the data increases and the agreement worsens. The data of Sage et al. [87] have a substantial dispersion and have the largest deviation from the calculated values of the studies being reviewed.

Overall, the combined literature data on the enthalpy of evaporation of propane are quite satisfactory in completeness, accuracy and the covered range.

Table 1.13 lists the experimental data on the specific heat c_σ of liquid propane in the state of liquid-vapor equilibrium. Dana et al. [31], whose study was mentioned at the beginning of this section, measured the specific heat c_σ for the temperature range 242–292 K. Later the c_σ values were measured over a broad temperature range by Kemp and Egan [62] and Goodwin [43]. Cutler and Morrison [29] studied mainly the properties of propane mixture and obtained a limited number of points widely scattered over a narrow temperature range.

Guigo et al. [6] measured the values of c_σ by using a Strelkov's calorimeter. The calorimeter was calibrated using the rather reliable data for freon-23; the values of c_v for pseudo-isochores were also measured. From these results, the values of c_v for the isochore $v = 3.1$ cm^3/g and c_σ for points lying on the saturation curve were calculated. The authors of [6] compared their results with those of Din [100] and Das [32]. They pointed out that a disagreement exists between different data, and that a maximum of c_σ in tables [100] contradicts the experimental data. The expression for $c_\sigma(T)$ taken from Goodwin [44] was used for the comparison. The present authors did not wish to analyze the quality of approximation for c_σ at this stage, but instead used the calculated values of c_σ to find their deviation from the experimental data. The results in Fig. 5a show that most of the data of different authors is in self-agreement within $\pm 0.5\%$. The exception is the data of Dana et al. [31] and Guigo et al. [6], which have a considerable scatter and deviate systematically (Fig. 5b).

The experimental velocities of sound w' in saturated liquid propane are given in two studies (Table 1.14). Rao [83] measured the velocity by using the impulse method in a sample containing up to 1% impurities. The uncertainty in the value of w' does not exceed 1% according to [83]. Younglove [108] measured the value of w' for a broader temperature range with a carefully purified propane sample. According to Younglove, his results have a high accuracy; however, a comparison shows that in the overlapping temperature range, the data of [108] exceed those of [83] by about 15%.

METHOD OF CONSTRUCTING A SINGLE
EQUATION OF STATE AND CALCULATING
THE TABLES OF THERMODYNAMIC PROPERTIES

Monographs [9, 11] reported in detail the method for deriving a single equation of state for the gaseous and liquid phases from the experimental thermodynamic and thermal data. The fundamental concepts of the method and the formulas for calculating the thermodynamic functions will be discussed below.

The equation of state for propane can be written in the form

$$z = 1 + \sum_{i=1}^{r} \sum_{j=0}^{s_i} b_{ij} \frac{\omega^i}{\tau^j} \tag{2.1}$$

where $z = pv/RT$ is the compressibility coefficient; $\omega = \rho/p_{crit}$ is the reduced density; and $\tau = T/T_{crit}$ is the reduced temperature.

The coefficients b_{ij} in the equation (2.1) were determined by the method of least squares from the p, v, T data, the second virial coefficient, and the isochoric specific heat. The minimized function also contained a term to satisfy

Maxwell's rule. The calculations and corrections of the relative weights of the experimental data were made according to the recommendations of [10, 11].

The program outputs a series of equations of state. This is achieved by eliminating from each successive equation the least significant coefficient, for which the ratio of its absolute value to the value of its uncertainty is the smallest [3, 11]. Simultaneously the remaining coefficients are readjusted. Hence, the accuracy of the next equation in the series is found to be almost the same as that of the previous equation. This procedure is correct until the number of coefficients is substantially reduced.

The tables of the thermodynamic functions of propane were calculated from a single equation of state. The coefficients of this equation were obtained by averaging the corresponding coefficients from the series of equations of state which are equivalent in the accuracy of approximating the experimental data. These equations were obtained by eliminating the least significant coefficients, and also by changing the relative weights of the data used in deriving the equation of state.

For convenience in programming, the following additional notation is introduced:

$$
\left.\begin{aligned}
A_0 &= \sum_{i=1}^{r} \sum_{j=0}^{s_i} b_{ij} \frac{\omega^i}{\tau^j} \\
A_1 &= \sum_{i=1}^{r} \sum_{j=0}^{s_i} (i+1) b_{ij} \frac{\omega^i}{\tau^j} \\
A_2 &= -\sum_{i=1}^{r} \sum_{j=0}^{s_i} (j-1) b_{ij} \frac{\omega^i}{\tau^j} \\
A_3 &= \sum_{i=1}^{r} \sum_{j=0}^{s_i} \frac{i+j}{i} b_{ij} \frac{\omega^i}{\tau^j} \\
A_4 &= \sum_{i=1}^{r} \sum_{j=0}^{s_i} \frac{j-1}{i} b_{ij} \frac{\omega^i}{\tau^j} \\
A_5 &= -\sum_{i=1}^{r} \sum_{j=0}^{s_i} \frac{j(j-1)}{i} b_{ij} \frac{\omega^i}{\tau^j}
\end{aligned}\right\} \tag{2.2}
$$

The thermodynamic functions for the single-phase region were calculated from the formulas obtained from the differential form of thermodynamic equations.

$$
\begin{aligned}
\rho &= p/RT(1+A_0) \\
h &= h_0 - RTA_3 \\
s &= s_0 - R\ln(\omega/\omega_0) + RA_4 \\
c_v &= c_{v_0} + RA_5 \\
c_p &= c_v + R(1+A_2)^2/(1+A_1) \\
w/w_0 &= \sqrt{1+A_1}
\end{aligned} \tag{2.3}
$$

$$\mu/\mu_0 = (A_2 - A_1)/(1 + A_1)$$
$$k/k_0 = (1 + A_1)/(1 + A_0)$$
$$\alpha/\alpha_0 = (1 + A_2)/(1 + A_1)$$
$$\gamma/\gamma_0 = (1 + A_2)/(1 + A_v)$$

(2.3)
(*Cont.*)

In the formulas (2.3) h_0, s_0, and c_{v_0} are the enthalpy, entropy and the idea gas state; and ω_0, w_0, μ_0, l_0, α_0, γ_0, are the thermodynamic normalizing functions given by

$$\omega_0 = p_{st}/(\rho_{crit} RT) \text{ at } p_{st} = 0.101325 \text{ MPa}$$

$$w_0 = \sqrt{RT c_p/c_v}$$
$$\mu_0 = 1/\rho c_p$$
$$k_0 = c_p/c_v$$
$$\alpha_0 = 1/T$$
$$\gamma_0 = 1/T$$

(2.4)

The properties of gaseous and liquid propane for points on the saturation line were determined from the equation of state (2.1) and the relationship (2.2) assuming Maxwell's rule. The specific heat of the liquid (c_s') and the vapor (c_s'') were calculated along the saturation curve from the following formulas

$$c_s' = c_p' - R \frac{a'}{\alpha_0} \cdot \frac{z_{crit}}{\omega'} \cdot \frac{d\pi_s}{d\tau_s}$$

(2.5)

$$c_s'' = c_p'' - R \frac{a''}{\alpha_0} : \frac{z_{crit}}{\omega''} \cdot \frac{d\pi_s}{d\tau_s}$$

(2.6)

where

$$\frac{d\pi_s}{d\tau_s} = -\frac{\omega' \cdot \omega''}{z_{crit}} \cdot \frac{A_s'' - A_s'}{\omega'' - \omega'}$$

The properties for the points on the melting curve have not been determined, because for propane this curve is outside the region of parameters covered by the tables given here.

THREE

EQUATIONS OF STATES AND TABLES FOR THE THERMODYNAMIC FUNCTIONS OF PROPANE

The majority of the equations of state and tables of the thermodynamic functions for propane describe the properties of the gaseous phase. A single equation of state for gaseous and liquid propane was obtained only recently [7, 44].

This book provides a series of single equations of state, describing the experimental thermodynamic and thermal data for propane, the average equation of state and the tables of the thermodynamic functions. The reliability of the calculated thermodynamic functions was determined by comparing them to the experimental results and the values calculated by other authors. The uncertainties in the calculated thermodynamic functions were also estimated from the series of the equations of state.

3.1 THERMODYNAMIC FUNCTIONS IN THE IDEAL GAS STATE

The thermodynamic functions for propane in the ideal gas state were calculated in a number of studies. Chao et al. [25] was one of the most reliable sources of information. They calculated the thermodynamic functions of propane using the "rigid rotator–harmonic oscillator" approximation, compared the calculated specific heat with the experimental data and the published tabulated data, and showed that a good agreement exists between their calculated data and the experimental values. Goodwin and Haynes [44] used the results of [25] for calculating the tables of the thermodynamic functions of propane.

Similar to [44], the authors of this book adopted from Chao et al. [25] the values of the thermodynamic functions for propane in the ideal gas state. For convenience of computer calculations, the values of the specific c_p^0 were approximated by a polynomial

$$c_p^0 = \sum_{i=0}^{9} a_i \left(\frac{T}{100} \right)^i \tag{3.1}$$

If specific heat c_p^0 is expressed in kJ/(kg·K), then the polynomial coefficients (3.1) have the following values:

$a_0 = \ \ \ 0.8929236 \cdot 10^0$ $\qquad\qquad$ $a_5 = -0.3613520 \cdot 10^0$

$a_1 = -0.1034171 \cdot 10^1$ $\qquad\qquad$ $a_6 = \ \ \ 0.6889628 \cdot 10^{-1}$

$a_2 = \ \ \ 0.2457288 \cdot 10^1$ $\qquad\qquad$ $a_7 = -0.7903528 \cdot 10^{-2}$

$a_3 = -0.2242043 \cdot 10^1$ $\qquad\qquad$ $a_8 = \ \ \ 0.5008473 \cdot 10^{-3}$

$a_4 = \ \ \ 0.1162396 \cdot 10^1$ $\qquad\qquad$ $a_9 = -0.1346702 \cdot 10^{-4}$

For the range from 50 to 800 K, the standard deviation of the approximation amounts to 0.002%; the maximum deviation, 0.07%. Exclude the value of c_p^0 at $T = 300$ K, which was in poor agreement with the rest of the data, and the deviation decreases to 0.01% and 0.03%, respectively. The calculated values of the thermodynamic functions for propane in the ideal gas state are compared with the data of Chao et al. [25] in Table 3.1.

According to [25], the $c_p^0(T)$ functions has a shape that is hard to approximate by a single expression. The $c_p^0(T)$ function consists of two segments of different shape at low and high temperatures. Goodwin and Haynes [44] could not express this function analytically with an acceptable precision below 1500 K, and found only the equation for the enthalpy. The c_p^0 values were obtained in [44] by differentiating the enthalpy equation. This caused a decrease in the accuracy and a deviation from [25] of up to 0.4%. Since practical applications do not require tabulated data above 700 K, the present authors approximated the values of c_p^0 only for the range 50–800 K. This restriction permitted to achieve a high precision in the analytical description by means of a regular polynomial.

The enthalpy and entropy values for the ideal gas state were calculated for the present tables from the following expressions

$$h_0 = \int_{T_0}^{T} c_p^0 \, dT + h_{00} + h_0^0 \tag{3.2}$$

Table 3.1 Thermodynamic Functions of Propane for the Ideal Gas State*

T, K	c_p^0, kJ/(kg·K)		$H^0-H_0^0$, kJ/kg		s_0, kJ/(kg·K)	
	[25]	Calcula-tion[†]	[25]	Calcula-tion[†]	[25]	Calcula-tion[†]
50	0,7723	0,7723	37,86	37,72	4,2146	4,2132
100	0,9365	0,9365	80,18	80,18	4,7953	4,7953
150	1,1063	1,1063	131,41	131,33	5,2081	5,2078
200	1,2714	1,2715	190,81	190,72	5,5487	5,5484
273,15	1,5589	1,5585	293,94	293,84	5,9861	5,9854
298,15	1,6690	1,6694	334,27	334,18	6,1275	6,1267
300	1,6766	1,6777	337,30	337,27	6,1370	6,1370
400	2,1320	2,1320	527,83	527,87	6,6825	6,6822
500	2,5533	2,5533	762,47	762,48	7,2044	7,2040
600	2,9186	2,9185	1036,49	1036,72	7,7025	7,7030
700	3,2355	3,2353	1344,57	1344,21	8,1769	8,1763
800	3,5097	3,5090	1682,16	1684,94	8,6276	8,6309

*In Russian decimal points are written as commas. Where original Russian tables have been incorporated into this volume, please read commas as decimal points.
[†]Results of calculations made for this book.

$$s_0 = \int_{T_0}^{T} \frac{c_p^0}{T} dT + s_{00} + s_0^0 \tag{3.3}$$

where h_{00} and s_{00} are the enthalpy and entropy at temperature T_0, h_0^0 is the enthalpy of sublimation at $T = 0$ K, and s_0^0 is a so-called origin constant (in this study $s_0^0 = 0$). The value for the enthalpy of sublimation of propane, $h_0^0 = 6238.49$ J/mol $= 141.47$ kJ/kg, was adopted from [25].

The calculations started at $T_0 = 100$ K; the enthalpy and entropy at this temperature were adopted from [25] and are given in Table 3.1.

3.2 EQUATIONS FOR CALCULATING THE THERMODYNAMIC PROPERTIES OF PROPANE

To derive an average equation of state for propane the six series of single equations were obtained from the most reliable experimental p, v, T data values of the second virial coefficient and the isochoric specific heats. Table 3.2 lists the p, v, T data, used for deriving the equations, and the uncertainties adopted in calculating the relative weights for the experimental data. The choice of the data was based on the accuracy claimed by the authors of the original results and on the agreement among the results of different investigations, which was estimated by the present authors. For better compatibility of all of the p, v, T data, the temperatures in studies [33, 36, 84] were recalculated to the IPTS-68 scale.

Table 3.2 Summary of the Experimental Data, Used to Derive the Equations of State for Propane, and the Standard Deviation $\delta\rho_{av}$ for the Average Equation of State

Year	Author	ΔT, K	Δp, MPa	Number of points	$\delta\rho$, %	$\delta\rho_{av}$, %
			Parameter range			
1949	Reamer [84]	311–511	0.1–69	306	0.15–0.20	0.21
1960	Dawson and McKetta [33]	243–348	0.05–0.2	18	0.05	0.05
1962	Dittmar et al. [36]	273–413	1.0–103	319	0.15	0.15
1971	Tomlinson [104]	278–328	1.1–14	40	0.05	0.05
1978	Warowny et al. [106]	373–423	0.3–5.4	48	0.15–0.20	0.22
1978	Teichmann [99]	323–573	2.8–61	141	0.20	0.19
1978	Ely and Kobayashi [38]	166–324	0.4–43	222	0.05	0.06
1978	Golovskii and Tsymarny [4]	88–272	1.5–61	155	0.05	0.07
1982	Thomas and Harrison [101]	258–623	0.6–40	489	0.10	0.10
1983	Haynes [47]	90–300	0.6–37	196	0.05	0.06
1984	Kratzke and Müller [67]	247–491	2.2–61	60	0.05	0.07
			Data on ρ'			
1971	Tomlinson [104]	278–313	0.6–1.4	8	0.10	0.09
1973	Rodosevich and Miller [86]	91–115	$1\cdot10^{-9}$–$1\cdot10^{-6}$	4	0.10	0.08
1976	McClune [74]	93–173	$3\cdot10^{-9}$–$3\cdot10^{-3}$	17	0.10	0.09
1977	Haynes and Miza [48]	100–289	$3\cdot10^{-8}$–0.7	16	0.05	0.04
1978	Ely and Kobayashi [38]	166–288	$1\cdot10^{-3}$–0.7	18	0.05	0.07
1978	Orrit and Laupretre [70]	87–244	$3\cdot10^{-10}$–0.3	31	0.05	0.06
1982	Thomas and Harrison [101]	258–365	0.3–3.9	19	0.10	0.13
1982	Goodwin and Haynes* [44]	85.5	$2\cdot10^{-10}$	1	0.10	0.08
			Data on ρ''			
1982	Thomas and Harrison [101]	323–369	1.7–4.2	11	0.25	0.43
1982	Goodwin and Haynes* [44]	85.5–300	$2\cdot10^{-10}$–1.0	23	0.10	0.09
			Data on ρ_λ			
1982	Goodwin and Haynes* [44]	85.5–92	$1\cdot10^{-2}$–70	20	0.10	0.09

*The present authors used the results of [44] because the density data at these temperatures are either absent or unreliable.

To satisfy Maxwell's relations, the chosen data included the values of ρ' from studies [38, 48, 74, 79], the values of ρ'' from [44, 101] and the values of p_s from 31 isotherms for the temperature range from the triple point (85.47 K) to 365.15 K.

The equilibrium vapor pressures (in bar) for the corresponding p, v, T data were calculated from the equation derived by Goodwin and Haynes

$$\ln p_s = a/\tau + b + c\tau + d\tau^2 + e\tau^3 + f(1-\tau)^\varepsilon \tag{3.4}$$

where $a = -8.722780$ $b = 19.20308$
      $c = -15.61064$ $d = 12.68579$
      $e = -3.806543$ $f = 1.883215$
      $\epsilon = 1.35$

The equation (3.4) was obtained on the basis of the experimental data of [21, 62, 67, 101] for the temperature range from 166.19 to 369.75 K. The equation describes the pressure data with no more than 0.21% of error. For the temperatures from 230 K to the triple point, the calculated values of p_s from [44] are based on the experimental data of Goodwin [43] for c_s', on the specific heat of the liquid in the state of saturation.

From the triple to critical points, the values of the equilibrium vapor pressure of propane vary over a wide range (from $1.7 \cdot 10^{-10}$ MPa to 4.2 MPa); in addition, below 166 K the experimental values of p_s are not reliable. Therefore the low temperature values of p_s, calculated from the equation (3.4), cannot be highly accurate. As a result different uncertainties δp_s were assumed during calculations of the relative weights of different terms included in the minimization function to satisfy Maxwell's rule. For the temperature intervals 365–300 K, 290–250 K and 240–200 K the corresponding uncertainties δp_s were assumed to be 0.1, 0.2 and 0.3%; on approaching the triple point, the uncertainty increased up to 5%. The variable values of δp_s guaranteed that Maxwell's relation was satisfied accurately at different temperatures, the accuracy in the calculated values corresponding to the accuracy of the experimental equilibrium vapor pressures at these temperatures.

Twenty-eight values of the second virial coefficient B from [14, 35, 37, 46, 59, 84, 89) were used for deriving all of the series of equation for the temperature range of 211–609 K. For calculating the relative weights of the values of B, the uncertainty δB was assumed to be 2% for the second and third series of equations and 1% for the other series. One hundred and twenty eight values of the isochoric specific heat for liquid propane in the single-phase region and for points on the saturation curve from study [43] were also used. These values cover temperatures from 86 to 337 K and densities from 495 to 733 kg/m^3 (to the maximum pressure of 30 MPa). For calculating the relative weight of the data [43], the values of the relative uncertainty δc_v were assumed to be 3% for the first and second series, 2% for the third, fourth and sixth series and 1% for the fifth series.

At the first stage of calculations, the deviation $\delta \rho$ of the experimental values of density from the calculated ones was determined with help of an auxiliary equation of state. During later calculations a weight of zero was automatically used for the points, the deviation of which exceeded triple standard deviation from the average of the corresponding data. Since these points have not been used in deriving the series of equations of state, they are not included in Table 3.2 and are not shown on the graphs of deviations.

By regression analysis of the p, v, T data for isotherms and isochores, equations were derived with the maximum power of coefficients equal to 10 and

7 for equations in terms of ω and τ, respectively; the total number of coefficients in the initial equations of each series was 50. In deriving the subsequent series of equations, the least important coefficients were discarded, and simultaneously other coefficients were corrected according to the algorithm given in [3]. In the six series of equations, the standard deviations from approximating the experimental data were similar. For example, for the data of [104] and [38], they were 0.05–0.06%; for the data of [4] and [48], 0.06–0.08%. This allows to assume that the equations are equivalent in the accuracy of representing the experimental data. The number of equations in series varied from 8 to 15.

From the series of 72 equations of state, the obtained average had the following form

$$z = 1 + \sum_{i=1}^{r} \sum_{j=0}^{s_i} b_{ij} \frac{\omega^i}{\tau^j} \tag{3.5}$$

The coefficients of the average equation of state are given below.

$b_{10} = \ \ 0.9436738 \cdot 10^0$
$b_{11} = -0.1055440 \cdot 10^1$
$b_{12} = -0.4766330 \cdot 10^1$
$b_{13} = \ \ 0.7482938 \cdot 10^1$
$b_{14} = -0.4625821 \cdot 10^1$
$b_{15} = \ \ 0.7046490 \cdot 10^0$
$b_{16} = \ \ 0.1033962 \cdot 10^0$
$b_{17} = \ \ 0.1177091 \cdot 10^{-2}$

$b_{20} = -0.4298275 \cdot 10^1$
$b_{21} = \ \ 0.1335508 \cdot 10^2$
$b_{22} = -0.7959928 \cdot 10^1$
$b_{23} = -0.6743008 \cdot 10^1$
$b_{24} = \ \ 0.8107017 \cdot 10^1$
$b_{25} = -0.2110645 \cdot 10^1$
$b_{26} = -0.8221815 \cdot 10^{-1}$

$b_{30} = \ \ 0.1004808 \cdot 10^2$
$b_{31} = -0.3262745 \cdot 10^2$
$b_{32} = \ \ 0.2818743 \cdot 10^2$
$b_{33} = \ \ 0.5316384 \cdot 10^0$
$b_{34} = -0.5747877 \cdot 10^1$
$b_{35} = \ \ 0.8858945 \cdot 10^0$

$b_{40} = -0.1122063 \cdot 10^2$
$b_{41} = \ \ 0.3629536 \cdot 10^2$
$b_{42} = -0.3349436 \cdot 10^2$
$b_{43} = -0.2882508 \cdot 10^{-1}$
$b_{44} = \ \ 0.4737234 \cdot 10^1$
$b_{45} = -0.1006505 \cdot 10^{-1}$

$b_{50} = \ \ 0.8504364 \cdot 10^1$
$b_{51} = -0.2229182 \cdot 10^2$
$b_{52} = \ \ 0.2137842 \cdot 10^2$
$b_{53} = \ \ 0.6621292 \cdot 10^0$
$b_{54} = -0.1749976 \cdot 10^1$

$b_{60} = -0.6078228 \cdot 10^1$
$b_{61} = \ \ 0.8995876 \cdot 10^1$
$b_{62} = -0.8168041 \cdot 10^1$
$b_{63} = -0.1685770 \cdot 10^1$
$b_{64} = \ \ 0.1396997 \cdot 10^0$

$b_{70} = \ \ 0.3993340 \cdot 10^1$
$b_{71} = -0.3007530 \cdot 10^1$
$b_{72} = \ \ 0.2447105 \cdot 10^1$
$b_{73} = \ \ 0.7423182 \cdot 10^0$

$b_{80} = -0.1854627 \cdot 10^1$
$b_{81} = \ \ 0.8573689 \cdot 10^0$
$b_{82} = -0.4134646 \cdot 10^0$
$b_{83} = -0.6480302 \cdot 10^{-1}$

$b_{90} = \ \ 0.5116171 \cdot 10^0$
$b_{91} = -0.2324079 \cdot 10^0$
$b_{92} = -0.2398793 \cdot 10^{-3}$

$b_{10,0} = -0.6146529 \cdot 10^{-1}$
$b_{10,1} = \ \ 0.3957138 \cdot 10^{-1}$

During calculations the following values of the critical parameters and the gas constant were adopted: T_{crit} = 369.85 K, p_{crit} = 4.2475 MPa, ρ_{crit} = 220.49 kg/m^3, R = 188.549 J(kg·K).

Table 3.2 lists the standard deviations, $\delta\rho_{av}$, of the experimental densities, used for deriving the equation, from the calculated values. This table shows that the equation of state describes with a high accuracy the data of [4, 33, 38, 67, 101, 104] for the single-phase region and the data of [38, 44, 48, 74, 79, 86, 104] for the saturation curve; the $\delta\rho_{av}$ values for these data lie between 0.04 and 0.1%. The equation describes reasonably all other data used in its derivation.

Since new reliable experimental data on the density of liquid propane for pressures of up to 61 MPa [4, 47, 67] were used for deriving the series of equations of state, at low temperatures the accuracy of the average equation of state was improved in comparison to the equation of [7, 44].

Table 3.3 lists $\delta\rho_{av}$ for the data which were only used for comparing them with values, calculated from the average equation of state. These data usually either have a lower accuracy than those, used for deriving the equation, or belong to the critical region, where the equation in the form of (3.5) cannot accurately describe the thermodynamic functions of substances. Therefore higher values of $\delta\rho_{av}$ are observed in Table 3.3. Nevertheless, the agreement between the experimental [14, 22, 26, 49, 53, 55, 57, 65, 72, 105, 109] and the calculated densities, and between the calculated pressures [15, 99, 101], and the experimental values describing the critical region, is satisfactory. The corresponding standard deviations, $\delta\rho_{av}$ and δp_{av}, lie in the 0.07–0.37% and 0.09–0.34% ranges, respectively.

A histogram of deviations was plotted using most of the $\delta\rho$ values, given in Table 3.2 (Fig. 6). The histogram shows that the approximation for the p, v, T data is good. For 23 points, taken from [84, 99, 106] and not shown on the histogram the deviations exceed 0.5%, however, the calculated standard deviation of $\delta\rho_{av}$ = 0.14% given there included all 2162 points. The accuracy of the analytical description for the experimental data of different authors can be estimated in depth from the graphs of deviations, given in section 3.3.

Figure 7 shows the deviations of the experimental equilibrium vapor pressures [49, 62, 66, 99, 101] and data used in calculations [7, 44] from the values, calculated by the average equation of state that allows for Maxwell's relation. As can be seen from the figure, in the range from 130 K to the critical temperature, the calculated values of p_s agree with the data of [44], used for deriving the equation of state, the discrepancies being limited to less than 0.2%. Below 130 K the discrepancies increase; however, even at T = 100 K they do not exceed 1.1%. Only after a further decrease in temperature; that is, in the region of very low pressures (1.7 · 10^{-10}–2.5 · 10^{-8} MPa) and accuracies, the differences between the compared values reach 3–4%. For the temperature range from 130 K to T_{crit}, the values of p_s, calculated by the equation of A. V. Kletskii et al. [7], agree with the values, obtained by the present authors, within the limits of 0.05 to +0.5%. At T < 130 K the difference increases, reaching

Table 3.3 Standard Deviations $\delta\rho_{av}$ and δp_{av}, Used for Comparison of Experimental Data and Calculated Values

Year	Author	Parameter range $\Delta T, K$	Δp, MPa	Number of points	$\delta\rho_{av}$, %	δp_{av}, %
1934	Sage et al. [89]	294–378	0.2–21	155	1.38	
1937	Beattie et al. [14]	370–548	2.4–31	105	0.27	
		370–373	4.3–4.6	5		0.20
1940	Burgoyne [14]	243–294	0.5–6.1	41	1.28	
1940	Deshner and Brown [35]	303–570	0.1–14	232	0.87	
1949	Cherney et al. [26]	323–398	1.1–5.0	25	0.13	
1962	Dittmar et al. [36]	363–413	3.9–15	18	2.32	
1966	Huang et al. [33]	173–273	6.9–35	30	0.37	
1978	Teichmann [99]	369–388	4.3–5.7	7		0.34
1982	Thomas and Harrison [101]	365–373	3.9–4.9	302		0.09
1984	Straty and Palavra [97]	363–598	0.2–34	142	0.19*	

Data on ρ'

Year	Author	Parameter range $\Delta T, K$	Δp, MPa	Number of points	$\delta\rho_{av}$, %	δp_{av}, %
1921	Moass and Wright	195–249	0.01–0.2	13	0.27	
1926	Dana et al. [31]	273–329	0.5–1.9	12	0.49	
1934	Sage et al. [89]	294–366	0.9–4.0	19	2.77	
1937	van der Vet [105]	283–323	0.6–1.7	8	0.13	
1940	Deshner and Brown [35]	303–368	1.1–4.1	14	1.37	
1942	Carney [22]	228–333	0.1–2.1	12	0.32	
1949	Reamer et al. [84]	315–363	1.4–3.8	8	0.63	
1955	Clegg and Rowlinson [28]	323–353	1.7–3.1	5	1.62	
1963	Seeman and Urban [109]	279–299	0.6–1.0	8	0.19	
1967	Helgeson and Sage [49]	278–371	0.5–3.6	16	0.33	
1968	Klosek and McKinley [65]	89–133	$6\cdot10^{-10}$ $3\cdot10^{-5}$	9	0.07	
1969	Jensen and Kurata [55]	93–133	$3\cdot10^{-9}$ $3\cdot10^{-5}$	5	0.16	
1969	Sliwinski [94]	283–369.6	0.6–4.2	14	0.45	
1973	Kahre [57]	277–328	0.5–1.8	5	0.16	
1982	Thomas and Harrison [101]	367–369	4.0–4.2	3		0.15

Data on ρ''

Year	Author	Parameter range $\Delta T, K$	Δp, MPa	Number of points	$\delta\rho_{av}$, %	δp_{av}, %
1926	Dana et al. [31]	290–323	0.8–1.7	5	1.08	
1934	Sage et al. [89]	293–366	0.8–4.0	19	2.70	
1940	Deshner and Brown [35]	303–368	1.1–4.1	14	4.93	
1949	Reamer et al. [84]	313–363	1.4–3.8	8	1.58	
1955	Clegg and Rowlinson [28]	323–368	1.7–4.1	8	1.01	
1967	Helgeson and Sage [49]	278–361	0.5–3.6	16	0.7	
1969	Sliwinski [94]	283–369.6	0.6–4.2	15	0.62	

The values of δp_{av} are shown only for points in the critical region.

*During calculation of the standard deviation two points were discarded on the isochore 170 kg/m^3 at $T = 373$ and 383 K. For these points, the deviations in density are equal to -3.7 and -1.3%, respectively, and deviations in pressure are 0.33 and 0.29%, respectively.

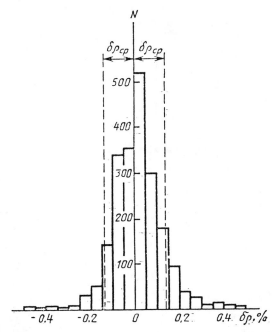

Figure 6 Histogram of deviations of the experimental density of propane from the values calculated from the average equation of state.

3% at the triple point. Figure 7 also shows that the average equation of state describes the experimental data of [49, 62, 66, 99, 101] with the uncertainty of less than 0.2% over the entire temperature range of experiments.

The results of [28, 31, 35, 77, 84, 89], which are not shown in Fig. 7, have a poorer agreement and a greater spread. The average equation of state and the equation (3.4) describe these data with the uncertainty that varies from -1.8 to $+1.1\%$. The results of [21, 24, 72, 102], which are systematically higher than the results of other investigators, deviate from the calculated values by 20–60%; this fact was also mentioned in [44].

Since the single average equation of state describes reliably the equilibrium vapor pressure (except for the region of the triple point), the thermal functions for the subcritical isotherms can be calculated for the entire single-phase region including the saturation curve. The values of p_s are calculated from the equation of state and Maxwell's relation. This guarantees consistency in the calculated properties for the single-phase region and for the state of saturation.

The values of the second virial coefficients, which were used for deriving the series of equations of state, are described by the average equation of state with a standard deviation of 0.08 cm^3/g. Figure 8 shows that the analytical description of data for the second virial coefficient is good.

The experimental data of Goodwin [43] for the isochoric specific heat at

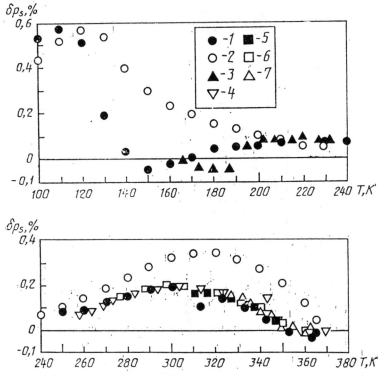

Figure 7 Deviations of the equilibrium vapor pressure, calculated by using the average equation of state, from the experimental data of: *1)* [44]; *2)* [7]; *3)* [62]; *4)* [101]; *5)* [66]; *6)* [49]; *7)* [99].

$T > 100$ K (110 points) are described by the average equation with a standard deviation of $\delta c_{v_{av}} = 2.4\%$. If one excludes four bad points (with deviation from -3.3 to $+11.5\%$) on the isochore 719 kg/m^3 belonging to the lowest temperatures (100.33–107.44 K), then the value of $\delta c_{v_{av}}$ decreases to 1.8%. Figure 9 shows the differences between the results of [43] and the values calculated by the present authors.

3.3 ESTIMATION OF THE RELIABILITY OF CALCULATED VALUES

The accuracy of the average equation of state (3.5) is primarily characterized by the precision of the analytical description of the experimental p, v, T data. This accuracy can be estimated from the standard deviation between the experimental values of density and the calculated ones, given in Tables 3.2 and 3.3. Figures 10–23 provide detailed information about the quality of analytical approximation through showing the graphs of deviations.

Figures 10–15 show that most of the experimental values of density [4, 33, 38, 47, 67, 104] deviates from the calculated values within ±0.1%. Only a small fraction of the experimental points of [4, 47, 67] deviated by 0.15–0.25%. Among them are 16 points from [47] on the isotherms 290 and 300 K. At these temperatures the equation of state describes the data of other authors [38, 67, 101] with the uncertainty of ±0.05%.

Greater deviations are typical for the data of [101, 106]. Figures 16 and 17 show that most of the experimental density values taken from these studies

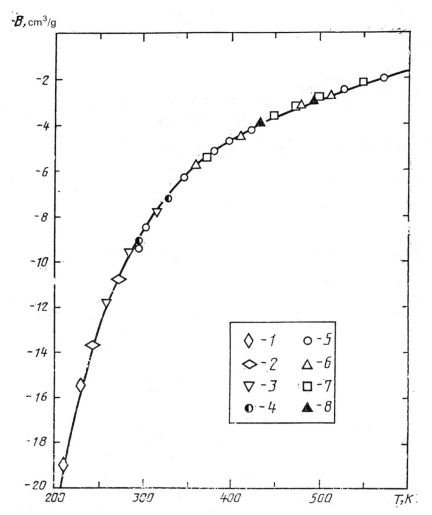

Figure 8 Deviations of the calculated values of the second virial coefficient from the experimental data of: *1*) [46]; *2*) [59]; *3*) [37]; *4*) [89]; *5*) [35]; *6*) [84]; *7*) [14]; *8*) [98].

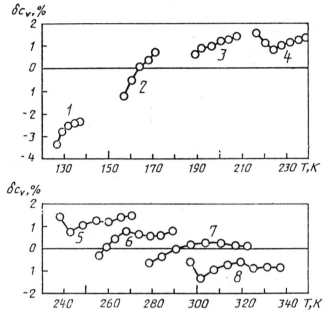

Figure 9 Deviations of the experimental isochoric specific heat, c_v, for the single-phase region [43] from the calculated values for the isochores: *1*) 0.69; *2*) 0.66; *3*) 0.63; *4*) 0.60; *5*) 0.57; *6*) 0.55; *7*) 0.52; *8*) 0.49 g/cm^3.

agree with the calculated ones with $\pm 0.2\%$. Only a few points deviated by 0.3–0.4%.

For most of the isochores the data of Teichmann [99] also agree with the calculated values within $\pm 0.2\%$ (Fig. 18). However, at the minimum density achieved in the experiment (~ 108 kg/m^3), for a number of points the deviations increase to 0.4–0.6%. For other isochores at temperatures close to the critical point the deviations amount to 0.4–2.4% for the ten points not shown in Fig. 18.

The data of Reamer et al. [84] have a greater scatter than those in the more recent investigations (Fig. 19). The deviations of these data lie within 0.4%, but for six points, not shown on Fig. 19, they are 0.6–1.8%.

The data of Dittmar et al. [36] (Fig. 20) describe a wider pressure range than the other investigations. For isotherms 273–343 K over the entire pressure range, and for all other isotherms at $p > 20$ MPa, the deviations of the experimental data from those, calculated by the present authors, do not exceed 0.2%. This confirms the accuracy of the obtained equation of state for the region of high pressures. Meanwhile, in the temperature range 363–413 K at pressures below 20 MPa, substantial discrepancies between the calculated values and the experimental values of [36] are observed. They increase with decrease in pressure and reach 1.6–5.2% for the starting points of the isotherms. This fact was noticed during the preliminary calculations. It can be explained for this range of parameters by the deviations of the data of [36] from the more reliable data of [84, 101]. Therefore 3 points, corresponding to the minimum pressures for the isotherms 363–413 K, have been discarded.

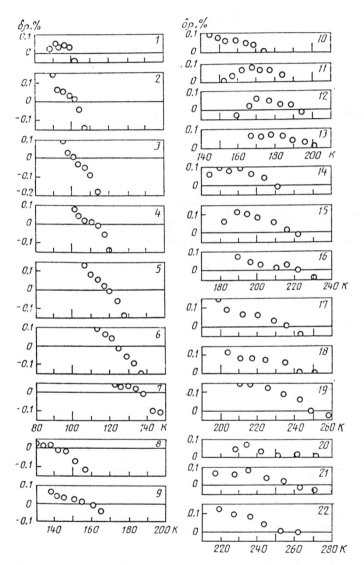

Figure 10 Deviations of the experimental density data of [4] from the calculated values for the isochores: *1*) 736.6; *2*) 732.0; *3*) 726.2; *4*) 719.7; *5*) 714.0; *6*) 706.1; *7*) 697.8; *8*) 690.0; *9*) 683.1; *10*) 675.7; *11*) 668.2; *12*) 659.5; *13*) 653.4; *14*) 646.6; *15*) 637.7; *16*) 631.0; *17*) 621.0; *18*) 615.6; *19*) 609.5; *20*) 699.2; *21*) 598.6; *22*) 597.1 kg/m^3.

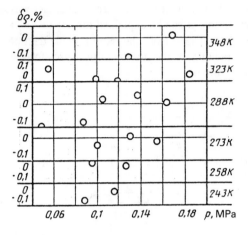

Figure 11 Deviations of the experimental density data of [33] from the calculated values.

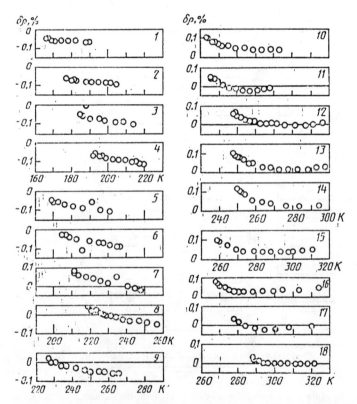

Figure 12 Deviations of the experimental density data of [38] from the calculated values for the isochores: *1*) 651; *2*) 640; *3*) 632; *4*) 624; *5*) 617; *6*) 611; *7*) 604; *8*) 596; *9*) 587; *10*) 581; *11*) 576; *12*) 563; *13*) 562; *14*) 559; *15*) 549; *16*) 536; *17*) 522; *18*) 508 kg/m^3.

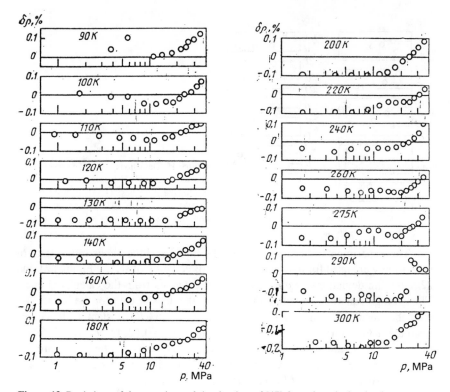

Figure 13 Deviations of the experimental density data of [47] from the calculated values.

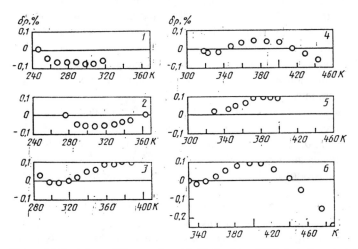

Figure 14 Deviations of the experimental density data of [67] from the calculated values for the isochores: *1*) 563; *2*) 527; *3*) 511; *4*) 473; *5*) 459; *6*) 443 kg/m³.

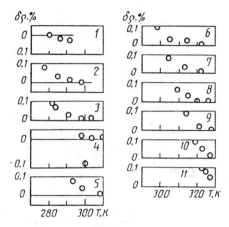

Figure 15 Deviations of the experimental density data of [104] from the calculated values for the isochores: *1*) 531; *2*) 524; *3*) 517; *4*) 509; *5*) 501; *6*) 493; *7*) 484; *8*) 477; *9*) 468; *10*) 460; *11*) 452 kg/m^3.

The experimental density values of the liquid and the vapor in the state of saturation were used for deriving the series of equation of states. Figures 21 and 22 show the deviations of these experimental values from the calculated ones. The average equation describes almost all of the density data for the liquid, ρ', with the uncertainty of $\pm 0.12\%$. Only at $T > 353$ K substantial deviations from the data of [101] are observed. They amount to 0.62, 0.79 and 0.90% for the temperatures 367.15, 368.15 and 369.15, respectively (the points are not shown in Fig. 21). The corresponding deviations in pressures amount to only -0.20, -0.15 and -0.06%, respectively.

For temperatures from the triple point to 280 K, the average equation describes the density data for the saturated vapor with the uncertainty of $+0.11\%$. For higher temperatures, the deviations from the data of [44] reach 0.51%, and from the data of [101], 0.76%. It should also be pointed out that the above deviation from the data of [101] is observed for the region in the vicinity of the critical point, where the pressure deviations do not exceed 0.15%. The deviations of the density values for the single-phase region, δp, and for the state of saturation are given in Table 3.4 for the points that are not shown in the illustrations.

The equation describes most of the data of [44] for the density of liquid propane on the freezing curve with the uncertainty of $\pm 0.09\%$ (Fig. 23). The deviations increase to 0.19 and 0.25% only for the pressures of 60 and 70 MPa.

Thus the average equation of state describes quite accurately the most reliable data on the density of liquid and gaseous propane along the curves of the phase equilibrium.

The present authors also compared the calculated values of the isochoric and isobaric specific heats and the velocity of sound at high pressures with the experimental data of other authors. Figure 9 shows that for the majority of the points in a single-phase region, the calculated values of c_v agree with [43] within $\pm 1.6\%$. For the isochore 0.69 g/cm^3, the deviations increase from

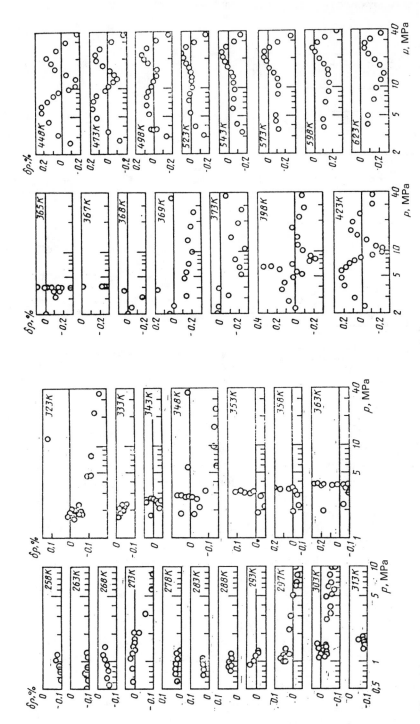

Figure 16 Deviations of the experimental density data of [101] from the calculated values.

53

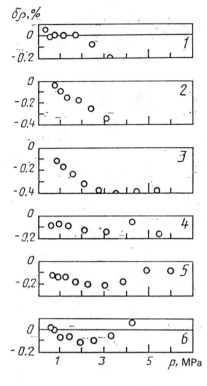

$\delta\rho, \%$

Figure 17 Deviations of the experimental density data of [106] from the calculated values for the isotherms; *1*) 373.15; *2*) 393.19; *3*) 407.50; *4*) 407.43; *5*) 422.97; *6*) 423.02 K.

−2.4 to −3.3%. At temperatures 100.3–107.4 K, for the isochore 0.72 g/cm³, which is not shown in the graphs, the deviations vary from −3.3 to 11.5%. For the curve of saturation, the deviations of the experimental values of [43] from the calculated ones lie within ±2% for the majority of the points (Fig. 24); however, for the temperature range 115–155 K, the deviations increase, reaching −5% at $T \approx$ 130 K.

For the temperature range of 116.5–422.0 K and the pressure range of 1.72–13.79 MPa, the average equation of state describes the majority of the experimental specific heat points [107] with the uncertainty of ±1% (Fig. 25). The deviations exceed 1% only for the two lowest temperatures 116.5 and 144.3 K, and for several points near the maximum of c_p. The experimental c_p values, obtained by Ernst and Busser [40] for T = 293.15–353.15 K and p = 0.05–1.37 MPa, are systematically below the calculated values. However, for most of the points, the deviations do not exceed 1.6% (Fig. 26), and only for two points near the saturation curve they reach −2.3%. The deviations are partially explained by the difference between the c_p values of the ideal gas, adopted in this book and in [40].

It is interesting to compare the calculated values of the velocity of sound in liquid and gaseous propane with the extensive experimental data. Figure 27

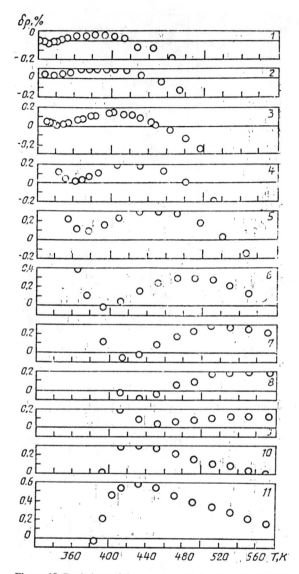

Figure 18 Deviations of the experimental density data of [99] from the calculated values for the isochores: *1*) 0.46; *2*) 0.45; *3*) 0.44; *4*) 0.40; *5*) 0.37; *6*) 0.33; *7*) 0.29; *8*) 0.26; *9*) 0.22; *10*) 0.16; *11*) 0.11 g/cm³.

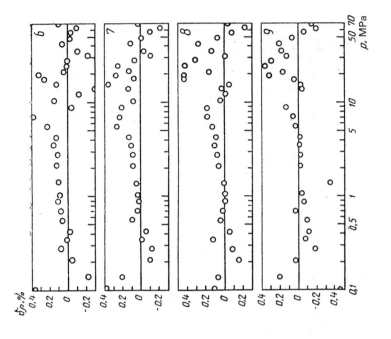

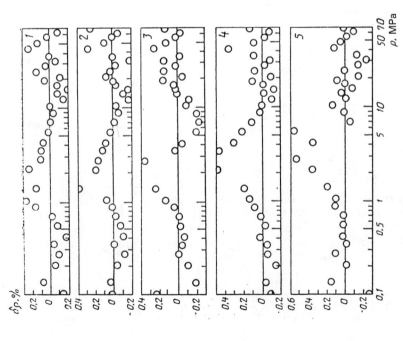

Figure 19 Deviations of the experimental density data of [84] from the calculated values for the isotherms: *1*) 310.9; *2*) 327.6; *3*) 344.3; *4*) 360.9; *5*) 377.6; *6*) 410.9; *7*) 444.3; *8*) 477.6; *9*) 510.9 K.

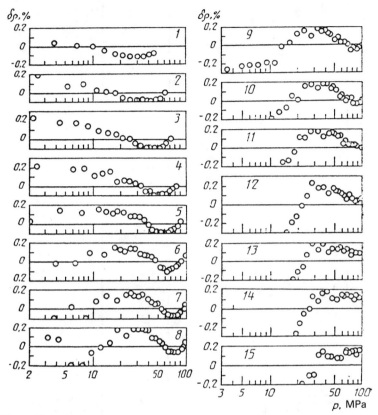

Figure 20 Deviations of the experimental density data of [36] from the calculated values for the isotherms: *1*) 273; *2*) 283; *3*) 293; *4*) 303; *5*) 313; *6*) 323; *7*) 333; *8*) 343; *9*) 353; *10*) 363; *11*) 373; *12*) 383; *13*) 393; *14*) 403; *15*) 413 K.

shows the deviations of the experimental data of Younglove [108] from those calculated by the present authors. At temperatures 160–300 K the deviations do not exceed 3%. The highest deviations are observed for the initial points of the isotherms, but they do not exceed 1.5% for the majority of the points. For the range of 90–140 K, the deviations of the values calculated by the present authors from the data of [108] increase as the temperature decreases. For the isotherms at 100 and 90 K, the deviations reach 14.4 and 36.0%. Eight points on these isotherms with deviations of greater than 10% are not shown in Fig. 27. It should be mentioned that most of the data of [108] systematically exceed the values calculated by the present authors and by the authors of [44]. For many points on the isotherms 160–300 K the deviations of the values of [44] from [108] are 2–3 times greater than the deviations of the results of the present authors. For the isotherms 90 and 100 K at low and moderate pressures the deviations are smaller and amount to 5–6%.

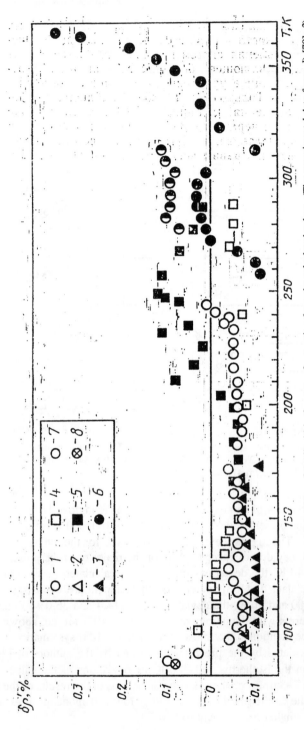

Figure 21 Deviations of the experimental densities of the liquid in the state of saturation from the calculated values. (The experimental data from: *1*) [79]; *2*) [86]; *3*) [74]; *4*) [47]; *5*) [38]; *6*) [101]; *7*) [104]; *8*) the triple point.)

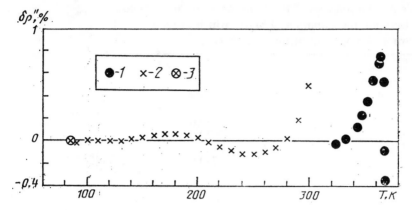

Figure 22 Deviations of the calculated density values for the saturated vapor from the experimental data of: *1*) [101]; *2*) [44]; *3*) triple point.

For the saturation curve in the temperature range 140–290 K, the values for the velocity of sound calculated by the present authors agree adequately with the data of [108]. However, at lower temperatures the deviations increase from 5.8% at $T = 130$ K to 46.9% at $T = 90$ K. For the saturated liquid at 140–230 K the less reliable data of Rao [83] are lower than the values calculated by the present authors by 9.5–12.2%. Similar deviation is observed between the calculated values of [44] and the data of [83]. Considering the substantial discrepancy between the experimental data of [108] and [83] above 120 K, the agreement between the calculated velocities of sound and the experimental data of [108] can be considered satisfactory.

Study [70] measured the velocity of sound at higher temperatures and pressures (498.15 K and 101.3 MPa). The deviations of the calculated values from most of the experimental data of [70] do not exceed 1.5% (Fig. 28). The

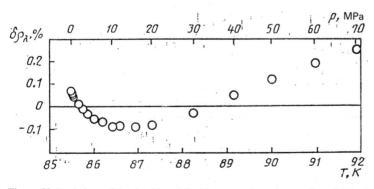

Figure 23 Deviations of the densities of liquid propane for points on the solidification curve [44] from the calculated values.

Table 3.4 Parameters of the Experimental Points Used for Deriving the Equation of State, which are not Shown on the Graphs of the Deviations

Author	p, MPa	T, K	$\delta\rho$, %
Dittmar [36]	5.88	363.15	−0.59
	7.85	363.15	−0.57
	9.12	363.15	−0.43
	8.23	373.15	−0.59
	10.6	373.15	−0.54
	12.1	373.15	−0.38
	10.7	383.15	−0.70
	13.3	383.15	−0.45
	13.1	393.15	−0.70
	16.1	393.16	−0.36
	15.1	403.16	−0.49
	17.9	413.16	−0.46
Reamer [84]	2.07	344.26	0.91
	2.76	360.93	1.03
	3.45	377.59	0.66
	1.03	510.93	1.78
	1.21	510.93	0.72
Teichmann [99]	50.2	403.29	0.46
	53.3	413.01	0.55
	59.4	432.82	0.58
	65.3	452.48	0.55
	71.4	472.83	0.47
	43.8	372.41	−1.50
	49.8	383.16	−0.45
	49.1	378.26	1.28
	56.5	387.62	0.80
	64.7	397.88	0.47
	46.0	373.37	2.36
	56.8	384.04	0.95
	42.6	369.35	1.35
	52.9	378.14	0.79
	59.4	548.96	−0.48
Data on ρ'			
Thomas and Harrison [101]	4.04	367.15	0.62
	4.12	368.15	0.79
	4.19	369.15	0.90

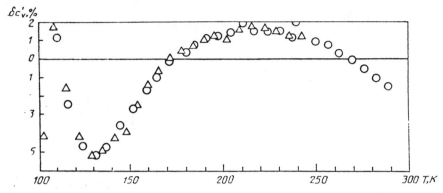

Figure 24 Deviations between two series of experimental data for the isochoric specific heat of liquid propane in the state of saturation [43] and the calculated values.

standard deviation of 200 points amounts to 1.76% and decreases to 1.4%, if to exclude one point ($T = 369.95$ K, $p = 4.36$ MPa) which is close to the critical point and has the deviation of -15.3%. For pressures above 70 MPa and for temperatures 369.95–498.15 K the deviations of the calculated values from the data of [70] increase to 2.5–4.1%; and the experimental data are systematically lower than the calculated ones. For a possible explanation, there is a disparity between the data of [70] and the p, v, T data of [36], which are the only data available for high temperatures and pressures, and which are not very accurate.

On the whole, the comparison of the experimental thermodynamic and thermal data shows that the average equation of state describes adequately the experimental data for a wide range of parameters, except for the temperature interval of 85–120 K, which is close to the triple point.

Tables 3.5–3.12 list standard deviations of the calculated values of the thermodynamic functions computed by the equation

$$\sigma_x = \sqrt{\sum_{k=1}^{N} (\bar{x} - x_k)^2 / (N-1)} \qquad (3.6)$$

This equation describes the standard deviation for an equation in the series of N equivalent equations of state. On the basis of the data listed in the tables, the expected deviations for the average values of the thermodynamic functions can be calculated by the method described in [9].

3.4 COMPARISON OF THE PREVIOUSLY PUBLISHED DATA

Tables of the thermodynamic properties of propane, calculated by Goodwin and Haynes [44] and Ershova and Kletskii [7], are the most recent and thorough

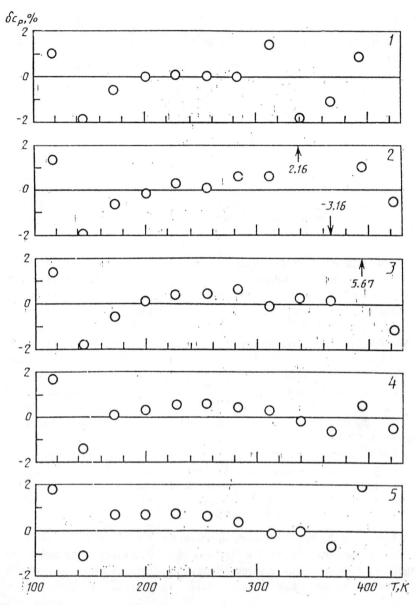

Figure 25 Deviations of the experimental data for the isobaric specific heat [107] from the calculated values. For the isobars: *1*) 1.72; *2*) 3.45; *3*) 6.89; *4*) 10.3; *5*) 13.8 MPa.

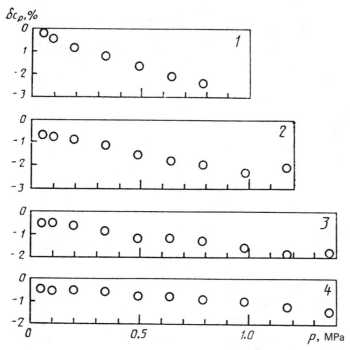

Figure 26 Deviations of the experimental isobaric specific heats [40] from the calculated values. For the isotherms: *1*) 293.15; *2*) 313.15; *3*) 333.15; *4*) 353.15 K.

among the known tables. These tables contain the data on the properties of gaseous and liquid propane for a wider range of parameters than the range used in similar previous publications (in [7, 44] the pressure varies from the saturation vapor pressure to 70 MPa, and the temperatures, from 85.5 to 700 K [44] and from 100 to 550 K [7]).

Tables of Goodwin and Haynes [44] were calculated using an equation of state, derived from the p, v, T data of [14, 26, 33, 36, 38, 47, 84, 101]. This equation of state describes the density of the above mentioned data with the standard deviations equal to 0.46, 0.27, 0.20, 0.21, 0.06, 0.07, 0.32, and 1.21%, respectively. For the subcritical and supercritical densities, the tables of Ershova and Kletskii [7] were calculated from two mutually consistent equations of state. These equations were obtained from the p, v, T data of [4, 14, 26, 33, 35, 36, 38, 84, 104] and the data on the heat of evaporation and the isochoric specific heat [6]. The equations describe the p, v, T data from the above nine sources with the standard deviations of $\delta\rho_{av}$ = 0.13, 0.40, 0.26, 0.10, 0.88, 0.49, 0.09, 0.27, and 0.10%, respectively. The authors of [7] pointed out that the deviations of the calculated heat of evaporation from the data of [6] do not exceed the experimental uncertainty (0.2–0.5%). In [7] the uncertainty in the heat capacity data was not specified.

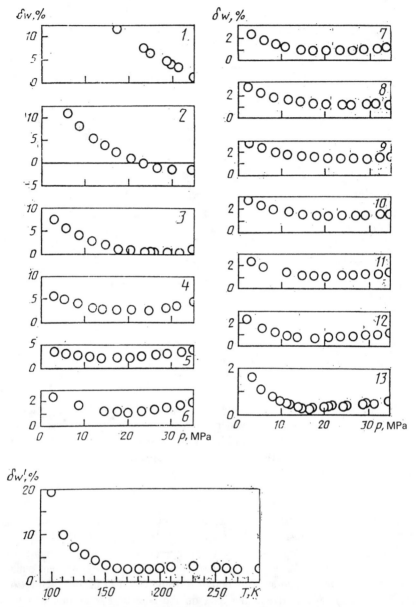

Figure 27 Deviations of the experimental data on velocity of sound in propane [108] from the calculated values for the isotherms: *1*) 90; *2*) 100; *3*) 110; *4*) 120; *5*) 140; *6*) 160; *7*) 180; *8*) 200; *9*) 220; *10*) 240; *11*) 260; *12*) 280; *13*) 300 K.

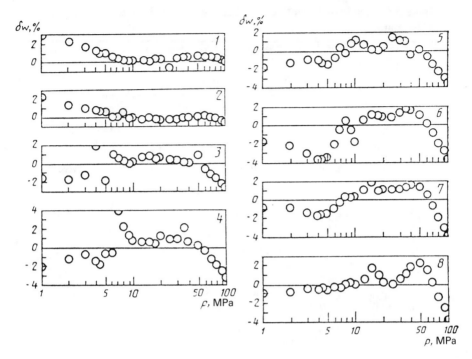

Figure 28 Deviations of the experimental data on the velocity of sound [70] from the calculated values for the isotherms: *1*) 298; *2*) 323; *3*) 369.95; *4*) 398; *5*) 423; *6) 448; 7*) 473; *8*) 498 K.

Since the tables [7, 44] summarize reliable p, v, T data which was published up to and including 1978, and since they are more accurate than the tables published previously, the present authors considered it quite sufficient to compare the thermodynamic properties of propane, calculated in this study, only with the data of [7, 44].

The density values in different tables are in reasonable agreement. Table 3.13 shows that the discrepancies for the majority of the points lie within ±0.2%, and for a number of points they do not exceed ±0.1%. Generally speaking, the data of the present authors agree better with the data of [44], particularly for many points at low temperatures, for the isotherm 350 K at $p = 2$ MPa (where the value of ρ in tables [7] is probably erroneous), for the isotherm 400 K at $p = 3$ and 5 MPa. In the critical region the deviations from the data of [44] reach −0.6%, and from the data of [7] −1.2%. For this region, a small difference in the configuration of the isobars leads to an appreciable difference in density. Above 450 K for the pressures 20–70 MPa, the discrepancy increases systematically reaching −1.3% at $T = 700$ K and $p = 70$ MPa. This fact could be expected because the data tabulated in [44]; from this were obtained for these temperatures and pressures through extrapolation.

To compare the enthalpy data and for a better compatibility between the

Table 3.5 Standard Deviations for the Calculated Values of the Density of Propane

T, K	Deviation, %, at pressure p, MPa								
	0,1	1,0	2,0	3,0	4,0	5,0	6,0	8,0	10
100	0,01	0,01	0,01	0,01	0,01	0,01	0,01	0,01	0,01
120	0,01	0,01	0,01	0,01	0,01	0,01	0,01	0,01	0,01
140	0,01	0,01	0,01	0,01	0,01	0,01	0,01	0,01	0,01
160	0,02	0,02	0,02	0,01	0,01	0,01	0,01	0,01	0,01
180	0,02	0,02	0,02	0,02	0,02	0,01	0,01	0,01	0,01
200	0,01	0,01	0,01	0,01	0,01	0,01	0,01	0,01	0,01
220	0,01	0,01	0,01	0,01	0,01	0,01	0,01	0,01	0,01
240	0,04	0,01	0,01	0,01	0,01	0,01	0,01	0,01	0,01
260	0,02	0,01	0,01	0,01	0,01	0,01	0,01	0,01	0,01
280	0,01	0,01	0,01	0,01	0,01	0,01	0,01	0,01	0,01
300	0,01	0 01	0,01	0,01	0,01	0,01	0,01	0,01	0,01
320	0,01	0,03	0,01	0,01	0,01	0,01	0,01	0,02	0,02
340	0,01	0,03	0,01	0,02	0,02	0,02	0,02	0,02	0,02
360	0,01	0,03	0,02	0,03	0,01	0,02	0,02	0,02	0,02
380	0,01	0,03	0,03	0,03	0,04	0,14	0,09	0,03	0,02
400	0,00	0,02	0,02	0,03	0,02	0,03	0,06	0,08	0,03
450	0,00	0,01	0,02	0,02	0,03	0,05	0,06	0,05	0,07
500	0,00	0,02	0,03	0,03	0,03	0,03	0,04	0,06	0,04
550	0,00	0,02	0,04	0,05	0,05	0,05	0,04	0,03	0,03
600	0,00	0,03	0,05	0,07	0,08	0,08	0,08	0,07	0,06
650	0,01	0,05	0,08	0,10	0,11	0,13	0,13	0,14	0,14
700	0,01	0,07	0,11	0,14	0,16	0,18	0,19	0,22	0,25

T, K	Deviation, %, at pressure p, MPa								
	20	30	40	50	60	70	80	90	100
100	0,01	0,01	0,02	0,03	0,06	0,09			
120	0,01	0,01	0,01	0,02	0,05	0,11	0,15		
140	0,01	0,01	0,01	0,02	0,04	0,08	0,15		
160	0,01	0,01	0,01	0,02	0,03	0,05	0,09	0,14	0,22
180	0,01	0,01	0,01	0,01	0,02	0,03	0,05	0,08	0,12
200	0,01	0,01	0,01	0,01	0,01	0,02	0,03	0,05	0,07
220	0,01	0,01	0,01	0,01	0,01	0,01	0,02	0,03	0,05
240	0,01	0,01	0,01	0,01	0,01	0,02	0,03	0,04	0,05
260	0,01	0,01	0,01	0,01	0,01	0,02	0,04	0,05	0,06
280	0,01	0,01	0,01	0,01	0,02	0,03	0,04	0,06	0,08
300	0,01	0,01	0,01	0,01	0,02	0 03	0,05	0,06	0,08
320	0,01	0,01	0,01	0,01	0,02	0,03	0,04	0,06	0,08
340	0,01	0,01	0,01	0,01	0,01	0,02	0,03	0,05	0,07
360	0,01	0,01	0,01	0,01	0,01	0,01	0,02	0,04	0,06
380	0,01	0,02	0,02	0,02	0,02	0,02	0,02	0,04	0,05
400	0,02	0,02	0,02	0,02	0,02	0,03	0,04	0,05	0,06
450	0,05	0,03	0,01	0,01	0,03	0,05	0,07	0,09	0,11
500	0,03	0,03	0,05	0,06	0,06	0,08	0,10	0,13	0,16
550	0,02	0,02	0,09	0,14	0,15	0,16	0,17	0,19	0,22
600	0,05	0,03	0,12	0,21	0,25	0,27	0,28	0,28	0,30
650	0,09	0,04	0,12	0,25	0,34	0,39	0,41	0,41	0,42
700	0,15	0,07	0,12	0,26	0,39	0,48	0,53	0,55	0,56

Table 3.6 Standard Deviations for the Calculated Values of the Enthalpy of Propane

T, K	Deviation, kJ/kg, at pressure p, MPa								
	0,1	1,0	2,0	3,0	4,0	5,0	6,0	8,0	10
100	1,15	1,15	1,15	1,15	1,15	1,16	1,16	1,16	1,16
120	0,29	0,29	0,29	0,29	0,29	0,29	0,29	0,29	0,29
140	0,27	0,27	0,27	0,27	0,27	0,27	0,27	0,27	0,27
160	0,38	0,38	0,38	0,38	0,38	0,38	0,38	0,38	0,38
180	0,27	0,27	0,27	0,27	0,27	0,27	0,27	0,27	0,26
200	0,25	0,25	0,25	0,25	0,25	0,25	0,25	0,24	0,24
220	0,26	0,26	0,25	0,25	0,25	0,25	0,25	0,25	0,25
240	0,20	0,21	0,21	0,21	0,21	0,21	0,21	0,21	0,21
260	0,09	0,18	0,18	0,18	0,18	0,18	0,18	0,18	0,18
280	0,07	0,23	0,23	0,23	0,23	0,22	0,22	0,22	0,22
300	0,05	0,27	0,27	0,27	0,27	0,27	0,27	0,27	0,26
320	0,03	0,26	0,26	0,26	0,26	0,26	0,26	0,26	0,26
340	0,02	0,16	0,22	0,19	0,20	0,20	0,20	0,20	0,21
360	0,02	0,13	0,16	0,15	0,15	0,14	0,14	0,14	0,14
380	0,02	0,15	0,20	0,17	0,15	0,28	0,14	0,14	0,15
400	0,02	0,18	0,26	0,27	0,24	0,21	0,23	0,19	0,20
450	0,02	0,17	0,29	0,37	0,42	0,45	0,45	0,38	0,37
500	0,02	0,15	0,26	0,35	0,41	0,47	0,52	0,58	0,56
550	0,03	0,22	0,38	0,49	0,58	0,65	0,71	0,81	0,87
600	0,04	0,33	0,59	0,78	0,93	1,05	1,14	1,29	1,40
650	0,05	0,44	0,80	1,09	1,32	1,51	1,67	1,91	2,09
700	0,06	0,53	0,98	1,36	1,68	1,95	2,18	2,54	2,82

T, K	Deviation, kJ/kg, at pressure p, MPa								
	20	30	40	50	60	70	80	90	100
100	1,16	1,16	1,17	1,19	1,22	1,28			
120	0,30	0,30	0,30	0,30	0,30	0,31	0,35		
140	0,28	0,28	0,28	0,29	0,29	0,31	0,33		
160	0,38	0,38	0,38	0,37	0,36	0,34	0,31	0,29	0,34
180	0,26	0,26	0,25	0,24	0,23	0,21	0,19	0,17	0,18
200	0,24	0,24	0,24	0,24	0,24	0,24	0,24	0,24	0,25
220	0,24	0,25	0,25	0,26	0,26	0,27	0,29	0,30	0,31
240	0,21	0,21	0,22	0,23	0,24	0,26	0,27	0,29	0,32
260	0,18	0,18	0,19	0,20	0,21	0,23	0,24	0,26	0,29
280	0,22	0,22	0,22	0,23	0,23	0,24	0,25	0,26	0,28
300	0,26	0,26	0,26	0,27	0,28	0,29	0,30	0,31	0,33
320	0,26	0,27	0,27	0,28	0,30	0,31	0,33	0,36	0,39
340	0,22	0,23	0,24	0,26	0,27	0,30	0,33	0,37	0,43
360	0,15	0,17	0,18	0,20	0,22	0,25	0,29	0,35	0,42
380	0,14	0,14	0,14	0,15	0,17	0,20	0,25	0,31	0,40
400	0,22	0,20	0,19	0,20	0,21	0,23	0,27	0,33	0,40
450	0,34	0,37	0,40	0,44	0,50	0,56	0,61	0,67	0,72
500	0,53	0,49	0,52	0,60	0,72	0,87	1,01	1,13	1,23
550	0,90	0,86	0,82	0,85	0,96	1,14	1,35	1,56	1,74
600	1,57	1,54	1,46	1,41	1,44	1,56	1,76	2,00	2,24
650	2,50	2,51	2,39	2,28	2,23	2,26	2,38	2,57	2,81
700	3,54	3,67	3,56	3,40	3,27	3,22	3,24	3,36	3,54

Table 3.7 Standard Deviations for the Calculated Values of the Entropy of Propane

T, K	Deviation, %, at pressure p, MPa								
	0,1	1,0	2,0	3,0	4,0	5,0	6,0	8,0	10
100	0,46	0,46	0,46	0,46	0,47	0,47	0,47	0,47	0,47
120	0,10	0,10	0,10	0,10	0,10	0,10	0,10	0,10	0,10
140	0,05	0,05	0,05	0,05	0,05	0,05	0,05	0,05	0,05
160	0,06	0,06	0,06	0,06	0,06	0,06	0,06	0,06	0,07
180	0,04	0,04	0,04	0,04	0,04	0,04	0,04	0,04	0,04
200	0,03	0,03	0,03	0,03	0,03	0,03	0,03	0,03	0,03
220	0,03	0,03	0,03	0,03	0,03	0,03	0,03	0,03	0,03
240	0,01	0,02	0,02	0,02	0,02	0,02	0,02	0,02	0,02
260	0,01	0,02	0,02	0,02	0,02	0,02	0,02	0,02	0,02
280	0,01	0,02	0,02	0,02	0,02	0,02	0,02	0,02	0,02
300	0,01	0,02	0,02	0,02	0,02	0,02	0,02	0,02	0,02
320	0,01	0,01	0,02	0,02	0,02	0,02	0,02	0,02	0,02
340	0,01	0,01	0,01	0,01	0,01	0,01	0,01	0,01	0,01
360	0,01	0,01	0,01	0,01	0,01	0,01	0,01	0,01	0,01
380	0,01	0,01	0,01	0,01	0,01	0,01	0,01	0,01	0,01
400	0,01	0,01	0,01	0,01	0,01	0,01	0,01	0,01	0,01
450	0,01	0,01	0,01	0,01	0,02	0,02	0,02	0,01	0,01
500	0,01	0,00	0,01	0,01	0,01	0,02	0,02	0,02	0,02
550	0,01	0,01	0,01	0,01	0,02	0,02	0,02	0,02	0,03
600	0,01	0,01	0,01	0,02	0,02	0,03	0,03	0,03	0,04
650	0,01	0,01	0,02	0,03	0,03	0,04	0,04	0,05	0,05
700	0,01	0,01	0,02	0,03	0,04	0,04	0,05	0,06	0,06

T, K	Deviation, %, at pressure p, MPa								
	20	30	40	50	60	70	80	90	100
100	0,48	0,48	0,49	0,50	0,52	0,55			
120	0,10	0,11	0,11	0,11	0,10	0,10	0,12		
140	0,05	0,05	0,05	0,05	0,06	0,06	0,07		
160	0,07	0,07	0,07	0,06	0,06	0,06	0,06	0,06	0,07
180	0,04	0,04	0,04	0,04	0,03	0,03	0,03	0,03	0,03
200	0,03	0,03	0,03	0,03	0,03	0,03	0,03	0,03	0,04
220	0,03	0,03	0,03	0,03	0,03	0,03	0,04	0,04	0,04
240	0,02	0,02	0,02	0,02	0,03	0,03	0,03	0,03	0,04
260	0,02	0,02	0,02	0,02	0,02	0,02	0,02	0,03	0,03
280	0,02	0,02	0,02	0,02	0,02	0,02	0,02	0,02	0,02
300	0,02	0,02	0,02	0,02	0,02	0,02	0,02	0,02	0,02
320	0,02	0,02	0,02	0,02	0,02	0,02	0,02	0,02	0,03
340	0,01	0,02	0,02	0,02	0,02	0,02	0,02	0,02	0,03
360	0,01	0,01	0,01	0,01	0,01	0,02	0,02	0,02	0,03
380	0,01	0,01	0,01	0,01	0,01	0,01	0,01	0,02	0,02
400	0,01	0,01	0,01	0,01	0,01	0,01	0,01	0,02	0,02
450	0,01	0,01	0,02	0,02	0,02	0,02	0,03	0,03	0,03
500	0,02	0,02	0,02	0,02	0,03	0,03	0,04	0,04	0,05
550	0,03	0,03	0,03	0,03	0,03	0,04	0,05	0,05	0,06
600	0,04	0,04	0,04	0,04	0,04	0,05	0,05	0,06	0,07
650	0,06	0,06	0,06	0,06	0,06	0,06	0,07	0,07	0,08
700	0,08	0,09	0,08	0,08	0,08	0,08	0,08	0,09	0,09

Table 3.8 Standard Deviations for the Calculated Values of the Isochoric Specific Heat of Propane

T, K	Deviation, %, at pressure p, MPa								
	0,1	1,0	2,0	3,0	4,0	5,0	6,0	8,0	10
140	1,64	1,61	1,58	1,56	1,54	1,52	1,51	1,49	1,48
160	0,39	0,39	0,40	0,40	0,41	0,42	0,43	0,44	0,45
180	0,82	0,83	0,84	0,85	0,86	0,87	0,87	0,89	0,90
200	0,53	0,53	0,54	0,55	0,56	0,56	0,57	0,58	0,59
220	0,25	0,25	0,26	0,26	0,27	0,28	0,29	0,31	0,32
240	0.94	0,27	0,27	0,28	0,29	0,30	0,31	0,33	0.35
260	0,30	0,29	0,28	0,27	0,27	0,28	0,29	0,32	0,34
280	0,09	0,30	0,26	0,23	0,21	0,21	0,21	0,23	0,25
300	0,06	0,41	0,34	0,28	0,23	0,20	0,17	0,15	0,16
320	0,06	0,33	0,49	0,41	0,34	0,29	0,25	0,19	0,17
340	0,05	0.34	0,33	0,57	0,46	0,39	0,33	0,27	0,23
360	0,03	0,27	0,39	0,31	0,62	0,47	0,39	0,31	0,27
380	0,02	0,19	0,32	0,36	0,31	0,54	0,40	0,30	0,27
400	0,01	0,13	0,22	0,29	0,31	0,29	0,26	0,26	0,27
450	0,01	0,09	0,14	0,17	0,19	0,22	0,24	0,27	0,30
500	0,01	0,12	0,21	0,26	0,29	0,31	0,33	0,35	0,37
550	0,02	0,13	0,24	0,31	0,37	0,41	0,44	0,49	0,52
600	0,01	0,13	0,24	0,32	0,39	0,45	0,49	0,56	0,60
650	0,01	0,12	0,23	0,31	0,38	0,44	0,49	0,58	0,64
700	0,01	0,11	0,21	0,29	0,36	0,42	0,47	0,56	0,62

T, K	Deviation, %, at pressure p, MPa								
	20	30	40	50	60	70	80	90	100
140	1,53	1,70	2,00	2,42	2,92	3,47			
160	0,44	0,33	0,27	0,44	0,70	0,98	1,23	1,41	1,52
180	0,88	0,82	0,75	0,73	0,79	0,91	1,06	1,22	1,39
200	0,57	0,51	0,48	0,52	0,62	0,76	0,91	1,05	1,18
220	0,33	0,29	0,29	0,38	0,51	0,67	0,82	0,96	1,09
240	0,40	0,39	0,39	0,43	0,52	0,64	0,77	0,91	1,06
260	0,42	0,42	0,41	0,41	0,45	0,52	0,63	0,76	0,92
280	0,35	0,36	0,34	0,31	0,31	0,35	0,44	0,56	0,72
300	0,25	0,27	0,25	0,21	0,20	0,24	0,33	0,45	0,61
320	0,20	0,21	0,20	0,20	0,23	0,29	0,39	0,50	0,64
340	0,21	0,20	0,21	0,25	0,31	0,39	0,49	0,61	0,74
360	0,23	0,23	0,25	0,30	0,38	0,47	0,58	0,69	0,82
380	0,24	0,25	0,29	0,35	0,43	0,52	0,63	0,74	0,87
400	0,24	0,25	0,31	0,38	0,46	0,55	0,65	0,76	0,88
450	0,25	0,26	0,35	0,44	0,51	0,58	0,66	0,75	0,85
500	0,36	0,35	0,42	0,50	0,56	0,61	0,66	0,71	0,78
550	0,53	0,50	0,53	0,58	0,62	0,64	0,67	0,69	0,73
600	0,67	0,65	0,64	0,66	0,68	0,69	0,69	0,70	0,71
650	0,76	0,75	0,73	0,72	0,73	0,73	0,72	0,71	0,71
700	0,79	0,80	0,78	0,77	0,76	0,75	0,74	0,72	0,71

Table 3.9 Standard Deviations for the Calculated Values of the Isobaric Specific Heat of Propane

T, K	Deviation, %, at pressure p, MPa								
	0,1	1,0	2,0	3,0	4,0	5,0	6,0	8,0	10
140	1,01	1,01	1,01	1,01	1,02	1,02	1,02	1,02	1,02
160	0,34	0,34	0,34	0,34	0,34	0,34	0,34	0,33	0,33
180	0,55	0,55	0,55	0,55	0,55	0,55	0,55	0,55	0,55
200	0,32	0,32	0,32	0,32	0,32	0,32	0,32	0,32	0,32
220	0,20	0,20	0,20	0,20	0,20	0,20	0,20	0,20	0,20
240	0,91	0,30	0,30	0,30	0,30	0,30	0,29	0,29	0,29
260	0,28	0,32	0,32	0,32	0,32	0,32	0,32	0,32	0,32
280	0,09	0,25	0,25	0,25	0,25	0,25	0,25	0,25	0,25
300	0,07	0,15	0,15	0,15	0,15	0,15	0,15	0,16	0,16
320	0,06	0,33	0,15	0,14	0,14	0,14	0,14	0,13	0,13
340	0,05	0,32	0,29	0,17	0,18	0,18	0,19	0,19	0,19
360	0,03	0,24	0,35	0,25	0,17	0,12	0,15	0,20	0,22
380	0,02	0,16	0,27	0,32	0,24	1,04	0,31	0,13	0,17
400	0,01	0,10	0,18	0,25	0,30	0,31	0,22	0,24	0,13
450	0,01	0,09	0,13	0,16	0,17	0,20	0,23	0,29	0,17
500	0,01	0,11	0,19	0,24	0,27	0,29	0,30	0,31	0,33
550	0,01	0,11	0,19	0,26	0,32	0,36	0,39	0,43	0,45
600	0,01	0,09	0,17	0,24	0,31	0,36	0,40	0,47	0,52
650	0,01	0,07	0,14	0,21	0,27	0,32	0,37	0,46	0,52
700	0,01	0,06	0,11	0,16	0,22	0,27	0,32	0,40	0,48

T, K	Deviation, %, at pressure p, MPa								
	20	30	40	50	60	70	80	90	100
140	1,04	1,04	1,03	1,01	0,96	0,89			
160	0,33	0,33	0,34	0,36	0,39	0,43	0,48	0,58	0,86
180	0,55	0,55	0,55	0,55	0,55	0,56	0,57	0,59	0,64
200	0,32	0,31	0,31	0,30	0,30	0,31	0,33	0,36	0,42
220	0,19	0,19	0,19	0,20	0,21	0,23	0,26	0,30	0,36
240	0,29	0,29	0,29	0,30	0,31	0,33	0,34	0,37	0,39
260	0,31	0,31	0,32	0,33	0,35	0,37	0,39	0,42	0,44
280	0,25	0,26	0,27	0,29	0,31	0,34	0,37	0,41	0,46
300	0,16	0,17	0,19	0,21	0,23	0,26	0,30	0,35	0,41
320	0,13	0,13	0,14	0,15	0,17	0,20	0,23	0,28	0,33
340	0,18	0,17	0,17	0,18	0,19	0,20	0,22	0,25	0,29
360	0,23	0,22	0,22	0,23	0,24	0,25	0,27	0,29	0,31
380	0,23	0,24	0,25	0,26	0,28	0,30	0,32	0,34	0,36
400	0,19	0,22	0,24	0,26	0,29	0,33	0,36	0,38	0,41
450	0,16	0,15	0,17	0,20	0,25	0,30	0,36	0,41	0,46
500	0,27	0,26	0,25	0,24	0,25	0,28	0,33	0,39	0,45
550	0,44	0,41	0,38	0,37	0,36	0,36	0,37	0,40	0,44
600	0,59	0,57	0,53	0,50	0,48	0,47	0,46	0,47	0,48
650	0.67	0,70	0,66	0,62	0,58	0,56	0,55	0,55	0,55
700	0,67	0,75	0,75	0,72	0,67	0,64	0,62	0,62	0,61

Table 3.10 Standard Deviations for the Calculated Values of the Velocity of Sound in Propane

T, K	Deviation, %, at pressure p, MPa								
	0,1	1,0	2,0	3,0	4,0	5,0	6,0	8,0	10
100	5,34	4,79	4,24	3,74	3,29	2,87	2,49	1,81	1,26
120	1,23	1,09	0,95	0,82	0,71	0,62	0,54	0,44	0,41
140	0,59	0,56	0,53	0,52	0,51	0,50	0,51	0,51	0,52
160	0,42	0,45	0,48	0,50	0,53	0,55	0,58	0,61	0,63
180	0,47	0,49	0,52	0,54	0,57	0,59	0,61	0,64	0,66
200	0,41	0,40	0,40	0,41	0,42	0,44	0,45	0,48	0,50
220	0,44	0,39	0,35	0,31	0,29	0,28	0,27	0,27	0,29
240	0,03	0,49	0,42	0,36	0,30	0,26	0,22	0,16	0,14
260	0,01	0,58	0,49	0,42	0,35	0,29	0,24	0,15	0,09
280	0,01	0,63	0,53	0,44	0,36	0,29	0,23	0,14	0,07
300	0,01	0,69	0,55	0,44	0,35	0,27	0,20	0,10	0,04
320	0,01	0,02	0,64	0,48	0,35	0,25	0,18	0,07	0,05
340	0,01	0,01	0,06	0,60	0,40	0,27	0,18	0,08	0,09
360	0,00	0,01	0,04	0,08	0,51	0,24	0,14	0,09	0,11
380	0,00	0,01	0,03	0,04	0,07	0,35	0,15	0,16	0,16
400	0,00	0,01	0,03	0,03	0,04	0,06	0,15	0,22	0,24
450	0,00	0,02	0,03	0,03	0,05	0,06	0,05	0,10	0,06
500	0,00	0,02	0,04	0,04	0,03	0,04	0,05	0,04	0,10
550	0,00	0,03	0,05	0,06	0,05	0.04	0,04	0,06	0,08
600	0,01	0,04	0,07	0,09	0,09	0,09	0,10	0,11	0,11
650	0,01	0,07	0,10	0,12	0,14	0,15	0,17	0,18	0,18
700	0,01	0,09	0,14	0,17	0,19	0,21	0,24	0,26	0,26

T, K	Deviation, %, at pressure p, MPa								
	20	30	40	50	60	70	80	90	100
100	1,02	2,00	2,68	3,42					
120	0,50	0,48	1,10	2,54					
140	0,50	0,61	1,25	2,36					
160	0,60	0,46	0,67	1,30	2,16	3,22			
180	0,64	0,48	0,37	0,64	1,13	1,71	2,34	3,00	3,67
200	0,52	0,39	0,27	0,40	0,72	1,09	1,48	1,87	2,23
220	0,32	0,24	0,17	0,32	0,58	0,86	1,13	1,40	1,65
240	0,16	0,13	0,17	0,35	0,57	0,81	1,03	1,25	1,45
260	0,09	0,10	0,20	0,38	0,59	0,81	1,03	1,25	1,46
280	0,10	0,10	0,18	0,35	0,55	0,78	1,01	1,25	1,49
300	0,14	0,11	0,12	0,27	0,46	0,69	0,93	1,19	1,45
320	0,17	0,14	0,08	0,18	0,36	0,57	0,80	1,06	1,34
340	0,18	0,15	0,08	0,12	0,27	0,46	0,67	0,92	1,19
360	0,16	0,13	0,09	0,13	0,24	0,40	0,58	0,80	1,05
380	0,13	0,10	0,10	0,16	0,27	0,40	0,55	0,73	0,94
400	0,10	0,07	0,11	0,20	0,32	0,44	0,58	0,73	0,90
450	0,12	0,18	0,21	0,28	0,42	0,58	0,73	0,87	1,00
500	0,09	0,27	0,36	0,39	0,50	0,69	0,89	1,06	1,20
550	0,09	0,30	0,50	0,54	0,59	0,76	0,99	1,22	1,41
600	0,08	0,29	0,60	0,73	0,75	0,83	1,05	1,32	1,57
650	0,10	0,26	0,63	0,88	0,96	0,97	1,10	1,36	1,65
700	0,13	0,24	0,61	0,97	1,15	1,18	1,21	1,38	1,67

Table 3.11 Standard Deviations for the Calculated Values of the Adiabatic Joule-Thomson Effect in Propane

T, K	Deviation, %, at pressure p, MPa								
	0,1	1,0	2,0	3,0	4,0	5,0	6,0	8,0	10
100	1,33	1,31	1,30	1,30	1,29	1,29	1,29	1,30	1,31
120	0,39	0,38	0,38	0,37	0,36	0,35	0,35	0,34	0,33
140	1,01	1,01	1,01	1,01	1,02	1,02	1,02	1,03	1,03
160	0,38	0,38	0,37	0,37	0,37	0,36	0,36	0,36	0,35
180	0,62	0,61	0,61	0,60	0,60	0,59	0,59	0,58	0,57
200	0,43	0,42	0,41	0,41	0,40	0,40	0,39	0,38	0.37
220	0,27	0,26	0,25	0,25	0,24	0,24	0,23	0,23	0,22
240	3,23	0,33	0,32	0,32	0,32	0,31	0,31	0,30	0,30
260	1,91	0,46	0,46	0,45	0,44	0,44	0,43	0,41	0,40
280	1,93	0,58	0,59	0,58	0,57	0,55	0,53	0,49	0,46
300	1,74	2,48	3,05	1,55	1,17	0,96	0,82	0,64	0,52
320	1,32	0,28	0,51	0,53	0,71	1,04	1,72	9,61	1,25
340	0,96	0,27	0,25	0'25	0,22	0,29	0,36	0,51	0,79
360	0,99	0,36	0,17	0,16	0,28	0,29	0,28	0,25	0,32
380	1,28	0,58	0,24	0,18	0,17	0,18	0,50	0,42	0,35
400	1,55	0,82	0,38	0,23	0,19	0,17	0,18	0,34	0,33
450	1,71	1,12	0,70	0,46	0,32	0,24	0,18	0,23	0,25
500	1,75	1,27	0,93	0,76	0,68	0,62	0,52	0,28	0,32
550	2,86	2,22	1,70	1,39	1,26	1,22	1,18	0,97	0,66
600	4,75	3,92	3,18	2,65	2,32	2,16	2,09	1,96	1,70
650	6,94	5,99	5,11	4,41	3,90	3,56	3,36	3,19	3,04
700	9,24	8,27	7,33	6,54	5,90	5,42	5,08	4,75	4,63

T, K	Deviation, %, at pressure p, MPa								
	20	30	40	50	60	70	80	90	100
100	1,49	1,79	2,22	2,77	3,46	4,38	5,71		
120	0,34	0,38	0,46	0,59	0,81	1,28			
140	1,04	1,03	1,02	1,00	0,99	1,09	1,55		
160	0,33	0,32	0,32	0,33	0,37	0,49	0,76	1,31	2,58
180	0,54	0,52	0,49	0,47	0,46	0,48	0,57	0,79	1,15
200	0,34	0,31	0,28	0,26	0,27	0,31	0,41	0,57	0,78
220	0,21	0,21	0,22	0,24	0,28	0,33	0,40	0,50	0,63
240	0,30	0,31	0,33	0,37	0,41	0,46	0,51	0,58	0,65
260	0,35	0,34	0,36	0,40	0,45	0,50	0,57	0,65	0,73
280	0,33	0,29	0,29	0,31	0,36	0,42	0,49	0,58	0,69
300	0,27	0,20	0,18	0,19	0,22	0,27	0,33	0,41	0,51
320	0,25	0,21	0,20	0,21	0,23	0,27	0,31	0,35	0,41
340	0,54	0,36	0,33	0,34	0,38	0,43	0,49	0,54	0,59
360	3,45	0,57	0,43	0,43	0,49	0,58	0,68	0,77	0,86
380	2,02	0,94	0,44	0,42	0,52	0,65	0,79	0,93	1,06
400	1,05	2,32	0,39	0,28	0,43	0,62	0,82	1,01	1,18
450	0,41	1,42	3,09	1,29	0,80	0,62	0,71	0,93	1,18
500	0,34	0,93	22,54	4,63	2,79	1,99	1,52	1,28	1,26
550	0,49	0,80	4,93	10,41	5,19	3,87	3,05	2,46	2,05
600	0,79	0,72	2,79	22,40	6,89	5,40	4,55	3,84	3,24
650	1,75	1,25	2,81	369,83	7,18	5,99	5,51	4,99	4,41
700	3,54	2,32	3,86	45,15	6,87	5,58	5,68	5,61	5,28

Table 3.12 Standard Deviations for the Calculated Values of the Adiabatic Index in Propane

T, K	Deviation, %, at pressure p, MPa								
	0,1	1,0	2,0	3,0	4,0	5,0	6,0	8,0	10
100	11,04	9,89	8,73	7,68	6,72	5,85	5,06	3,67	2,54
120	2,49	2,20	1,91	1,65	1,43	1,24	1,09	0,88	0,81
140	1,18	1,12	1,07	1,04	1,02	1,01	1,01	1,03	1,05
160	0,86	0,91	0,96	1,02	1,07	1,11	1,16	1,23	1,27
180	0,95	1,00	1,04	1,09	1,14	1,18	1,22	1,28	1,32
200	0,82	0,81	0,81	0,83	0,85	0,88	0,91	0,96	1,00
220	0,88	0,78	0,70	0,63	0,59	0,56	0,55	0,55	0,58
240	0,05	0,99	0,84	0,72	0,61	0,52	0,44	0,33	0,28
260	0,03	1,17	1,00	0,84	0,70	0,58	0,48	0,31	0,19
280	0,02	1,28	1,07	0,88	0,72	0,58	0,47	0,28	0,14
300	0,01	1,40	1,12	0,88	0,69	0,53	0,39	0,19	0,08
320	0,01	0,04	1,29	0,95	0,69	0,50	0,34	0,12	0,10
340	0,01	0,04	0,13	1,21	0,79	0,52	0,34	0,17	0,19
360	0,01	0,03	0,09	0,14	1,02	0,47	0,27	0,19	0,24
380	0,01	0,03	0,08	0,09	0,13	0,69	0,31	0,34	0,32
400	0,01	0,03	0,06	0,07	0,09	0,13	0,30	0,49	0,49
450	0,01	0,03	0,05	0,06	0,07	0,09	0,07	0,18	0,15
500	0,01	0,03	0,05	0,06	0,05	0,07	0,09	0,11	0,22
550	0,01	0,04	0,07	0,08	0,07	0,07	0,09	0,13	0,17
600	0,01	0,06	0,09	0,11	0,12	0,13	0,14	0,17	0,17
650	0,01	0,08	0,13	0,15	0,18	0,20	0,22	0,24	0,23
700	0,01	0,11	0,16	0,20	0,23	0,27	0,30	0,33	0,31

T, K	Deviation, %, at pressure p, MPa								
	20	30	40	50	60	70	80	90	100
100	2,05	3,96	5,26	6,66					
120	1,00	0,96	2,21	5,12					
140	1,01	1,22	2,51	4,76					
160	1,20	0,92	1,34	2,60	4,34	6,47			
180	1,28	0,94	0,73	1,28	2,26	3,41	4,65	5,93	7,21
200	1,02	0,78	0,53	0,80	1,43	2,17	2,93	3,68	4,38
220	0,63	0,47	0,35	0,65	1,16	1,70	2,25	2,76	3,24
240	0,32	0,26	0,35	0,71	1,14	1,60	2,05	2,48	2,87
260	0,18	0,21	0,40	0,75	1,17	1,61	2,04	2,47	2,87
280	0,20	0,20	0,35	0,69	1,09	1,54	1,99	2,46	2,92
300	0,28	0,23	0,24	0,53	0,92	1,35	1,83	2,33	2,85
320	0,34	0,27	0,15	0,35	0,70	1,12	1,58	2,08	2,63
340	0,36	0,29	0,15	0,24	0,54	0,90	1,33	1,80	2,33
360	0,33	0,26	0,18	0,26	0,49	0,79	1,15	1,57	2,05
380	0,25	0,19	0,19	0,33	0,54	0,79	1,09	1,44	1,84
400	0,19	0,15	0,21	0,40	0,62	0,86	1,12	1,41	1,75
450	0,22	0,37	0,43	0,55	0,82	1,12	1,40	1,65	1,89
500	0,16	0,54	0,69	0,73	0,95	1,31	1,68	1,99	2,24
550	0,18	0,58	0,92	0,97	1,07	1,42	1,88	2,29	2,62
600	0,16	0,56	1,09	1,26	1,28	1,49	1,96	2,49	2,93
650	0,19	0,53	1,16	1,53	1,60	1,64	1,98	2,54	3,10
700	0,27	0,50	1,13	1,69	1,92	1,93	2,06	2,50	3,12

Table 3.13 Deviations of the Tabulated Data of Density, δp, [44] (line 1) and [7] (line 2) from the Calculated Values

T, K	Deviation δp, %, at pressure p, MPa									
	0,5	1	2	3	5	10	20	30	50	70
100	−0,08	−0,08	−0,09	−0,09	−0,10	−0,11	−0,10	−0,07	0,02	0,07
	−0,09	−0,11	−0,11	−0,13	−0,14	−0,18	−0,22	−0,22	−0,22	−0,25
120*	−0,08	−0,08	−0,08	−0,08	−0,08	−0,08	−0,06	−0,05	−0,06	−0,20
150	−0,06	−0,06	−0,06	−0,05	−0,05	−0,03	0	0,03	0,03	−0,04
	−0,06	−0,05	−0,06	−0,06	−0,07	−0,08	−0,08	−0,08	−0,14	−0,28
200	−0,04	−0,04	−0,04	−0,04	−0,04	−0,03	0	0,03	0,08	0,11
	−0,10	−0,10	−0,11	−0,12	−0,12	−0,12	−0,11	−0,08	−0,06	−0,08
250	0,03	0,03	0,02	0,02	0,01	0	0,01	0,01	0,01	0
	−0,05	−0,05	−0,06	−0,06	−0,07	−0,10	−0,10	−0,09	−0,08	−0,12
300	0,21	−0,02	−0,03	−0,04	−0,05	−0,03	0,01	0,03	0,03	0,01
	0,23	−0,01	−0,02	−0,04	−0,07	−0,08	−0,07	−0,04	−0,01	−0,03
350	0,35	0,28	0,01	0,13	−0,17	−0,16	−0,06	−0,01	0,04	0,06
	0,06	0	−2,92	0,17	−0,07	−0,13	−0,23	−0,08	0,02	0,08
400	0,31	0,27	0,01	−0,33	−0,57	−0,50	0,01	−0,02	−0,06	0
	−0,02	−0,11	−0,36	−0,74	−1,20	0,05	−0,10	−0,17	−0,17	0
450	0,28	0,24	0,03	−0,24	−0,54	−0,13	0,11	0,11	−0,09	−0,21
	−0,03	−0,10	−0,21	−0,38	−0,65	0,63	0,14	−0,13	−0,37	−0,24
500	0,26	0,22	0,07	−0,14	−0,44	0,21	0,09	0,23	−0,28	−0,48
	−0,02	−0,06	−0,05	−0,07	−0,08	−0,48	0,26	0,05	−0,48	−0,51
540*	0,26	0,24	0,12	−0,05	−0,34	0	0,35	0,29	−0,32	−0,71
550**	0	0,01	0,04	0,13	0,31	−0,12	0,37	0,18	−0,40	−0,69
620*	0,26	0,29	0,25	0,14	−0,08	−0,19	0,74	0,53	−0,34	−1,09
700*	0,27	0,34	0,38	0,33	0,16	−0,15	0,53	0,76	−0,24	−1,30

*For the data of [44].
**For the data of [7].

values of h at $p = 1$ bar and $T = 300$ K, the present authors introduced a correction to the values given in [44] ($+127.4$ kJ/kg) and in [7] ($+24.8$ kJ/kg). The comparison of the enthalpy values (Table 3.14) showed a satisfactory agreement. The deviations for the majority of the points lie within ± 2 kJ/kg, and for many points they do not exceed ± 1 kJ/kg. As in the case of density, for most of the isotherms a better agreement is observed with the results of [44]. This is especially obvious at 100, 350, and 500 K. In the critical region the deviations are of about the same magnitude as for the neighboring isotherms and isobars. Meanwhile for temperatures above 450 K and pressures 10–70 MPa, the deviations between the values calculated by the present authors and the data of [44] and [7] are visibly increasing. At the pressure of 70 MPa and temperatures of 550 K and 700 K the deviation reaches 5.9 kJ/kg for the data of [7] and 4.2 kJ/kg for the data of [44], respectively. These deviations can be explained by the fact that the equations of state are extrapolated, and by possible errors in the data of [7] for the isotherms 500 and 550 K (because the agreement with the data of [44] at $T = 500$ and 540 K is quite satisfactory).

To compare the entropy data, a correction of 1.276 kJ/(kg·K) was introduced to the data of [7] for a better compatibility at $p = 0.1$ MPa and $T = 300$

Table 3.14 Deviations of the Tabulated Enthalpy Data, Δh, [44] (line 1) and [7] (line 2) from the Calculated Values

T, K	Δh, kJ/kg at p, MPa									
	0,5	1	2	3	5	10	20	30	50	70
100	−0,7	−0,7	−0,7	−0,7	−0,7	−0,7	−0,7	−0,7	−0,8	−1,2
	−1,6	−1,6	−1,6	−1,6	−1,6	−1,6	−1,5	−1,5	−1,5	−1,7
120*	1,5	1,6	1,6	1,6	1,5	1,6	1,6	1,7	1,8	1,7
150	−0,1	−0,1	0	−0,1	0	0	0	0	0,1	0,3
	0,6	00,0	0,7	0,6	0,6	0,6	0,7	1,0	0,8	1,1
200	−0,5	−0,5	−0,4	−0,4	−0,4	−0,4	−0,4	−0,5	−0,6	−0,7
	0,2	0,2	0,3	0,2	0,3	0,3	0,2	0,2	0,2	0,1
250	1,3	1,3	1,3	1,3	1,3	1,3	1,3	1,4	1,4	1,3
	0,6	0,6	0,6	0,7	0,7	0,7	0,7	0,9	0,9	1,0
300	1,0	2,2	2,2	2,2	2,1	2,1	1,9	1,9	1,9	2,1
	−0,5	−0,1	0	−0,1	0	−0,1	−0,1	−0,1	0	0,3
350	−0,1	−0,1	0	−0,9	−0,9	−0,6	−0,7	−0,8	−1,0	−1,1
	−0,8	−1,3	−1,9	−3,7	−3,5	−3,4	−3,5	−4,6	−3,8	−3,9
400	−0,6	−0,7	−0,8	−0,7	−0,6	−0,9	−1,4	−1,1	−1,2	−1,7
	−0,8	−1,0	−1,0	−0,6	−0,2	−1,9	−1,4	−1,3	−1,9	−2,4
450	−0,7	−0,9	−0,8	−0,6	−0,5	−0,4	0,8	1,0	1,1	0,4
	−0,6	−0,6	−0,1	0,5	1,9	0,5	1,6	2,1	1,9	1,0
500	−0,8	−0,8	−0,6	−0,3	0,1	−0,5	1,6	1,8	2,0	1,2
	−0,3	0,8	0,3	1,1	2,6	5,3	4,0	4,4	5,1	4,5
540*	−0,9	−0,8	−0,5	−0,1	0,6	0,5	1,2	2,1	2,4	1,6
550**	−0,3	0	0,4	1,0	2,3	5,4	4,4	4,5	5,6	5,9
620*	−0,9	−0,8	−0,2	0,4	1,4	2,3	1,2	2,2	3,1	2,7
700*	−0,2	0,1	0,7	1,4	2,6	4,2	3,0	2,6	3,9	4,2

*For the data of [44].

**For the data of [7].

K between the data of [7] and the values, calculated by the present authors. No correction was made to the data of [44]. The entropy values calculated by the present authors agree quite reasonably with the data of [7, 14]. Table 3.15 shows that for the majority of the points the deviations do not exceed ± 0.01 kJ/(kg·K), and for many points they lie within ± 0.005 kJ/(kg·K). Greater deviations from the data of [44] are observed for the 120 K isotherm and from the data of [7] for the 100 and 350 K isotherms. When the temperatures and pressures increase (at $T \geq 500$ K and $p \geq 10$ MPa), the deviations increase. This

Table 3.15 Deviations of the Tabulated Entropy Data, Δs, [44] (line 1) and [7] (line 2) from the Calculated Values

T, K	Δs, J/(kg·K) at p, MPa									
	0,5	1	2	3	5	10	20	30	50	70
100	−4	−3	−3	−3	−3	−4	−4	−4	−5	−9
	−13	−13	−13	−13	−13	−13	−13	−12	−13	−17
120	13	13	13	13	13	13	13	13	14	14
150	1	1	0	0	1	1	1	1	2	3
	3	3	3	2	3	3	3	3	4	6
200	−1	−2	−2	−1	−2	−2	−2	−2	−3	−2
	1	1	1	1	0	0	0	0	−1	0
250	6	6	6	6	6	6	6	6	6	8
	2	2	3	3	2	3	3	3	3	3
300	5	9	9	9	9	8	8	8	8	8
	−1	0	0	0	0	−1	0	−5	10**	
350	2	2	2	0	1	1	0	0	0	−1
	−2	−4	−5	−12	−10	−10	−11	−12	−12	−12
400	0	1	0	0	1	−1	−2	−2	−2	−3
	−3	−3	−3	−2	−2	−6	−4	−5	−6	−8
450	0	1	0	0	1	1	3	3	4	2
	−2	−1	−1	0	3	0	2	3	3	0
500	0	0	1	1	1	0	4	5	6	3
	−1	−1	0	1	4	9	7	8	9	8
540	0	1	1	2	3	2	4	5	6	5
550*	−1	0	1	2	3	9	8	8	10	10
620	0	0	1	3	4	6	4	6	8	6
700	0	2	3	4	6	8	6	6	9	8

*For the data of [7].

**In tables [7] a mistake was probably made ($s = 3.063$ instead of 3.053).

Table 3.16 Deviations of the Tabulated Data of the Specific Heat, δc_v, [44] (line 1) and [7] (line 2) from the Calculated Values

T, K	δc_v, %, at pressure p, MPa									
	0,5	1	2	3	5	10	20	30	50	70
150	—1,24	—1,14	—1,00	—0,86	—0,66	—0,27	—0,04	—0,31	—1,46	—2,41
	—3,30	—3,16	—2,95	—2,74	—2,38	—1,66	—1,01	—1,00	—2,16	—3,50
200	1,01	1,04	1,18	1,26	1,37	1,61	1,60	1,30	0,50	—0,06
	—1,50	—1,43	—1,22	—1,07	—0,79	—0,14	0,43	0,42	—0,28	—1,09
250	—0,63	—0,53	—0,41	—0,28	—0,03	0.37	0,52	0,44	—0,03	—0,39
	—1,12	—0,92	—0,73	—0,53	—0,13	0,53	1,19	1,38	1,04	0,45
300	—1,57	—2,21	—1,86	—1,61	—1,17	—0,56	—0,12	—0,14	—0,28	—0,13
	—1,16	—4,96	—4,63	—4,34	—3,89	—3,26	—2,61	—2,42	—2,40	—2,39
350	—0,77	—1,22	—1,26	—1,86	—0,39	1,00	1,45	1,24	0,89	1,03
	—0,29	—0,11	0,27	—3,36	—2,43	—1,74	—1,54	—1,69	—1,83	—1,62
400	—0,28	—0,36	—0,45	—0,42	0,04	3,67	4,60	4,03	3,13	3,03
	0	0,31	0,90	1,57	3,03	2,87	2,69	2,20	1,70	1,83
450	—0,10	—0,01	0,14	0,29	0,47	0,67	1,85	1,16	—0,18	—0,58
	0,23	0,42	0,83	1,28	2,11	3,25	3,13	2,67	1,98	2,05
500	—0,08	0,04	0,24	0,40	0,64	0,81	1,15	0,60	—0,84	—1,52
	0,30	0,42	0,59	0,76	1,09	1,64	1,43	1,26	0,88	0,91
540	—0,14	0,01	0,18	0,33	0,56	0,72	0,79	0,42	—0,94	—1,72
550*	0,20	0,27	0,27	0,31	0,35	0,23	—0,28,	—0 53	—0,60	—0,48
620	0,24	0,31	0,48	0,58	0,74	0,86	0,75	0,50	—0,47	—1,25
700	0,14	0,18	0,29	0,38	0,47	0,55	0,37	0,19	—0,44	—1,06

*For the data of [7].

tendency can be explained in the same way as the corresponding tendency for the density and enthalpy. However for the above range of parameters the deviations in the entropy do not exceed 0.01 kJ/(kg·K).

Table 3.16 shows that the calculated values of the isochoric specific heat agree well with the data of [44]; the deviations of more than 2% are observed for only six points, and they do not exceed 5%. The agreement is somewhat worse for the data of [7]. The deviations of more than 2% occur for 26 points; however, the maximum deviation does not exceed 5% in this case as well.

The calculated values of the isobaric specific heat agree with the majority of the data of [44] within ±1% (Table 3.17). Only for the 150 K isotherm and for seven points on some other isotherms, are higher deviations observed; at 400 K and high pressures they reach 3%. At lower temperatures (150, 200 K), a better agreement is observed with the data of [7] than with the data of [44]. However for other isotherms, the data of [7] have worse agreement with the c_p values calculated by the present authors than do the dat of [44].

A comparison of the calculated values of the velocity of sound in gaseous and liquid propane with the data of [7, 44] shows, that for the temperature range of 300–700 K and for the pressures up to and including 50 MPa, a good agreement is observed, and the deviations for the majority of the points do not exceed 1% (Table 3.18). At lower temperatures and the pressure of 70 MPa, the deviations increase and reach a maximum at $T = 150$ K.

On the whole, the calculated properties of propane, presented in this book and in [7, 44], agree quite reasonably. For the temperature range 100–140 K the

Table 3.17 Deviations of the Tabulated Data of the Specific Heat, δc_p, [44] (line 1) and [7] (line 2) from the Calculated Values

T, K	δc_p, %, at pressure p, MPa									
	0,5	1	2	3	5	10	20	30	50	70
150	−1,72	−1,75	−1,76	−1,77	−1,78	−1,78	−1,84	−1,90	−2,06	−2,20
	−0,74	−0,79	−0,79	−0,79	−0,79	−0,84	−0,89	−1,00	−1,15	−1,30
200	0,34	0,31	0,36	0,33	0,36	0,39	0,42	0,41	0,29	0,03
	0,09	0,09	0,14	0,09	0,14	0,19	0,19	0,19	0,15	0,10
250	0,13	0,13	0,13	0,11	0,07	0,07	0,05	0,08	0,27	0,48
	0,30	0,30	0,30	0,30	0,26	0,31	0,31	0,41	0,65	0,89
300	−1,73	0,11	0,12	0,10	0,05	−0,04	−0,16	−0,29	−0,39	−0,34
	−0,77	−1,67	−1,66	−1,67	−1,74	−1,88	−2,03	−2,23	−2,42	−2,45
350	−0,76	−1,10	−0,61	−0,54	0,43	0,51	0,80	0,92	0,81	0,52
	−0,10	0,19	0,70	−1,37	−1,07	−1,03	−0,91	−0,98	−1,36	−1,82
400	−0,24	−0,28	−0,29	−0,30	−0,13	3,29	2,44	2,83	3,14	2,91
	0,05	0,32	0,77	1,02	2,32	−4,49	1,68	2,09	2,21	1,84
450	−0,04	0,01	0,13	0,21	0,37	−1,48	0,84	0,69	0,88	0,76
	0,21	0,33	0,53	0,63	0,82	4,23	1,84	2,09	2,66	2,67
500	−0,09	0	0,16	0,26	0,45	0,71	0,04	0,39	0,41	0,39
	0,23	0,31	0,34	0,26	0,14	0,96	0,83	0,81	1,36	1,81
540*	−0,14	−0,06	0,06	0,18	0,33	0,75	−0,35	0,16	0,27	0,33
550**	0,15	0,13	0,10	−0,03	−1,20	−0,56	−0,13	−0,55	−0,48	−0,20
620*	0,17	0,20	0,28	0,35	0,44	0,73	0,43	0,05	0,34	0,57
700*	0,10	0,11	0,13	0,14	0,20	0,39	0,52	−0,01	−0,03	0,29

*For the data of [44].
**For the data of [7].

Table 3.18 Deviations of the Tabulated Data of the Velocity of Sound, δw, [44] (line 1) and [7] (line 2) from the Calculated Values

T, K	δw, %, at pressure p, MPa									
	0,5	1	2	3	5	10	20	30	50	70
120*	1,18	0,95	0,50	0,06	−0,66	−2,03	−2,96	−2,11	3,96	—
150	−1,52	−1,64	−1,81	−2,05	−2,39	−2,87	−3,07	−2,35	0,83	5,86
	3,26	3,10	2,78	2,47	1,93	0,98	0,30	0,88	4,71	11,24
200	−0,36	−0,45	−0,60	−0,81	−1,17	−1,63	−2,04	−1,96	−1,30	−0,77
	2,39	2,20	1,90	1,61	1,03	0,07	−0,88	−0,94	0,16	1,86
250	1,59	1,38	1,17	0,96	0,56	−0,09	−0,42	−0,31	0,22	0,47
	5,05	2,72	2,32	2,00	1,44	0,38	−0,54	−0,69	−5,67**	1,26
300	−0,38	2,36	1,84	1,30	0,53	−0,31	−0,67	−0,38	0,17	0,46
	0	3,80	3,20	2,72	1,96	0,78	−0,18	−0,34	0,03	0,82
350	−0,31	−0,36	0,14	4,53	1,02	−0,72	−1,04	−0,74	−0,48	−0,68
	0,08	0,24	0,82	3,25	1,48	0,60	0,08	−0,21	−0,61	−0,72
400	−0,39	−0,11	0,08	0,42	−0,30	−2,47	−1,22	−0,21	−0,13	−1,14
	0,07	0,15	0,39	0,50	−0,05	−0,73	0,16	0,35	−0,55	−1,74
450	−0,44	−0,34	−0,07	0,11	0,08	0,12	−1,64	0,49	1,43	−0,15
	0,03	0,03	0,11	0,07	−0,08	−2,20	−0,16	1,00	0,58	−1,44
500	−0,38	−0,10	−0,03	0,27	−0,07	−0,91	−1,23	0,21	2,59	1,11
	−0,03	0	−0,10	−0,20	−0,39	−0,18	−0,08	0,81	1,78	−0,36
540*	−0,24	−0,31	−0,19	0,13	0,26	−1,07	−0,45	−0,02	3,00	2,26
550***	−0,03	−0,06	−0,19	−0,35	−0,58	0,43	0,27	0,15	2,38	0,92
620*	−0,48	−0,31	−0,37	−0,23	0,09	−0,49	−0,84	0,17	3,13	3,75
700*	−0,45	−0,29	−0,51	−0,24	−0,08	−0,22	−1,09	−0,02	3,12	4,76

*For the data of [44].

In tables [7] a mistake was probably made (w = 1306.5 instead of 1386.5).

***For the data of [7].

deviations of the specific heats δc_v and δc_p, and the velocity of sound from the data of [7, 44], which are not shown in Tables 3.16–3.18, increase and reach −9.9%, 4.1% and 19.9%, respectively. It should be mentioned that during calculations that use a single equation of state at low temperatures, a negligible uncertainty in small values of p_s can generate a considerable error in the calculated thermal properties of the liquid. Therefore, in the final tables, the values of δc_v, δc_p and w are not given for the lowest range of temperatures. Generally the current tables describe a greater number of thermodynamic properties than do those of [7, 44], covering a wider pressure range (up to 100 MPa) than the tables of [7, 44] (up to 70 MPa).

TABLES OF THERMODYNAMIC PROPERTIES OF PROPANE

BASIC NUMERICAL VALUES AND PRINCIPAL CONSTANTS

Molar mass	$\mu = 44.0972$
Gas constant	$R = 188.549$ J/(kg·K)
Temperature at the triple point	$T_{tr} = 85.47$ K
Pressure at the triple point	$p_{tr} = 1.6895 \cdot 10^{-10}$ MPa
Normal boiling point	$T_b = 231.07$ K
Temperature at the critical point	$T_{crit} = 369.85$ K
Density at the critical point	$\rho_{crit} = 220.49$ kg/m^3
Pressure at the critical point	$p_{crit} = 4.2475$ MPa
Heat of sublimation at 0 K	$h_0^0 = 141.47$ kJ/kg

DEFINITION OF SYMBOLS AND UNITS

T — temperature, K

p — pressure, MPa

p_s — saturation vapor pressure, MPa

ρ — density, kg/m^3

h — enthalpy, kJ/kg

s — entropy, kJ/(kg·K)

c_v — isochoric specific heat, kJ/(kg·K)

c_p — isobaric specific heat, kJ/(kg·K)

w — velocity of sound, m/s

μ — adiabatic Joule-Thomson coefficient, K/MPa

k — adiabatic index

α/α_0 — volume expansion coefficient

γ/γ_0 — thermal coefficient of pressure

c_s — specific heat on the saturation curve, kJ/(kg·K)

c_λ — specific heat on the freezing curve, kJ/(kg·K)

$(')$ — property for points on the boiling curve

$('')$ — property for points on the condensation curve

Table II.1 Thermodynamic Properties of Propane on the Boiling and Condensation Curves (for the Indicated Temperatures)*

T	p	ρ'	ρ''	h'	h''	s'	s''
100	$2,485 \cdot 10^{-8}$	718,94	$1,318 \cdot 10^{-6}$	156,5	704,1	2,185	7,665
102	4,390	716,95	2,283	159,8	706,0	2,222	7,577
104	7,573	714,95	3,862	163,3	707,8	2,257	7,492
106	$1,278 \cdot 10^{-7}$	712,93	6,392	167,0	709,7	2,292	7,412
108	2,111	710,90	$1,037 \cdot 10^{-5}$	170,6	711,7	2,326	7,335
110	3,419	708,86	1,648	174,3	713,6	2,359	7,262
112	5,435	706,82	2,574	178,0	715,5	2,393	7,192
114	8,487	704,78	3,949	181,7	717,5	2,426	7,126

*In Russian decimal points are written as commas. Where original Russian tables have been incorporated into this volume, please read commas as decimal points.

Table II.1 (*Continued*)

T	p	ρ'	ρ''	h'	h''	s'	s''
116	$1,303 \cdot 10^{-6}$	702,73	5,959	185,6	719,5	2,459	7,062
118	1,970	700,69	8,856	189,5	721,5	2,492	7,001
120	2,933	698,65	$1,296 \cdot 10^{-4}$	193,4	723,5	2,525	6,943
122	4,303	696,61	1,871	197,3	725,5	2,558	6,887
124	6,228	694,57	2,664	201,3	727,6	2,590	6,834
126	8,897	692,53	3,745	205,3	729,6	2,622	6,783
128	$1,255 \cdot 10^{-5}$	690,49	5,201	209,3	731,7	2,654	6,735
130	1,750	688,45	7,140	213,3	733,7	2,685	6,688
132	2,413	686,42	9,694	217,1	735,9	2,715	6,644
134	3,291	684,38	$1,302 \cdot 10^{-3}$	221,4	737,9	2,746	6,601
136	4,442	682,35	1,732	225,5	740,1	2,776	6,560
138	5,938	680,31	2,282	229,7	742,4	2,804	6,521
140	7,865	678,28	2,979	233,6	744,4	2,835	6,483
142	$1,032 \cdot 10^{-4}$	676,25	3,856	237,6	746,5	2,863	6,447
144	1,344	674,21	4,949	241,7	748,7	2,892	6,413
146	1,735	672,18	6,301	245,7	750,8	2,920	6,380
148	2,222	670,14	7,962	249,8	753,1	2,947	6,348
150	$2,824 \cdot 10^{-4}$	668,10	$9,987 \cdot 10^{-3}$	253,9	755,3	2,975	6,318
152	3,565	666,06	$1,244 \cdot 10^{-2}$	257,9	757,5	3,002	6,288
154	4,469	664,02	1,539	262,0	759,7	3,028	6,260
156	5,566	661,98	1,893	266,1	762,0	3,054	6,234
158	6,888	659,93	2,313	270,1	764,2	3,080	6,208
160	8,472	657,88	2,810	274,2	766,5	3,106	6,183
162	$1,036 \cdot 10^{-3}$	655,83	3,394	278,3	768,8	3,131	6,159
164	1,260	653,77	4,078	282,4	771,1	3,156	6,136
166	1,524	651,72	4,874	286,5	773,4	3,181	6,115
168	1,834	649,65	5,797	290,6	775,7	3,206	6,094
170	2,196	647,59	6,861	294,6	778,0	3,230	6,073
172	2,617	645,52	8,084	298,7	780,3	3,254	6,054
174	3,104	643,44	9,482	302,9	782,7	3,278	6,035
176	3,666	641,36	$1,107 \cdot 10^{-1}$	306,9	785,0	3,301	6,017
178	4,310	639,28	1,288	311,1	787,4	3,324	6,000
180	5,048	637,18	1,493	315,2	789,8	3,347	5,984
182	5,888	635,09	1,723	319,4	792,1	3,370	5,968
184	6,842	632,99	1,981	323,5	794,5	3,393	5,953
186	7,921	630,88	2,271	327,6	796,8	3,415	5,938
188	9,138	628,76	2,593	331,8	799,2	3,438	5,924
190	$1,051 \cdot 10^{-2}$	626,64	2,952	336,0	801,6	3,460	5,910
192	1,204	624,51	3,350	340,1	804,0	3,482	5,897
194	1,375	622,37	3,790	344,4	806,3	3,503	5,885
196	1,566	620,22	4,276	348,6	808,7	3,525	5,873
198	1,777	618,07	4,810	352,8	811,1	3,546	5,861
200	$2,012 \cdot 10^{-2}$	615,91	$5,396 \cdot 10^{-1}$	357,0	813,5	3,568	5,850
202	2,271	613,73	6,038	362,1	815,8	3,589	5,839
204	2,556	611,55	6,739	365,5	818,2	3,610	5,829
206	2,870	609,36	7,502	369,8	820,6	3,631	5,819
208	3,215	607,17	8,333	374,1	823,0	3,651	5,810
210	3,592	604,96	9,235	378,4	825,4	3,672	5,800
212	4,003	602,74	$1,021 \cdot 10^{0}$	382,7	827,8	3,692	5,792
214	4,451	600,51	1,127	387,1	830,2	3,713	5,783
216	4,939	598,27	1,241	391,4	832,5	3,733	5,775
218	5,468	596,02	1,364	395,7	834,9	3,753	5,767
220	6,041	593,75	1,496	400,1	837,3	3,773	5,760
222	6,661	591,48	1,637	404,5	839,7	3,793	5,753

Table II.1 (*Continued*)

T	p	ρ'	ρ''	h'	h''	s'	s''
224	7,329	589,19	1,789	408,9	842,0	3,812	5,746
226	8,050	586,90	1,952	413,4	844,5	3,832	5,739
228	8,824	584,59	2,126	417,8	846,8	3,851	5,733
230	9,656	582,26	2,312	422,3	849,2	3,871	5,727
232	1,055·10⁻¹	579,93	2,510	426,7	851,5	3,890	5,721
234	1,150	577,58	2,721	431,3	853,8	3,910	5,715
236	1,252	575,22	2,945	435,8	856,2	3,929	5,710
238	1,361	572,84	3,183	440,3	858,5	3,948	5,705
240	1,477	570,45	3,436	444,9	860,8	3,967	5,700
242	1,601	568,05	3,703	449,6	863,3	3,986	5,695
244	1,732	565,63	3,987	454,1	865,5	4,005	5,691
246	1,872	563,20	4,287	458,8	867,9	4,024	5,687
248	2,020	560,74	4,604	463,4	870,2	4,042	5,682
250	2,177·10⁻¹	558,28	4,938	468,1	872,5	4,061	5,679
252	2,342	555,80	5,291	472,8	874,8	4,080	5,675
254	2,518	553,29	5,663	477,6	877,1	4,098	5,671
256	2,703	550,78	6,055	482,3	879,4	4,117	5,668
258	2,898	548,24	6,467	487,1	881,7	4,135	5,665
260	3,103	545,69	6,901	491,8	883,9	4,154	5,662
262	3,319	543,11	7,357	496,7	886,2	4,172	5,659
264	3,547	540,52	7,835	501,5	888,4	4,190	5,656
266	3,785	537,90	8,338	506,4	890,7	4,208	5,653
268	4,036	535,26	8,865	511,4	893,0	4,227	5,651
270	4,299	532,60	9,418	516,3	895,2	4,245	5,648
272	4,574	529,92	9,998	521,3	897,4	4,263	5,646
274	4,863	527,21	10,61	526,2	899,6	4,281	5,644
276	5,164	524,48	11,24	531,3	901,8	4,299	5,642
278	5,480	521,72	11,91	536,3	904,0	4,317	5,640
280	5,809	518,93	12,60	541,5	906,2	4,335	5,638
282	6,161	516,12	13,33	546,6	908,4	4,353	5,636
284	6,511	513,27	14,10	551,7	910,4	4,371	5,634
286	6,885	510,39	14,89	556,9	912,5	4,389	5,633
288	7,274	507,48	15,73	562,2	914,7	4,407	5,631
290	7,680	504,54	16,60	567,5	916,8	4,425	5,630
292	8,101	501,56	17,51	572,7	918,8	4,443	5,628
294	8,540	498,54	18,46	578,1	920,8	4,461	5,627
296	8,995	495,48	19,46	583,5	922,8	4,479	5,626
298	9,469	492,38	20,50	588,9	924,8	4,497	5,624
300	9,960·10⁻¹	489,24	21,59	594,4	926,8	4,515	5,623
302	1,047·10⁰	486,05	22,72	599,9	928,7	4,533	5,622
304	1,100	482,81	23,91	605,5	930,6	4,551	5,621
306	1,155	479,51	25,16	611,1	932,5	4,569	5,619
308	1,211	476,17	26,46	616,8	934,3	4,587	5,618
310	1,270	472,76	27,82	622,5	936,1	4,605	5,617
312	1,331	469,30	29,24	628,2	937,8	4,623	5,616
314	1,394	465,77	30,73	634,1	939,5	4,641	5,614
316	1,459	462,16	32,30	639,9	941,2	4,660	5,613
318	1,526	458,49	33,93	645,9	942,8	4,678	5,612
320	1,596	454,74	35,65	651,8	944,3	4,696	5,610
322	1,668	450,90	37,45	657,9	945,8	4,715	5,609
324	1,742	446,97	39,35	664,0	947,3	4,733	5,607
326	1,819	442,95	41,34	670,2	948,6	4,752	5,606
328	1,898	438,82	43,43	676,5	949,9	4,770	5,604
330	1,979	434,58	45,63	682,8	951,1	4,789	5,602

Table II.1 (*Continued*)

T	p	ρ'	ρ''	h'	h''	s'	s''
332	2,064	430,22	47,96	689,3	952,2	4,808	5,600
334	2,151	425,73	50,42	695,8	953,2	4,827	5,597
336	2,240	421,09	53,02	702,4	954,1	4,846	5,595
338	2,332	416,29	55,77	709,1	954,9	4,865	5,592
340	2,427	411,33	58,69	716,0	955,6	4,885	5,589
342	2,525	406,17	61,81	722,9	956,1	4,904	5,586
344	2,626	400,79	65,13	730,0	956,5	4,924	5,583
346	2,730	395,17	68,69	737,2	956,7	4,944	5,579
348	2,837	389,28	72,52	744,6	956,7	4,965	5,574
350	2,948	383,07	76,66	752,2	956,4	4,986	5,569
352	3,061	376,49	81,16	760,0	955,9	5,007	5,564
354	3,178	369,47	86,08	768,0	955,0	5,029	5,557
356	3,298	361,91	91,53	776,3	953,8	5,051	5,550
358	3,422	353,68	97,62	784,9	952,1	5,075	5,541
360	3,550	344,58	104,55	794,1	949,7	5,099	5,531
362	3,682	334,29	112,63	803,8	946,5	5,125	5,519
364	3,818	322,23	122,40	814,5	942,0	5,153	5,504
365	3,888	315,21	128,23	820,4	939,1	5,169	5,494
366	3,959	307,20	134,99	826,7	935,5	5,185	5,483
367	4,031	297,71	143,14	833,9	930,9	5,204	5,469
368	4,105	285,75	153,61	842,4	924,7	5,227	5,460
369	4,180	268,48	169,01	853,8	915,0	5,257	5,423

Table II.2 Thermodynamic Properties of Propane on the Boiling and Condensation Curves (for the Indicated Temperatures)

T	c_v'	c_v''	c_p'	c_p''	c_s'	c_s''	w'	w''
100		0,748		0,937		−4,542	1712	153,6
102		0,755		0,944		−4,411	1744	155,0
104		0,762		0,951		−4,285	1766	156,4
106		0,769		0,958		−4,163	1781	157,7
108		0,776		0,965		−4,044	1788	159,1
110		0,783		0,972		−3,929	1791	160,4
112		0,790		0,979		−3,818	1790	161,7
114		0,797		0,986		−3,710	1788	163,0
116		0,804		0,993		−3,606	1783	164,3
118		0,811		1,000		−3,504	1777	165,6
120		0,818		1,006		−3,407	1771	166,9
122		0,825		1,013		−3,312	1764	168,1
124		0,831		1,020		−3,219	1756	169,4
126		0,838		1,026		−3,128	1749	170,6
128		0,845		1,033		−3,041	1741	171,8
130	1,415	0,851	2,014	1,039	2,018	−2,957	1733	173,0
132	1,415	0,858	2,019	1,046	2,023	−2,877	1724	174,2
134	1,414	0,864	2,023	1,052	2,027	−2,800	1716	175,4
136	1,412	0,870	2,025	1,058	2,029	−2,724	1707	176,6
138	1,409	0,877	2,027	1,065	2,030	−2,648	1698	177,8
140	1,407	0,883	2,029	1,071	2,031	−2,573	1688	178,9
142	1,403	0,889	2,030	1,077	2,032	−2,500	1679	180,1

Table II.2 (*Continued*)

T	c_v'	c_v''	c_p'	c_p''	c_s'	c_s''	w'	w''
144	1,400	0,895	2,030	1,083	2,032	−2,429	1669	181,2
146	1,397	0,901	2,031	1,089	2,033	−2,361	1659	182,4
148	1,394	0,907	2,032	1,095	2,033	−2,295	1649	183,5
150	1,391	0,913	2,032	1,100	2,034	−2,232	1638	184,6
152	1,388	0,919	2,033	1,106	2,035	−2,170	1627	185,7
154	1,386	0,925	2,034	1,112	2,036	−2,109	1616	186,8
156	1,383	0,930	2,035	1,118	2,037	−2,050	1605	187,9
158	1,381	0,936	2,036	1,124	2,038	−1,994	1594	189,0
160	1,380	0,942	2,038	1,129	2,039	−1,940	1582	190,1
162	1,378	0,948	2,040	1,135	2,041	−1,886	1570	191,1
164	1,377	0,953	2,042	1,141	2,043	−1,833	1558	192,2
166	1,377	0,959	2,044	1,147	2 045	−1,782	1546	193,2
168	1,376	0,965	2,047	1,152	2,048	−1,732	1533	194,3
170	1,376	0,971	2,049	1,158	2,050	−1,683	1521	195,3
172	1,376	0,977	2,052	1,164	2,053	−1,636	1508	196,3
174	1,376	0,982	2,056	1,170	2,056	−1,590	1495	197,3
176	1,377	0,988	2,059	1,176	2,059	−1,545	1482	198,2
178	1,378	0,994	2,063	1,183	2,063	−1,501	1469	199,2
180	1,379	1,000	2,067	1,189	2,067	−1,458	1456	200,2
182	1,380	1,006	2,071	1,196	2,071	−1,416	1443	201,1
184	1,382	1,013	2,075	1,202	2,075	−1,376	1430	202,0
186	1,383	1,019	2,080	1,209	2,080	−1,337	1417	202,9
188	1,385	1,025	2,085	1,216	2,085	−1,299	1403	203
190	1,387	1,032	2,090	1,223	2,090	−1,262	1390	204,/
192	1,390	1,039	2,095	1,231	2,095	−1,226	1377	205,6
194	1,392	1,046	2,101	1,238	2,101	−1,191	1364	206,4
196	1,395	1,053	2,106	1,246	2,106	−1,157	1350	207,2
198	1,397	1,060	2,112	1,254	2,112	−1,123	1337	208,0
200	1,400	1,067	2,118	1,263	2,118	−1,091	1324	208,8
202	1,403	1,075	2,125	1,271	2,124	−1,059	1311	209,6
204	1,406	1,082	2,131	1,280	2,130	−1,028	1298	210,4
206	1,410	1,090	2,138	1,289	2,137	−0,997	1284	211,1
208	1,413	1,098	2,145	1,299	2,144	−0,967	1271	211,8
210	1,417	1,106	2,152	1,308	2,151	−0,939	1258	212,5
212	1,421	1,115	2,160	1,318	2,158	−0,911	1245	213,2
214	1,424	1,123	2,167	1,329	2,165	−0,884	1232	213,8
216	1,428	1,132	2,175	1,339	2,173	−0,857	1219	214,5
218	1,433	1,141	2,183	1,350	2,181	−0,831	1206	215,1
220	1,437	1,150	2,191	1,361	2,189	−0,805	1194	215,7
222	1,442	1,159	2,199	1,373	2,197	−0,779	1181	216,2
224	1,446	1,169	2,208	1,384	2,205	−0,754	1168	216,8
226	1,451	1,178	2,217	1,396	2,214	−0,730	1155	217,3
228	1,456	1,188	2,226	1,409	2,223	−0,707	1143	217,8
230	1,461	1,198	2,235	1,422	2,232	−0,684	1130	218,3
232	1,466	1,209	2,245	1,435	2,241	−0,662	1117	218,7
234	1,471	1,219	2,254	1,448	2,250	−0,640	1105	219,1
236	1,477	1,230	2,264	1,461	2,260	−0,618	1092	219,5
238	1,482	1,240	2,274	1,475	2,269	−0,597	1080	219,9
240	1,488	1,251	2,284	1,490	2,279	−0,576	1067	220,2
242	1,494	1,262	2,295	1,504	2,289	−0,556	1055	220,6
244	1,500	1,274	2,306	1,519	2,299	−0,536	1042	220,8
246	1,506	1,285	2,317	1,534	2,310	−0,517	1030	221,1
248	1,512	1,291	2,328	1,550	2,320	−0,498	1017	221,3
250	1,519	1,300	2,340	1,565	2,331	−0,479	1005	221,5

Table II.2 (*Continued*)

T	c_v'	c_v''	c_p'	c_p''	c_s'	c_s''	w'	w''
252	1,525	1,320	2,351	1,582	2,342	—0,461	992,3	221,7
254	1,532	1,332	2,363	1,598	2,353	—0,443	979,9	221,8
256	1,538	1,344	2,376	1,615	2,365	—0,425	967,5	221,9
258	1,545	1,356	2,388	1,632	2,377	—0,408	955,1	222,0
260	1,552	1,368	2,401	1,649	2,389	—0,392	942,7	222,0
262	1,559	1,381	2,414	1,667	2,401	—0,376	930,3	222,0
264	1,567	1,393	2,428	1,685	2,414	—0,360	917,9	222,0
266	1,574	1,406	2,441	1,703	2,427	—0,345	905,5	221,9
268	1,581	1,418	2,456	1,722	2,440	—0,330	893,0	221,8
270	1,589	1,431	2,470	1,742	2,453	—0,316	880,5	221,6
272	1,596	1,444	2,485	1,761	2,466	—0,303	868,1	221,5
274	1,604	1,457	2,500	1,781	2,480	—0,290	855,6	221,2
276	1,612	1,470	2,516	1,802	2,495	—0,278	843,0	221,0
278	1,620	1,483	2,532	1,822	2,510	—0,267	830,5	220,7
280	1,628	1,496	2,549	1,844	2,525	—0,256	817,9	220,3
282	1,636	1,509	2,566	1,866	2,540	—0,245	805,3	220,0
284	1,644	1,522	2,584	1,888	2,555	—0,235	792,6	219,5
286	1,653	1,535	2,602	1,911	2,570	—0,226	779,9	219,1
288	1,661	1,549	2,621	1,935	2,586	—0,218	767,2	218,6
290	1,670	1,562	2,640	1,959	2,603	—0,211	754,4	218,0
292	1,678	1,576	2,660	1,984	2,621	—0,204	741,6	217,4
294	1,687	1,589	2,681	2,009	2,610	—0,199	728,7	216,8
296	1,696	1,602	2,703	2,036	2,658	—0,194	715,8	216,1
298	1,705	1,616	2,725	2,063	2,677	—0,190	702,8	215,4
300	1,714	1,630	2,748	2,092	2,696	—0,187	689,8	214,6
302	1,723	1,643	2,772	2,121	2,716	—0,185	676,7	213,7
304	1,732	1,657	2,798	2,152	2,737	—0,185	663,5	212,9
306	1,741	1,671	2,824	2,183	2,758	—0,185	650,3	211,9
308	1,751	1,684	2,851	2,217	2,781	—0,187	637,8	210,9
310	1,760	1,698	2,880	2,251	2,804	—0,190	623,6	209,9
312	1,770	1,712	2,911	2,288	2,828	—0,195	610,2	208,8
314	1,780	1,726	2,942	2,326	2,853	—0,202	596,7	207,6
316	1,790	1,740	2,976	2,366	2,879	—0,210	583,1	206,4
318	1,800	1,754	3,011	2,409	2,906	—0,220	569,4	205,2
320	1,810	1,768	3,049	2,455	2,934	—0,232	555,6	203,8
322	1,820	1,781	3,089	2,503	2,964	—0,246	541,7	202,5
324	1,831	1,797	3,131	2,555	2,995	—0,263	527,7	201,0
326	1,842	1,811	3,177	2,610	3,028	—0,282	513,6	199,5
328	1,853	1,826	3,226	2,670	3,063	—0,304	499,4	197,9
330	1,864	1,841	3,278	2,736	3,100	—0,329	485,1	196,3
332	1,876	1,856	3,335	2,807	3,139	—0,358	470,7	194,6
334	1,887	1,871	3,398	2,884	3,181	—0,392	456,1	192,8
336	1,899	1,886	3,466	2,970	3,227	—0,430	441,4	190,9
338	1,912	1,901	3,541	3,066	3,276	—0,473	426,5	189,0
340	1,924	1,917	3,624	3,173	3,328	—0,522	411,5	187,0
342	1,937	1,933	3,718	3,294	3,385	—0,579	396,2	184,9
344	1,951	1,950	3,824	3,432	3,447	—0,645	380,8	182,7
346	1,965	1,966	3,946	3,592	3,517	—0,722	365,2	180,4
348	1,980	1,983	4,088	3,780	3,597	—0,812	349,3	178,0
350	1,195	2,001	4,256	4,003	3,686	—0,918	333,2	175,5
352	2,011	2,019	4,460	4,275	3,791	—1,045	316,8	172,9
354	2,028	2,038	4,712	4,613	3,915	—1,203	300,1	170,2
356	2,046	2,058	5,035	5,047	4,068	—1,401	283,0	167,3
358	2,065	2,079	5,465	5,626	4,260	—1,652	265,5	164,3

Table II.2 (*Continued*)

T	c_v'	$c_v^{\bullet}$	c_p'	$c_p^{\bullet}$	c_s'	$c_s^{\bullet}$	w'	$w^{\bullet}$
360	2,086	2,101	6,072	6,438	4,514	−1,993	217,5	161,1
362	2,109	2,124	6,996	7,666	4,875	−2,175	229,0	157,8
364	2,135	2,150	8,582	9,743	5,437	−3,225	209,7	151,3
365	2,149	2,165	9,894	11,43	5,864	−3,778	199,8	152,5
366	2,165	2,180	11,93	14,01	6,418	−4,600	189,7	150,6
367	2,183	2,197	15,49	18,41	7,315	−5,792	179,3	148,8
368	2,204	2,217	23,15	27,55	9,270	−7,935	168,6	146,9
369	2,231	2,240	49,30	57,08	13,59	−12,88	157,6	145,2

Table II.3 Thermodynamic Properties of Propane on the Boiling and Condensation Curves (for the Indicated Temperatures)

T	μ'	$\mu^{\bullet}$	k'	$k^{\bullet}$	α'/α_0	$\alpha^{\bullet}/\alpha_0$	γ'/γ_0	$\gamma^{\bullet}/\gamma_0$
100	−0,63		$8,48\cdot10^{10}$	1,25	0,138	1,000	$9,05\cdot10^{9}$	1,000
102	−0,65		4,97	1,25	0,148	1,000	5,53	1,000
104	−0,65	−6659	2,94	1,25	0,146	1,000	3,17	1,000
106	−0,65	−5750	1,77	1,25	0,151	1,000	1,94	1,000
108	−0,65	−4971	1,08	1,24	0,154	1,000	1,20	1,000
110	−0,64	−4303	$6,65\cdot10^{9}$	1,24	0,158	1,000	$7,53\cdot10^{8}$	1,000
112	−0,63	−3727	4,17	1,24	0,162	1,000	4,82	1,000
114	−0,62	−3231	2,65	1,24	0,165	1,000	3,12	1,000
116	−0,62	−2804	1,71	1,23	0,169	1,000	2,06	1,000
118	−0,61	−2434	1,12	1,23	0,172	1,000	1,37	1,000
120	−0,60	−2113	$7,47\cdot10^{8}$	1,23	0,175	1,000	$9,27\cdot10^{7}$	1,000
122	−0,60	−1836	5,04	1,23	0,179	1,000	6,39	1,000
124	−0,59	−1594	3,44	1,23	0,182	1,000	4,42	1,000
126	−0,59	−1385	2,38	1,22	0,185	0,999	3,11	1,000
128	−0,59	−1202	1,67	1,22	0,189	0,999	2,22	0,999
130	−0,58	−1044	1,18	1,22	0,192	0,999	1,59	0,999
132	−0,58	−905,1	$8,65\cdot10^{7}$	1,22	0,196	0,999	1,16	0,999
134	−0,58	−784,2	4,98	1,22	0,199	0,999	$8,52\cdot10^{6}$	0,999
136	−0,58	−678,6	4,52	1,22	0,203	0,999	6,34	0,999
138	−0,58	−586,3	3,22	1,21	0,206	0,999	4,73	0,999
140	−0,57	−505,6	2,57	1,21	0,210	0,998	3,58	0,998
142	−0,57	−434,9	2,05	1,21	0,214	0,998	2,73	0,998
144	−0,57	−373,1	1,47	1,21	0,217	0,998	2,09	0,998
146	−0,57	−318,9	1,05	1,21	0,221	0,998	1,62	0,998
148	−0,57	−271,4	$8,56\cdot10^{6}$	1,21	0,225	0,998	1,26	0,998
150	−0,57	−229,8	$6,36\cdot10^{6}$	1,21	0,229	0,997	$9,95\cdot10^{5}$	0,997
152	−0,57	−193,4	4,96	1,20	0,233	0,997	7,87	0,997
154	−0,57	−161,4	3,85	1,20	0,237	0,997	6,27	0,997
156	−0,56	−133,4	3,08	1,20	0,241	0,997	5,02	0,997
158	−0,56	−108,9	2,44	1,20	0,245	0,997	4,05	0,997
160	−0,56	−87,41	1,95	1,20	0,249	0,997	3,28	0,997
162	−0,56	−68,63	1,57	1,20	0,254	0,997	2,68	0,997
164	−0,56	−52,20	1,25	1,20	0,258	0,998	2,19	0,997
166	−0,55	−37,86	1,02	1,19	0,262	0,998	1,80	0,997
168	−0,55	−25,35	$8,30\cdot10^{5}$	1,19	0,267	0,998	1,49	0,997
170	−0,55	−14,45	6,82	1,19	0,272	0,999	1,25	0,997
172	−0,55	−4,980	5,60	1,19	0,276	1,000	1,04	0,998

Table II.3 (*Continued*)

T	μ'	μ^*	k'	k^*	a'/a_0	a^*/a_0	τ'/τ_0	τ^*/τ_0
174	−0,54	+3,236	4,63	1,19	0,281	1,000	8,72·10⁴	0,998
176	−0,54	10,34	3,84	1,19	0,286	1,000	7,35	0,999
178	−0,54	16,47	3,20	1,19	0,291	1,003	6,22	0,999
180	−0,53	21,74	2,68	1,18	0,296	1,004	5,28	1,000
182	−0,53	26,24	2,25	1,18	0,301	1,005	4,51	1,001
184	−0,53	30,07	1,89	1,18	0,306	1,007	3,86	1,003
186	−0,52	33,31	1,60	1,18	0,312	1,009	3,31	1,004
188	−0,52	36,02	1,35	1,18	0,317	1,011	2,85	1,005
190	−0,52	38,27	1,15	1,18	0,323	1,014	2,47	1,007
192	−0,51	40,12	9,83·10⁴	1,18	0,328	1,017	2,14·10⁴	1,009
194	−0,51	41,61	8,41	1,17	0,334	1,020	1,86	1,011
196	−0,51	42,78	7,22	1,17	0,340	1,023	1,63	1,013
198	−0,50	43,68	6,22	1,17	0,346	1,026	1,42	1,016
200	−0,50	44,34	5,37·10⁴	1,17	0,352	1,030	1,25·10⁴	1,018
202	−0,49	44,78	4,64	1,17	0,358	1,034	1,10	1,021
204	−0,49	45,05	4,03	1,17	0,365	1,039	9,70·10³	1,024
206	−0,48	45,15	3,50	1,16	0,371	1,044	8,58	1,028
208	−0,48	45,11	3,05	1,16	0,378	1,049	7,60	1,031
210	−0,47	44,95	2,67	1,16	0,385	1,054	6,76	1,035
212	−0,47	44,68	2,33	1,16	0,392	1,060	6,02	1,039
214	−0,46	44,33	2,05	1,16	0,399	1,066	5,37	1,044
216	−0,46	43,90	1,80	1,16	0,406	1,073	4,81	1,048
218	−0,45	43,41	1,59	1,15	0,414	1,080	4,31	1,053
220	−0,44	42,86	1,40	1,15	0,421	1,087	3,87	1,058
222	−0,44	42,27	1,24	1,15	0,429	1,095	3,48	1,063
224	−0,43	41,65	1,10	1,15	0,437	1,103	3,14	1,069
226	−0,43	41,00	9,73·10³	1,15	0,445	1,112	2,84	1,074
228	−0,42	40,32	8,65	1,14	0,453	1,121	2,56	1,081
230	−0,41	39,63	7,70	1,14	0,462	1,130	2,32	1,087
232	−0,41	38,92	6,86	1,14	0,471	1,140	2,11	1,093
234	−0,40	38,21	6,13	1,14	0,480	1,151	1,92	1,100
236	−0,39	37,50	5,48	1,13	0,489	1,161	1,75	1,107
238	−0,39	36,78	4,90	1,13	0,498	1,173	1,59	1,115
240	−0,38	36,07	4,40	1,13	0,508	1,185	1,45	1,122
242	−0,37	35,37	3,95	1,13	0,517	1,197	1,33	1,130
244	−0,36	34,67	3,55	1,12	0,528	1,210	1,22	1,139
246	−0,35	33,98	3,19	1,12	0,538	1,223	1,12	1,147
248	−0,35	33,31	2,87	1,12	0,549	1,238	1,02	1,156
250	−0,34	32,64	2,59·10³	1,11	0,560	1,252	9,41·10²	1,165
252	−0,33	31,99	2,34	1,11	0,571	1,268	8,65	1,174
254	−0,32	31,36	2,11	1,11	0,583	1,284	7,97	1,184
256	−0,31	30,74	1,91	1,10	0,595	1,301	7,35	1,194
258	−0,30	30,14	1,73	1,10	0,607	1,318	6,78	1,204
260	−0,29	29,55	1,56	1,10	0,620	1,336	6,26	1,215
262	−0,28	28,98	1,42	1,09	0,633	1,355	5,79	1,226
264	−0,27	28,43	1,28	1,09	0,647	1,375	5,36	1,238
266	−0,26	27,89	1,17	1,08	0,661	1,396	4,96	1,249
268	−0,25	27,38	1,06	1,08	0,676	1,418	4,60	1,262
270	−0,24	26,88	9,61·10²	1,08	0,691	1,441	4,27	1,274
272	−0,22	26,40	8,73	1,07	0,707	1,465	3,96	1,287
274	−0,21	25,93	7,94	1,07	0,723	1,490	3,68	1,301
276	−0,20	25,49	7,22	1,06	0,740	1,516	3,42	1,315
278	−0,18	25,06	6,57	1,06	0,758	1,544	3,19	1,329
280	−0,17	24,64	5,98	1,05	0,777	1,573	2,97	1,344

Table II.3 (*Continued*)

T	μ'	μ^*	k'	k^*	a'/a_0	a^*/a_0	τ'/τ_0	τ^*/τ_0
282	−0,15	21,25	5,44	1,05	0,796	1,603	2,76	1,359
284	−0,14	23,87	4,95	1,04	0,817	1,635	2,57	1,375
286	−0,12	23,51	4,51	1,04	0,838	1,669	2,40	1,392
288	−0,10	23,16	4,11	1,03	0,861	1,705	2,24	1,409
290	−0,09	22,83	3,74	1,03	0,884	1,742	2,09	1,427
292	−0,07	22,51	3,40	1,02	0,909	1,782	1,95	'6
294	−0,05	22,21	3,10	1,02	0,935	1,824	1,82	1,465
296	−0,03	21,93	2,82	1,01	0,963	1,869	1,70	1,485
298	−0,01	21,65	2,57	1,00	0,992	1,916	1,59	1,506
300	0,02	21,40	$2,34 \cdot 10^2$	1,00	1,023	1,966	$1,49 \cdot 10^2$	1,528
302	0,04	21,15	2,13	0,99	1,056	2,019	1,39	1,551
304	0,07	20,92	1,93	0,99	1,090	2,076	1,30	1,575
306	0,09	20,70	1,76	0,98	1,127	2,137	1,22	1,600
308	0,12	20,50	1,60	0,97	1,167	2,202	1,14	1,626
310	0,15	20,31	1,44	0,96	1,209	2,272	1,07	1,653
312	0,19	20,13	1,31	0,96	1,254	2,346	1,00	1,682
314	0,22	19,96	1,19	0,95	1,302	2,427	$9,37 \cdot 10^1$	1,712
316	0,25	19,80	1,08	0,94	1,354	2,513	8,77	1,743
318	0,30	19,66	$9,74 \cdot 10^1$	0,94	1,410	2,607	8,21	1,776
320	0,34	19,52	8,80	0,93	1,471	2,709	7,68	1,811
322	0,39	19,40	7,93	0,92	1,537	2,819	7,18	1,848
324	0,43	19,29	7,15	0,91	1,608	2,939	6,72	1,887
326	0,49	19,19	6,43	0,90	1,687	3,071	6,28	1,928
328	0,55	19,10	5,77	0,90	1,773	3,215	5,87	1,971
330	0,61	19,02	5,16	0,89	1,867	3,374	5,49	2,017
332	0,68	18,95	4,62	0,88	1,972	3,550	5,12	2,066
334	0,75	18,89	4,12	0,87	2,089	3,746	4,78	2,118
336	0,84	18,84	3,66	0,86	2,220	3,966	4,46	2,173
338	0,93	18,80	3,24	0,85	2,368	4,213	4,15	2,233
340	1,03	18,77	2,87	0,85	2,535	4,494	3,86	2,296
342	1,15	18,74	2,52	0,84	2,728	4,814	3,59	2,365
344	1,27	18,73	2,21	0,83	2,951	5,185	3,33	2,438
346	1,42	18,72	1,93	0,82	3,212	5,617	3,09	2,519
348	1,78	18,72	1,44	0,80	3,523	6,129	2,86	2,606
350	1,780	18,72	14,42	0,801	3,899	6,743	$2,64 \cdot 10^1$	2,702
352	2,005	18,73	12,33	0,793	4,363	7,495	2,43	2,807
354	2,271	18,74	10,46	0,784	4,949	8,437	2,23	2,925
356	2,591	18,74	8,779	0,777	5,715	9,652	2,04	3,057
358	2,982	18,73	7,276	0,770	6,755	11,28	1,86	3,209
360	3,472	18,69	5,939	0,764	8,250	13,57	1,69	3,386
362	4,100	18,59	4,753	0,761	10,57	17,04	1,52	3,597
364	4,939	18,39	3,706	0,763	14,62	22,90	1,35	3,861
365	5,474	18,23	3,232	0,767	18,01	27,67	1,27	4,023
366	6,117	17,98	2,787	0,774	23,32	34,92	1,19	4,215
367	6,911	17,62	2,369	0,786	32,69	47,28	1,10	4,451
368	7,926	17,06	1,975	0,808	53,00	72,85	1,01	4,761
369	9,336	16,09	1,590	0,854	122,6	154,4	$8,97 \cdot 10^0$	5,231

Table II.4 Thermodynamic Properties of Propane on the Boiling and Condensation Curves (for the Indicated Pressures)

p	T	ρ'	ρ''	h'	h''	s'	s''
$0,5 \cdot 10^{-7}$	102,48	716,55	$2,5877 \cdot 10^{-6}$	160,3	706,4	2,228	6,556
$0,1 \cdot 10^{-6}$	105,06	713,94	5,0483	165,0	708,8	2,273	7,449
0,2	107,79	711,17	9,8411	170,0	711,5	2,320	7,343
0,3	109,45	709,47	$1,4537 \cdot 10^{-5}$	173,1	713,1	2,348	7,282
0,4	110,67	708,22	1,9169	175,4	714,3	2,369	7,238
0,5	111,64	707,23	2,3754	177,2	715,2	2,385	7,205
0,6	112,44	706,41	2,8301	178,7	716,0	2,399	7,177
0,8	113,73	705,09	3,7306	181,1	717,3	2,420	7,134
$0,1 \cdot 10^{-5}$	114,76	704,04	4,6216	183,1	718,3	2,437	7,101
0,2	118,08	700,65	8,9833	189,5	721,6	2,493	6,996
0,4	121,61	697,03	$1,7444 \cdot 10^{-4}$	196,5	725,1	2,551	6,898
0,6	123,80	694,80	2,5704	200,8	727,4	2,586	6,839
0,8	125,40	693,17	3,3835	204,0	729,0	2,612	6,798
$0,1 \cdot 10^{-4}$	126,67	691,87	4,1867	206,5	730,3	2,632	6,767
0,2	130,82	687,64	8,1077	214,9	734,6	2,697	6,669
0,4	135,29	683,08	$1,5680 \cdot 10^{-3}$	223,9	739,3	2,765	6,574
0,6	138,07	680,26	2,3046	229,6	742,3	2,806	6,519
0,8	140,12	678,17	3,0278	233,8	744,5	2,836	6,481
$0,1 \cdot 10^{-3}$	141,76	676,50	3,7411	237,1	746,3	2,860	6,451
0,2	147,14	671,03	7,2090	248,0	752,1	2,935	6,361
0,3	150,51	667,59	$1,0572 \cdot 10^{-2}$	254,9	755,8	2,981	6,310
0,4	153,01	665,04	1,3867	260,0	758,6	3,015	6,274
0,5	155,01	662,99	1,7111	264,0	760,9	3,041	6,247
0,6	156,70	661,27	2,0314	267,5	762,8	3,063	6,224
0,7	158,15	659,78	2,3484	270,4	764,4	3,082	6,206
0,8	159,44	658,47	2,6624	273,0	765,9	3,099	6,190
0,9	160,59	657,27	2,9739	275,4	767,2	3,113	6,176
$0,1 \cdot 10^{-2}$	161,64	656,21	$3,2832 \cdot 10^{-2}$	277,5	768,4	3,127	6,163
0,2	168,95	648,68	6,2870	292,5	776,8	3,217	6,084
0,3	173,59	643,87	9,1850	302,0	782,2	3,273	6,039
0,4	177,07	640,25	$1,2015 \cdot 10^{-1}$	309,2	786,3	3,314	6,008
0,5	179,87	637,32	1,4794	315,0	789,6	3,346	5,985
0,6	182,24	634,84	1,7532	319,9	792,4	3,373	5,966
0,7	184,30	632,67	2,0238	324,1	794,8	3,396	5,950
0,8	186,13	630,74	2,2915	327,9	797,0	3,417	5,937
0,9	187,78	628,99	2,5568	331,4	798,9	3,435	5,925
$0,1 \cdot 10^{-1}$	189,28	627,40	2,8199	334,5	800,7	3,452	5,915
0,15	195,33	620,94	4,1095	347,2	807,9	3,518	5,876
0,2	199,90	616,02	5,3668	356,8	813,4	3,567	5,850
0,25	203,62	611,98	6,6007	364,7	817,8	3,606	5,831
0,3	206,77	608,52	7,8162	371,5	821,6	3,639	5,815
0,35	209,53	605,48	9,0168	377,4	824,9	3,667	5,802
0,4	211,98	602,76	$1,0205 \cdot 10^{0}$	382,7	827,8	3,692	5,792
0,45	214,20	600,28	1,1382	387,5	830,4	3,715	5,782
0,5	216,24	598,00	1,2550	391,9	832,9	3,735	5,774
0,55	218,11	595,89	1,3710	396,0	835,1	3,754	5,767
0,6	219,86	593,91	1,4862	399,9	837,2	3,771	5,760
0,65	221,49	592,06	1,6007	403,4	839,1	3,787	5,754
0,7	223,03	590,30	1,7147	406,8	840,9	3,803	5,749
0,75	224,48	588,64	1,8281	410,0	842,6	3,817	5,744
0,8	225,86	587,05	1,9410	413,1	844,3	3,831	5,740
0,85	227,18	585,54	2,0534	416,0	845,8	3,843	5,735
0,9	228,43	584,09	2,1654	418,8	847,3	3,856	5,731

Table II.4 (*Continued*)

p	T	ρ'	ρ''	h'	h''	s'	s''
0,95	229,63	582,69	2,2770	421,5	848,7	3,867	5,728
0,10	230,78	581,35	2,3882	424,1	850,1	3,878	5,724
0,11	232,96	578,80	2,6097	429,0	852,6	3,900	5,718
0,12	234,99	576,42	2,8299	433,5	855,0	3,919	5,713
0,13	236,89	574,17	3,0490	437,9	857,3	3,937	5,708
0,14	238,68	572,03	3,2672	441,9	859,4	3,954	5,703
0,15	240,37	570,01	3,4845	445,8	861,3	3,970	5,699
0,16	241,98	568,07	3,7011	449,5	863,2	3,986	5,695
0,17	243,51	566,22	3,9170	453,0	865,0	4,000	5,692
0,18	244,98	564,44	4,1322	456,4	866,7	4,014	5,689
0,19	246,38	562,72	4,3469	459,7	868,3	4,027	5,686
0,20	247,73	561,07	4,5611	462,8	869,9	4,040	5,683
0,22	250,29	557,92	4,9881	468,8	872,8	4,064	5,678
0,24	252,68	554,91	5,4136	474,4	875,6	4,086	5,673
0,26	254,90	552,16	5,8379	479,7	878,1	4,106	5,670
0,28	257,01	549,49	6,2612	484,7	880,5	4,126	5,666
0,30	259,01	546,95	6,6837	489,5	882,8	4,144	5,663
0,32	260,91	544,52	7,1055	494,1	885,0	4,162	5,660
0,34	262,72	542,18	7,5269	498,5	887,0	4,178	5,657
0,36	264,45	539,92	7,9479	502,7	889,0	4,194	5,655
0,38	266,12	537,74	8,3687	506,8	890,9	4,209	5,653
0,40	267,71	535,64	8,7894	510,7	892,6	4,224	5,651
0,42	269,25	533,59	9,2100	514,5	894,4	4,238	5,649
0,44	270,74	531,61	9,6308	518,2	896,0	4,252	5,647
0,46	272,18	529,67	10,052	521,7	897,6	4,265	5,646
0,48	273,57	527,79	10,473	525,2	899,1	4,277	5,644
0,50	274,92	525,95	10,894	528,6	900,6	4,289	5,643
0,55	278,12	521,54	11,950	536,7	904,1	4,318	5,639
0,60	281,12	517,35	13,008	544,3	907,4	4,345	5,637
0,65	283,93	513,36	14,071	551,6	910,4	4,371	5,634
0,70	286,60	509,52	15,138	558,5	913,2	4,395	5,632
0,75	289,12	505,83	16,211	565,2	915,8	4,417	5,630
0,80	291,52	502,26	17,290	571,5	918,3	4,439	5,628
0,85	293,82	498,81	18,375	577,6	920,6	4,459	5,627
0,90	296,02	495,45	19,468	583,6	922,9	4,479	5,625
0,95	298,13	492,17	20,568	589,3	925,0	4,498	5,624
1,0	300,16	488,98	21,676	594,9	926,9	4,516	5,623
1,05	302,11	485,85	22,792	600,3	928,8	4,534	5,622
1,10	304,00	482,79	23,917	605,5	930,6	4,551	5,620
1,15	305,83	479,78	25,052	610,6	932,3	4,567	5,619
1,20	307,60	476,82	26,196	615,6	933,9	4,583	5,618
1,25	309,32	473,91	27,350	620,5	935,5	4,599	5,617
1,30	310,99	471,05	28,515	625,3	937,0	4,614	5,616
1,35	312,61	468,22	29,691	630,0	938,4	4,629	5,615
1,40	314,19	465,42	30,879	634,6	939,7	4,643	5,614
1,45	315,72	462,65	32,078	639,1	941,0	4,657	5,613
1,50	317,22	459,91	33,290	643,5	942,2	4,671	5,612
1,6	320,11	454,51	35,753	652,2	944,4	4,697	5,610
1,7	322,87	449,18	38,270	660,6	946,5	4,723	5,608
1,8	325,51	443,92	40,848	668,7	948,3	4,747	5,606
1,9	328,05	438,69	43,489	676,7	949,9	4,771	5,604
2,0	330,49	433,50	46,199	684,4	951,4	4,793	5,601
2,1	332,84	428,32	48,983	692,0	952,7	4,816	5,599
2,2	335,11	423,15	51,847	699,5	953,7	4,837	5,596
2,3	337,30	417,96	54,798	706,8	954,7	4,858	5,593

Table II.4 (*Continued*)

p	T	ρ'	ρ''	h'	h''	s'	s''
2,4	339,43	412,74	57,844	714,0	955,4	4,879	5,590
2,5	341,48	407,48	60,993	721,1	956,0	4,899	5,587
2,6	343,48	402,17	64,256	728,2	956,4	4,919	5,583
2,7	345,42	396,79	67,644	735,1	956,6	4,938	5,580
2,8	347,31	391,32	71,171	742,0	956,7	4,958	5,576
2,9	349,14	385,74	74,853	748,9	956,6	4,977	5,571
3,0	350,93	380,02	78,711	755,8	956,2	4,995	5,567
3,1	352,67	374,14	82,768	762,6	955,7	5,014	5,562
3,2	354,37	368,07	87,055	769,5	954,9	5,033	5,556
3,3	356,02	361,77	91,611	776,4	953,8	5,052	5,550
3,4	357,64	355,17	96,485	783,4	952,4	5,070	5,543
3,5	359,22	348,20	101,74	790,4	950,7	5,089	5,535
3,6	360,76	340,78	107,48	797,7	948,6	5,109	5,527
3,7	362,26	332,75	113,83	805,2	946,0	5,128	5,517
3,8	363,73	323,89	121,00	813,0	942,7	5,149	5,506
3,9	365,17	313,85	129,32	821,4	938,6	5,171	5,492
4,0	366,57	301,93	139,45	830,7	933,0	5,196	5,475
4,1	367,93	286,54	152,85	841,7	925,2	5,225	5,452
4,2	369,25	261,83	174,92	857,7	911,4	5,267	5,413

Table II.5 Thermodynamic Properties of Propane on the Boiling and Condensation Curves (for the Indicated Pressures)

p	c_v'	c_v''	c_p'	c_p''	c_s'	c_s''	w'	w''
$0,5 \cdot 10^{-7}$		0,757		0,945	—	—4,383	1752,9	155,4
$0,1 \cdot 10^{-6}$		0,766		0,955	—	—4,221	1776,7	157,1
0,2		0,776		0,964	—	—4,009	1789,0	158,9
0,3		0,782		0,970	—	—3,963	1791,7	160,0
0,4		0,786		0,974	—	—3,895	1792,1	160,9
0,5		0,789		0,978	—	—3,841	1791,5	161,5
0,6		0,792		0,981	—	—3,798	1790,7	162,0
0,8		0,796		0,985	—	—3,728	1788,6	162,8
$0,1 \cdot 10^{-5}$		0,800		0,989	—	—3,674	1786,4	163,5
0,2		0,811		1,000	—	—3,505	1777,3	165,6
0,4		0,823		1,012	—	—3,334	1765,3	167,9
0,6	1,410	0,831	1,993	1,019	1,993	—3,233	1757,2	169,2
0,8	1,413	0,836	2,002	1,024	2,002	—3,160	1751,0	170,2
$0,1 \cdot 10^{-4}$	1,415	0,840	2,007	1,029	2,007	—3,104	1746,0	171,0
0,2	1,418	0,854	2,020	1,042	2,020	—2,927	1729,0	173,5
0,4	1,415	0,868	2,028	1,056	2,028	—2,749	1709,7	176,2
0,6	1,412	0,877	2,030	1,065	2,030	—2,643	1697,1	177,8
0,8	1,408	0,883	2,031	1,071	2,031	—2,568	1687,5	179,0
$0,1 \cdot 10^{-3}$	1,406	0,888	2,032	1,076	2,032	—2,509	1679,7	180,0
0,2	1,397	0,904	2,033	1,092	2,033	—2,326	1652,9	183,0
0,3	1,392	0,914	2,034	1,102	2,034	—2,218	1635,1	184,9
0,4	1,388	0,922	2,035	1,109	2,035	—2,141	1621,5	186,3
0,5	1,386	0,928	2,036	1,115	2,036	—2,081	1610,4	187,4
0,6	1,384	0,932	2,037	1,120	2,037	—2,032	1600,8	188,3
0,7	1,382	0,937	2,038	1,124	2,038	—1,991	1592,4	189,1

Table II.5 (*Continued*)

p	c_v'	c_v''	c_p'	c_p''	c_s'	c_s''	w'	w''
0,8	1,381	0,940	2,039	1,128	2,039	$-1,955$	1584,9	189,8
0,9	1,380	0,944	2,040	1,131	2,040	$-1,923$	1578,1	190,4
$0,1 \cdot 10^{-2}$	1,380	0,947	2,041	1,134	2,040	$-1,895$	1571,8	191,0
0,2	1,377	0,968	2,049	1,155	2,049	$-1,708$	1526,9	194,7
0,3	1,377	0,981	2,056	1,169	2,055	$-1,598$	1497,5	197,1
0,4	1,378	0,991	2,062	1,180	2,061	$-1,521$	1475,0	198,8
0,5	1,379	1,000	2,067	1,189	2,067	$-1,461$	1456,7	200,1
0,6	1,381	1,007	2,072	1,196	2,072	$-1,412$	1441,2	201,2
0,7	1,382	1,014	2,076	1,203	2,076	$-1,370$	1427,6	202,2
0,8	1,384	1,019	2,081	1,209	2,080	$-1,335$	1415,5	203,0
0,9	1,385	1,025	2,085	1,215	2,084	$-1,303$	1404,6	203,7
$0,1 \cdot 10^{-1}$	1,387	1,030	2,088	1,221	2,088	$-1,275$	1394,7	204,4
0,15	1,394	1,050	2,105	1,244	2,104	$-1,168$	1354,6	206,9
0,2	1,400	1,067	2,118	1,262	2,118	$-1,092$	1324,4	208,8
0,25	1,406	1,081	2,130	1,278	2,129	$-1,034$	1299,9	210,2
0,3	1,411	1,093	2,141	1,293	2,140	$-0,986$	1279,2	211,4
0,35	1,416	1,104	2,151	1,306	2,149	$-0,946$	1261,1	212,3
0,4	1,421	1,114	2,160	1,318	2,158	$-0,911$	1245,2	213,2
0,45	1,425	1,124	2,168	1,330	2,166	$-0,881$	1230,7	213,9
0,5	1,429	1,133	2,176	1,340	2,174	$-0,854$	1217,6	214,5
0,55	1,433	1,141	2,183	1,351	2,181	$-0,829$	1205,5	215,1
0,6	1,437	1,149	2,190	1,360	2,188	$-0,807$	1194,3	215,6
0,65	1,440	1,157	2,197	1,370	2,195	$-0,786$	1183,8	216,1
0,7	1,444	1,164	2,204	1,379	2,201	$-0,767$	1174,0	216,5
0,75	1,447	1,171	2,210	1,387	2,208	$-0,749$	1164,7	216,9
0,8	1,451	1,178	2,216	1,396	2,213	$-0,733$	1155,9	217,3
0,85	1,454	1,184	2,222	1,404	2,219	$-0,717$	1147,6	217,6
0,9	1,457	1,191	2,228	1,412	2,225	$-0,702$	1139,6	217,9
0,95	1,460	1,197	2,233	1,419	2,230	$-0,689$	1132,0	218,2
$0,1 \cdot 10^{0}$	1,463	1,202	2,239	1,427	2,235	$-0,675$	1124,7	218,5
0,11	1,469	1,214	2,249	1,441	2,245	$-0,651$	1111,0	218,9
0,12	1,474	1,224	2,259	1,454	2,255	$-0,629$	1098,2	219,3
0,13	1,479	1,234	2,268	1,468	2,264	$-0,608$	1086,3	219,7
0,14	1,484	1,244	2,278	1,480	2,273	$-0,590$	1075,1	220,0
0,15	1,489	1,253	2,286	1,492	2,281	$-0,572$	1064,5	220,3
0,16	1,494	1,262	2,295	1,504	2,289	$-0,556$	1054,5	220,5
0,17	1,498	1,271	2,303	1,515	2,297	$-0,540$	1044,9	220,8
0,18	1,503	1,279	2,311	1,526	2,304	$-0,526$	1035,8	221,0
0,19	1,507	1,287	2,319	1,537	2,312	$-0,512$	1027,0	221,1
0,20	1,511	1,295	2,326	1,548	2,319	$-0,500$	1018,7	221,3
0,22	1,520	1,310	2,341	1,568	2,333	$-0,476$	1002,8	221,5
0,24	1,527	1,324	2,355	1,587	2,346	$-0,454$	988,0	221,7
0,26	1,535	1,337	2,369	1,605	2,359	$-0,435$	974,2	221,8
0,28	1,542	1,350	2,382	1,623	2,371	$-0,416$	961,1	221,9
0,30	1,549	1,362	2,395	1,641	2,383	$-0,400$	948,7	222,0
0,32	1,555	1,374	2,407	1,657	2,394	$-0,384$	936,9	222,0
0,34	1,562	1,385	2,419	1,673	2,406	$-0,370$	925,7	222,0
0,36	1,568	1,396	2,431	1,689	2,417	$-0,357$	914,9	221,9
0,38	1,574	1,406	2,442	1,705	2,427	$-0,344$	904,6	221,9
0,40	1,580	1,417	2,454	1,720	2,438	$-0,333$	894,6	221,8
0,42	1,586	1,426	2,465	1,734	2,448	$-0,322$	885,0	221,7
0,44	1,592	1,436	2,476	1,749	2,458	$-0,312$	875,8	221,6
0,46	1,597	1,445	2,486	1,763	2,467	$-0,302$	866,8	221,4
0,48	1,602	1,454	2,497	1,777	2,477	$-0,293$	858,1	221,3

Table II.5 (Continued)

p	c_v'	c_v''	c_p'	c_p''	c_s'	c_s''	w'	w''
0	1,608	1,463	2,508	1,790	2,487	—0,285	849,7	221,1
0,55	1,620	1,484	2,533	1,824	2,510	—0,266	829,6	220,7
0,6	1,632	1,503	2,558	1,856	2,532	—0,250	810,7	220,1
0,65	1,644	1,522	2,583	1,887	2,554	—0,236	792,9	219,6
0,7	1,655	1,539	2,607	1,918	2,575	—0,224	776,0	218,9
0,75	1,666	1,556	2,631	1,948	2,596	—0,214	759,9	218,3
0,8	1,676	1,572	2,655	1,978	2,617	—0,206	744,5	217,6
0,85	1,686	1,588	2,679	2,007	2,637	—0,199	729,8	216,8
0,9	1,696	1,603	2,703	2,036	2,658	—0,194	715,6	216,1
0,95	1,705	1,617	2,726	2,065	2,678	—0,190	701,9	215,3
1,0	1,714	1,631	2,750	2,094	2,697	—0,187	688,6	214,5
1,05	1,723	1,644	2,774	2,123	2,717	—0,185	675,8	213,7
1,1	1,732	1,657	2,798	2,152	2,737	—0,185	663,4	212,8
1,15	1,740	1,669	2,822	2,181	2,757	—0,185	651,3	212,0
1,2	1,749	1,682	2,846	2,210	2,776	—0,187	639,5	211,1
1,25	1,757	1,693	2,870	2,239	2,796	—0,190	628,1	210,2
1,3	1,765	1,705	2,895	2,269	2,816	—0,193	616,9	209,3
1,35	1,771	1,716	2,920	2,299	2,835	—0,197	606,0	208,4
1,4	1,781	1,727	2,945	2,330	2,855	—0,203	595,3	207,5
1,45	1,788	1,738	2,971	2,361	2,875	—0,209	584,8	206,6
1,5	1,796	1,749	2,997	2,392	2,895	—0,216	574,6	205,7
1,6	1,811	1,769	3,051	2,457	2,936	—0,233	554,7	203,8
1,7	1,825	1,789	3,107	2,525	2,978	—0,253	535,5	201,8
1,8	1,839	1,808	3,166	2 597	3,020	—0,277	516,9	199,9
1,9	1,853	1,826	3,227	2,672	3,064	—0,305	498,9	197,9
2,0	1,867	1,844	3,292	2,752	3,110	—0,336	481,5	195,9
2,1	1,880	1,862	3,361	2,838	3,157	—0,372	464,4	193,8
2,2	1,894	1,879	3,435	2,931	3,206	—0,412	447,8	191,8
2,3	1,907	1,896	3,514	3,031	3,257	—0,457	431,6	189,7
2,4	1,921	1,913	3,599	3,141	3,311	—0,508	415,7	187,6
2,5	1,934	1,929	3,693	3,261	3,369	—0,564	400,0	185,4
2,6	1,947	1,945	3,795	3,394	3,430	—0,627	384,7	183,3
2,7	1,961	1,961	3,909	3,543	3,496	—0,699	369,6	181,1
2,8	1,974	1,977	4,036	3,711	3,568	—0,779	354,7	178,9
2,9	1,988	1,993	4,181	3,902	3,646	—0,870	340,0	176,6
3,0	2,002	2,009	4,346	4,123	3,733	—0,975	325,5	174,3
3,1	2,016	2,026	4,538	4,380	3,831	—1,096	311,1	172,0
3,2	2,031	2,042	4,765	4,685	3,942	—1,236	296,9	169,6
3,3	2,046	2,058	5,039	5,053	4,070	—1,403	282,7	167,2
3,4	2,061	2,075	5,377	5,508	4,222	—1,602	268,6	164,8
3,5	2,077	2,092	5,807	6,084	4,406	—1,847	254,5	162,3
3,6	2,094	2,109	6,373	6,840	4,635	—2,154	240,4	159,8
3,7	2,112	2,128	7,155	7,876	4,933	—2,554	226,4	157,3
3,8	2,131	2,147	8,307	9,384	5,341	—3,097	212,2	154,7
3,9	2,152	2,167	10,17	11,78	5,945	—3,890	198,0	152,1
4,0	2,175	2,190	13,66	16,16	6,953	—5,175	183,7	149,6
4,1	2,203	2,215	22,33	26,59	9,050	—7,721	169,3	147,1
4,2	2 239	2,247	69,22	78,72	16,88	—16,37	154,5	144,8

Table II.6 Thermodynamic Properties of Propane on the Boiling and Condensation Curves (for the Indicated Pressures)

p	μ'	μ''	k'	k''	$\alpha' a/\alpha_0$	α''/α_0	γ'/γ_0	γ''/γ_0
$0,5 \cdot 10^{-7}$	—0,65	—7453	$4,40 \cdot 10^{10}$	1,25	0,144	1,000	$4,75 \cdot 10^9$	1,000
$0,1 \cdot 10^{-6}$	—0,65	—6161	2,25	1,25	0,149	1,000	2,46	1,000
0,2	—0,65	—5049	1,14	1,24	0,154	1,000	1,27	1,000
0,3	—0,64	—4475	$7,59 \cdot 10^9$	1,24	0,158	1,000	$8,61 \cdot 10^8$	1.000
0,4	—0,64	—4100	5,69	1,24	0,160	1,000	6,51	1.000
0,5	—0,63	—3825	4,54	1,24	0,161	1,000	5,24	1.000
0,6	—0,63	—3612	3,78	1,24	0,163	1,000	4,39	1,000
0,8	—0,62	—3294	2,82	1,24	0,165	1,000	3,32	1,000
$0,1 \cdot 10^{-5}$	—0,62	—3062	2,25	1,24	0,167	1,000	2,67	1,000
0,2	—0,61	—2421	1,11	1,23	0,172	1,000	1,36	1,000
0,4	—0,60	—1886	$5,43 \cdot 10^8$	1,23	0,178	1,000	$6,86 \cdot 10^7$	1,000
0,6	—0,59	—1618	3,58	1,23	0,182	1,000	4,60	1.000
0,8	—0,59	—1445	2,66	1,23	0,185	0,999	3,46	1,000
$0,1 \cdot 10^{-4}$	—0,59	—1321	2,11	1,22	0,187	0,999	2,78	0,999
0,2	—0,58	—984,3	1,03	1,22	0,194	0,999	1,40	0,999
0,4	—0,58	—714,4	$4,99 \cdot 10^7$	1,22	0,202	0,999	$7,02 \cdot 10^6$	0,999
0,6	—0,57	—583,3	3,27	1,21	0,207	0,999	4,69	0,999
0,8	—0,57	—501,0	2,41	1,21	0,210	0,998	3,52	0,998
$0,1 \cdot 10^{-3}$	—0,57	—442,9	1,91	1,21	0,213	0,998	2,82	0,998
0,2	—0,57	—291,0	$9,17 \cdot 10^6$	1,21	0,223	0,998	1,41	0,998
0,3	—0,57	—220,1	5,95	1,21	0,230	0,997	$9,36 \cdot 10^5$	0,997
0,4	—0,57	—176,7	4,37	1,20	0,235	0,997	7,01	0,997
0,5	—0,56	—146,8	3,44	1,20	0,239	0,997	5,59	0,997
0,6	—0,56	—124,5	2,82	1,20	0,242	0,997	4,65	0,997
0,7	—0,56	—107,2	2,39	1,20	0,245	0,997	3,98	0,997
0,8	—0,56	—93,17	2,07	1,20	0,248	0,997	3,48	0,997
0,9	—0,56	—81,59	1,82	1,20	0,250	0,997	3,08	0,997
$0,1 \cdot 10^{-2}$	—0,56	—71,82	$1,62 \cdot 10^6$	1,20	0,253	0,997	$2.77 \cdot 10^5$	0,997
0,2	—0,55	—19,97	$7,56 \cdot 10^5$	1,19	0,269	0,999	1,37	0,997
0,3	—0,54	1,66	4,81	1,19	0,280	1,000	$9,03 \cdot 10^4$	0,998
0,4	—0,54	13,73	3,48	1,19	0,289	1,002	6,72	0,999
0,5	—0,53	21,43	2,70	1,18	0,296	1,00	5,33	1,000
0,6	—0,53	26,74	2,20	1,18	0,302	1,006	4,42	1,002
0,7	—0,53	30,60	1,84	1,18	0,307	1,007	3,77	1,003
0,8	—0,52	33,50	1,58	1,18	0,312	1,009	3,28	1,004
0,9	—0,52	35,75	1,38	1,18	0,316	1,011	2,90	1,005
$0,1 \cdot 10^{-1}$	—0,52	37,52	1,22	1,18	0,321	1,013	2,60	1,006
0,15	—0,51	42,42	$7,60 \cdot 10^4$	1,17	0,338	1,022	1,70	1,012
0,2	—0,50	44,31	5,40	1,17	0,352	1,030	1,26	1,018
0,25	—0,49	45,01	4,14	1,17	0,364	1,038	$9,93 \cdot 10^3$	1,024
0,3	—0,48	45,15	3,32	1,16	0,374	1,046	8,18	1,029
0,35	—0,47	45,00	2,75	1,16	0,383	1,053	6,95	1,034
0,4	—0,47	44,69	2,34	1,16	0,392	1,060	6,02	1,039
0,45	—0,46	44,29	2,02	1,16	0,400	1,067	5,31	1,044
0,5	—0,46	43,85	1,77	1,16	0,407	1,074	4,74	1,049
0,55	—0,45	43,38	1,57	1,15	0,414	1,080	4,28	1,053
0,6	—0,45	42,91	1,41	1,15	0,421	1,087	3,90	1,058
0,65	—0,44	42,43	1,28	1,15	0,427	1,093	3,57	1,062
0,7	—0,44	41,96	1,16	1,15	0,433	1,099	3,30	1,066
0,75	—0,43	41,49	1,06	1,15	0,439	1,105	3,06	1,070
0,8	—0,43	41,04	$9,80 \cdot 10^3$	1,15	0,445	1,111	2,85	1,074
0,85	—0,42	40,60	9,07	1,14	0,450	1,117	2,67	1,078

Table II.6 (*Continued*)

p	μ'	μ''	k'	k''	α'/α_0	α''/α_0	γ'/γ_0	γ''/γ_0
0,9	−0,42	40,17	8,43	1,14	0,455	1,123	2,51	1,082
0,95	−0,41	39,76	7,86	1,14	0,460	1,128	2,37	1,086
$0{,}1 \cdot 10^0$	−0,41	39,35	$7{,}35 \cdot 10^3$	1,14	0,465	1,134	$2{,}24 \cdot 10^3$	1,089
0,11	−0,40	38,58	6,49	1,14	0,475	1,145	2,01	1,097
0,12	−0,40	37,86	5,79	1,13	0,484	1,156	1,83	1,104
0,13	−0,39	37,18	5,21	1,13	0,493	1,166	1,68	1,111
0,14	−0,38	36,54	4,72	1,13	0,501	1,177	1,54	1,117
0,15	−0,38	35,94	4,31	1,13	0,510	1,187	1,43	1,124
0,16	−0,37	35,38	3,95	1,13	0,517	1,197	1,33	1,130
0,17	−0,36	34,84	3,64	1,12	0,525	1,207	1,24	1,137
0,18	−0,36	34,34	3,36	1,12	0,533	1,217	1,17	1,143
0,19	−0,35	33,85	3,12	1,12	0,540	1.226	1,10	1,149
0,20	−0,35	33,40	2,91	1,12	0,547	1,236	1,04	1,155
0,22	−0,34	32,55	2,55	1,11	0,561	1,255	$9{,}29 \cdot 10^2$	1,166
0,24	−0,33	31,78	2,26	1,11	0,575	1,273	8,41	1,178
0,26	−0,32	31,08	2,02	1,11	0,588	1,291	7,68	1,189
0,28	−0,30	30,43	1,81	1,10	0,601	1,309	7,05	1,199
0,30	−0,30	29,84	1,64	1,10	0,613	1,327	6,51	1,210
0,32	−0,29	29,29	1,49	1,09	0,626	1,345	6,04	1,220
0,34	−0,28	28,78	1,37	1,09	0,638	1,363	5,63	1,230
0,36	−0,27	28.31	1,26	1,09	0,650	1,380	5,26	1,240
0,38	−0,26	27,86	1,16	1,08	0,662	1,397	4,94	1,250
0,40	−0,25	27,45	1,07	1,08	0,674	1,415	4,65	1,260
0,42	−0,24	27,06	$9{,}95 \cdot 10^2$	1,08	0,685	1,432	4,39	1.270
0,44	−0,23	26,70	9,27	1,07	0,697	1,450	4,15	1,279
0,46	−0,22	26,36	8,65	1,07	0,708	1,467	3,94	1,288
0,48	−0,21	26,03	8,10	1,07	0,720	1,484	3,74	1,298
0,50	−0,20	25,73	7,59	1,07	0,731	1,502	3,56	1,307
0,55	−0,18	25,03	$6{,}53 \cdot 10^2$	1,06	0,759	1,546	$3{,}17 \cdot 10^2$	1,330
0,6	−0,16	24,42	5,67	1,05	0,788	1,590	2,85	1,353
0,65	−0,14	23,88	4,97	1,04	0,816	1,634	2,58	1,375
0,7	−0,12	23,40	4,38	1,04	0,845	1,680	2,35	1,397
0,75	−0,09	22,97	3,89	1,03	0,874	1,726	2,15	1,419
0,8	−0,07	22,59	3,48	1,02	0,903	1,772	1,98	1,441
0,85	−0,05	22,24	3,13	1,02	0,933	1,820	1,84	1,464
0,9	−0,03	21,92	2,82	1,01	0,963	1,869	1,70	1,486
0,95	0	21,64	2,55	1,00	0,994	1,919	1,59	1,508
1,0	0,02	21,38	2,32	1,00	1,026	1,970	1,48	1,530
1,05	0,04	21,14	2,11	0,99	1,058	2,023	1,39	1,553
1,1	0,07	20,92	1,93	0,99	1,091	2,077	1,30	1,575
1,15	0,09	20,72	1,77	0,98	1,124	2,132	1,23	1,598
1,2	0,12	20,54	1,62	0,97	1,159	2,189	1,16	1,621
1,25	0,14	20,37	1,50	0,97	1,194	2,248	1,09	1,644
1,3	0,17	20,22	1,38	0,96	1,231	2,308	1,03	1,667
1,35	0,20	20,08	1,27	0,96	1,268	2,371	$9{,}81 \cdot 10^1$	1,691
1,4	0,22	19,95	1,18	0,95	1,307	2,435	9,31	1,715
1,45	0,25	19,83	1,09	0,94	1,347	2,501	8,85	1,739
1,5	0,28	19,71	1,01	0,94	1,388	2,570	8,42	1,763
1,6	0,34	19,52	$8{,}74 \cdot 10^1$	0,93	1,475	2,715	7,65	1,813
1,7	0,41	19,35	7,58	0,92	1,568	2,870	6,98	1,865
1,8	0,48	19,22	6,59	0,91	1,668	3,038	6,39	1,918
1,9	0,55	19,10	5.75	0,90	1.776	3,219	5,86	1,972
2,0	0,63	19,00	5.02	0,89	1,893	3.416	5,39	2,029
2,1	0,71	18,92	$4{,}40 \cdot 10^1$	0,88	2,021	3,631	$4{,}97 \cdot 10^1$	2,088

Table II.6 (*Continued*)

p	μ'	μ''	k'	k''	α'/α_0	α''/α_0	γ'/γ_0	γ''/γ_0
2,2	0,80	18,86	3,86	0,87	2,161	3,866	4,56	2,148
2,3	0,90	18,81	3,38	0,86	2,315	4,125	4,25	2,212
2,4	1,00	18,77	2,97	0,85	2,486	4,411	3,94	2,278
2,5	1,11	18,75	2,61	0,84	2,677	4,729	3,66	2,347
2,6	1,24	18,73	2,29	0,83	2,891	5,085	3,40	2,419
2,7	1,38	18,72	2,01	0,82	3,134	5,487	3,16	2,495
2,8	1,53	18,72	1,76	0,81	3,411	5,944	2,93	2,575
2,9	1,69	18,72	1,54	0,81	3,731	6,468	2,73	2,660
3,0	1,88	18,72	1,34	0,80	4,105	7,076	2,54	2,750
3,1	2,09	18,73	1,17	0,79	4,547	7,790	2,36	2,846
3,2	2,33	18,74	1,01	0,78	5,079	8,642	2,19	2,949
3,3	2,60	18,74	$8,76 \cdot 10^3$	0,78	5,731	9,674	2,04	3,060
3,4	2,91	18,73	7,53	0,77	6,549	10,95	1,89	3,181
3,5	3,27	18,71	6,44	0,77	7,604	12,58	1,75	3,314
3,6	3,69	18,66	5,47	0,76	9,014	14,72	1,62	3,462
3,7	4,20	18,57	4,61	0,76	10,99	17,65	1,49	3,630
3,8	4,81	18,43	3,84	0,76	13,94	21,92	1,37	3,823
3,9	5,57	18,19	3,16	0,77	18,78	28,71	1,25	4,054
4,0	6,54	17,80	2,55	0,78	27,99	41,11	1,13	4,343
4,1	7,84	17,11	2,00	0,81	51,19	70,54	1,01	4,739
4,2	9,82	15,70	1,49	0,87	178,9	217,1	$8,62 \cdot 10^0$	5,416

Table II.7 Thermodynamic Properties of Propane in the Single-Phase Region

T	ρ	h	s	c_v	c_p
			$p=0,01$		
100	718,94	156,5	2,185	—	—
110	708,91	174,1	2,358	—	—
120	698,68	193,3	2,524	—	—
130	688,48	213,3	2,684	1,418	2,018
140	678,30	233,5	2,834	1,409	2,031
150	668,12	253,8	2,974	1,392	2,034
160	657,89	274,2	3,106	1,381	2,039
170	647,60	294,6	3,230	1,377	2,050
180	637,19	315,2	3,347	1,379	2,067
190	0,2809	801,6	5,919	1,033	1,224
200	0,2667	814,1	5,983	1,075	1,267
210	0,2538	826,9	6,046	1,115	1,307
220	0,2421	840,2	6,108	1,154	1,346
230	0,2315	853,8	6,169	1,193	1,385
240	0,2217	867,9	6,228	1,233	1,424
250	0,2127	882,3	6,287	1,274	1,465
260	0,2045	897,2	6,346	1,315	1,506
270	0,1968	912,5	6,403	1,358	1,548
280	0,1898	928,2	6,460	1,401	1,592
290	0,1832	944,3	6,517	1,445	1,635
300	0,1771	960,9	6,573	1,490	1,680
310	0,1713	977,9	6,629	1,536	1,725
320	0,1659	995,4	6,684	1,581	1,771

Table II.7 (*Continued*)

T	ρ	h	s	c_v	c_p
			$p=0,01$		
330	0,1609	1013,3	6,740	1,627	1,816
340	0,1561	1031,7	6,795	1,673	1,862
350	0,1517	1050,6	6,849	1,719	1,908
360	0,1475	1069,9	6,904	1,764	1,953
370	0,1434	1089,6	6,958	1,810	1,999
380	0,1397	1109,8	7,012	1,855	2,044
390	0,1361	1130,5	7,065	1,899	2,088
400	0,1327	1151,6	7,119	1,944	2,133
410	0,1294	1173,2	7,172	1,988	2,177
420	0,1263	1195,1	7,225	2,031	2,220
430	0,1234	1217,6	7,278	2,074	2,263
440	0,1206	1240,4	7,330	2,117	2,306
450	0,1179	1263,7	7,382	2,159	2,348
460	0,1153	1287,4	7,434	2,201	2,390
470	0,1129	1311,5	7,486	2,243	2,432
480	0,1105	1336,0	7,538	2,284	2,473
490	0,1083	1360,9	7,589	2,325	2,513
500	0,1061	1386,3	7,641	2,365	2,554
520	0,1020	1438,1	7,742	2,444	2,632
540	0,09824	1491,5	7,843	2,520	2,709
560	0,0947	1546,5	7,943	2,593	2,782
580	0,0914	1602,8	8,042	2,663	2,852
600	0,0884	1660,5	8,140	2,730	2,919
620	0,0855	1719,5	8,236	2,793	2,982
640	0,0828	1779,8	8,332	2,855	3,044
660	0,0803	1841,3	8,427	2,916	3,105
680	0,0780	1904,0	8,520	2,979	3,168
700	0,0757	1968,0	8,613	3,047	3,236
			$p=0,05$		
100	718,96	156,6	2,185	—	—
110	708,93	174,2	2,357	—	—
120	698,70	193,3	2,524	—	—
130	688,50	213,3	2,684	1,418	2,018
140	678,32	233,6	2,834	1,408	2,031
150	668,14	253,9	2,974	1,392	2,034
160	657,92	274,2	3,106	1,381	2,039
170	647,62	294,7	3,230	1,377	2,050
180	637,22	315,3	3,347	1,379	2,067
190	626,67	336,1	3,460	1,388	2,090
200	515,93	357,1	3,568	1,400	2,119
210	604,97	378,1	3,672	1,417	2,152
220	1,2320	837,9	5,797	1,151	1,358
230	1,1749	851,7	5,859	1,195	1,401
240	1,1231	865,9	5,919	1,238	1,441
250	1,0759	880,6	5,979	1,280	1,481
260	1,0326	895,6	6,038	1,322	1,522
270	0,9929	911,0	6,096	1,364	1,562
280	0,9563	926,8	6,153	1,407	1,604
290	0,9223	943,1	6,210	1,451	1,647
300	0,8907	959,8	6,267	1,495	1,690
310	0,8614	976,9	6,323	1,539	1,734
320	0,8339	994,4	6,379	1,585	1,778

Table II.7 (*Continued*)

T	p	h	s	c_v	c_p
			$p=0,05$		
330	0,8081	1012,4	6,434	1,630	1,823
340	0,7839	1030,9	6,489	1,675	1,868
350	0,7612	1049,8	6,544	1,721	1,913
360	0,7398	1069,2	6,599	1,766	1,958
370	0,7195	1089,6	6,653	1,811	2,003
380	0,7004	1109,2	6,707	1,856	2,047
390	0,6823	1129,9	6,761	1,900	2,091
400	0,6650	1151,0	6,814	1,944	2,135
410	0,6487	1172,6	6,868	1,988	2,179
420	0,6331	1194,6	6,921	2,032	2,222
430	0,6183	1217,0	6,973	2,075	2,265
440	0,6041	1239,9	7,026	2,117	2,308
450	0,5905	1263,2	7,078	2,159	2,350
460	0,5776	1286,9	7,130	2,201	2,392
470	0,5653	1311,0	7,182	2,243	2,433
480	0,5534	1335,6	7,234	2,284	2,474
490	0,5421	1360,5	7,285	2,325	2,515
500	0,5312	1385,9	7,336	2,365	2,555
520	0,5106	1437,7	7,438	2,444	2,633
540	0,4916	1491,2	7,539	2,520	2,710
560	0,4740	1546,1	7,639	2,593	2,783
580	0,4576	1602,5	7,738	2,663	2,853
600	0,4423	1660,2	7,836	2,730	2,919
620	0,4280	1719,2	7,932	2,793	2,983
640	0,4146	1779,5	8,028	2,855	3,044
660	0,4020	1841,0	8,123	2,916	3,106
680	0,3901	1903,7	8,216	2,979	3,169
700	0,3789	1967,8	8,309	3,047	3,236
			$p=0,1$		
100	718,99	156,7	2,185	—	—
110	708,95	174,2	2,357	—	—
120	698,72	193,4	2,524	—	—
130	688,52	213,4	2,684	1,418	2,018
140	678,35	233,6	2,834	1,408	2,031
150	668,17	253,9	2,974	1,392	2,034
160	657,95	274,3	3,106	1,381	2,039
170	647,66	294,7	3,230	1,377	2,050
180	637,26	315,3	3,347	1,379	2,067
190	626,71	336,1	3,460	1,388	2,090
200	615,98	357,1	3,567	1,400	2,118
210	605,02	378,5	3,672	1,417	2,152
220	593,80	400,2	3,773	1,437	2,191
230	582,27	422,3	3,871	1,461	2,235
240	2,2853	863,4	5,781	1,245	1,465
250	2,1839	878,3	5,842	1,288	1,504
260	2,0922	893,5	5,901	1,330	1,542
270	2,0085	909,1	5,960	1,372	1,581
280	1,9316	925,1	6,018	1,414	1,621
290	1,8609	941,5	6,076	1,457	1,661
300	1,7955	958,3	6,133	1,500	1,703
310	1,7348	975,6	6,190	1,544	1,745
320	1,6784	993,2	6,246	1,589	1,788

Table II.7 (*Continued*)

T	ρ	h	s	c_v	c_p
			$p = 0,1$		
330	1,6255	1011,3	6,301	1,633	1,831
340	1,5761	1029,9	6,357	1,678	1,875
350	1,5297	1048,8	6,412	1,723	1,919
360	1,4860	1068,2	6,466	1,768	1,964
370	1,4448	1088,1	6,521	1,813	2,008
380	1,4060	1108,4	6,575	1,857	2,052
390	1,3691	1129,1	6,629	1,902	2,095
400	1,3342	1150,3	6,682	1,945	2,139
410	1,3011	1171,9	6,736	1,989	2,182
420	1,2696	1193,9	6,789	2,032	2,225
430	1,2396	1216,4	6,842	2,075	2,268
440	1,2109	1239,3	6,894	2,118	2,310
450	1,1837	1262,6	6,947	2,160	2,352
460	1,1576	1286,3	6,999	2,202	2,394
470	1,1326	1310,5	7,051	2.243	2,435
480	1,1088	1335,0	7,102	2,284	2,476
490	1,0859	1360,0	7,154	2,325	2,516
500	1,0639	1385,4	7,205	2,365	2,556
520	1,0226	1437,3	7,307	2,444	2,635
540	0,9844	1490,7	7,408	2,520	2,711
560	0,9490	1545,7	7,508	2,593	2,784
580	0,9160	1602,1	7,607	2,663	2,854
600	0,8853	1659,8	7,704	2,730	2,920
620	0,8565	1718,9	7,801	2,793	2,984
640	0,8296	1779,2	7,897	2,855	3,045
660	0,8043	1840,7	7,992	2,916	3,106
680	0,7805	1903,4	8,085	2,979	3,169
700	0,7581	1967,5	8,178	3,047	3,237
			$p = 0,101325$		
100	718,99	156,7	2,185	—	—
110	708,95	174,2	2,357	—	—
120	698,72	193,4	2,524	—	—
130	688,52	213,4	2,684	1,418	2,018
140	678,35	233,6	2,834	1,408	2,031
150	668,17	253,9	2,974	1,302	2,034
160	657,95	274,3	3,106	1,381	2,039
170	647,66	294,7	3,230	1,377	2,050
180	637,26	315,3	3,347	1,379	2,067
190	626,71	336,1	3,460	1,388	2,090
200	615,98	357,1	3,567	1,400	2,118
210	605,02	378,5	3,672	1,417	2,152
220	593,80	400,2	3,773	1,437	2,191
230	582,27	422,3	3,871	1,461	2,235
240	2,3165	863,3	5,778	1,245	1,465
250	2,2138	878,2	5,839	1,288	1,504
260	2,1205	893,4	5,899	1,330	1,543
270	2,0357	909,0	5,958	1,372	1,581
280	1,9578	925,1	6,016	1,414	1,621
290	1,8861	941,5	6,073	1,457	1,661
300	1,8197	958,3	6,130	1,501	1,703
310	1,7581	975,5	6,187	1,545	1,745
320	1,7010	993,2	6,243	1,589	1,788

Table II.7 (*Continued*)

T	p	h	s	c_v	c_p
		$p=0,101325$			
330	1,6474	1011,3	6,299	1,634	1,832
340	1,5972	1029,8	6,354	1,678	1,876
350	1,5501	1048,8	6,409	1,723	1,920
360	1,5059	1068,2	6,464	1,768	1,964
370	1,4641	1088,1	6,518	1,813	2,008
380,	1,4247	1108,4	6,572	1,857	2,052
390	1,3874	1129,1	6,626	1,902	2,096
400	1,3520	1150,3	6,680	1,945	2,139
410	1,3184	1171,9	6,773	1,989	2,182
420	1,2865	1193,9	6,786	2,032	2,225
430	1,2560	1216,4	6,839	2,075	2,268
440	1,2271	1239,3	6,892	2,118	2,310
450	1,1995	1262,6	6,944	2,160	2,352
460	1,1731	1286,3	6,996	2,202	2,394
470	1,1478	1310,5	7,048	2,243	2,435
480	1,1235	1335,0	7,100	2,284	2,476
490	1,1004	1360,0	7,151	2,325	2,516
500	1,0781	1385,3	7,203	2,365	2,556
520	1,0362	1437,3	7,304	2,444	2,635
540	0,9975	1490,7	7,405	2,520	2,711
560	0,9616	1545,7	7,505	2,593	2,784
580	0,9281	1602,1	7,604	2,663	2,854
600	0,8970	1659,8	7,702	2,730	2,920
620	0,8679	1718,9	7,799	2,793	2,984
640	0,8406	1779,1	7,894	2,855	3,045
660	0,8150	1840,7	7,989	2,916	3,106
680	0,7908	1903,4	8,083	2,979	3,169
700	0,7682	1967,5	8,176	3,047	3,237
		$p=0,15$			
100	719,01	157,7	2,185	—	—
110	708,97	174,3	2,357	—	—
120	698,75	193,4	2,524	—	—
130	688,54	213,4	2,684	1,418	2,018
140	678,37	233,7	2,834	1,408	2,031
150	668,19	254,0	2,974	1,392	2,034
160	657,98	274,4	3,106	1,381	2,039
170	647,69	294,8	3,229	1,377	2,050
180	637,29	315,4	3,347	1,379	2,067
190	626,75	336,2	3,459	1,388	2,090
200	616,02	357,2	3,567	1,400	2,118
210	605,07	378,5	3,671	1,417	2,152
220	593,85	400,2	3,772	1,437	2,191
230	582,33	422,4	3,871	1,461	2,235
240	570,45	445,0	3,967	1,488	2,284
250	3,3276	875,9	5,758	1,296	1,528
260	3,1807	891,3	5,819	1,338	1,564
270	3,0480	907,2	5,879	1,380	1,601
280	2,9273	923,4	5,938	1,422	1,638
290	2,8168	939,9	5,996	1,464	1,676
300	2,7151	956,9	6,053	1,506	1,716
310	2,6211	974,2	6,110	1,549	1,756
320	2,5340	992,0	6,167	1,593	1,798

Table II.7 (*Continued*)

T	ρ	h	s	c_v	c_p
			$p=0,15$		
330	2,4527	1010,2	6,223	1,637	1,840
340	2,3767	1028,8	6,278	1,681	1,883
350	2,3057	1047,9	6,333	1,726	1,926
360	2,2390	1067,3	6,388	1,770	1,969
370	2,1760	1087,2	6,443	1,815	2,013
380	2,1168	1107,6	6,497	1,859	2,056
390	2,0607	1128,4	6,551	1,903	2,099
400	2,0077	1149,6	6,605	1,946	2,143
410	1,9574	1171,2	6,658	1,990	2,185
420	1,9094	1193,3	6,711	2,033	2,228
430	1,8641	1215,8	6,764	2,076	2,270
440	1,8206	1238,7	6,817	2,118	2,312
450	1,7795	1262,0	6,869	2,160	2,354
460	1,7399	1285,8	6,921	2,202	2,396
470	1,7022	1309,9	6,973	2,243	2,437
480	1,6661	1334,5	7,025	2,284	2,477
490	1,6316	1359,5	7,077	2,325	2,518
500	1,5983	1384,9	7,128	2,365	2.558
520	1,5359	1436,8	7,230	2,444	2,636
540	1,4783	1490,3	7,331	2,520	2,712
560	1,4249	1545,3	7,431	2,593	2,785
580	1,3751	1601,7	7,530	2,663	2,855
600	1,3288	1659,4	7,628	2,730	2,921
620	1,2856	1718,5	7,724	2,793	2,985
640	1,2450	1778,8	7,820	2,855	3,046
660	1,2069	1840,3	7,915	2,916	3,107
680	1,1712	1903,1	8,008	2,979	3,170
700	1,1375	1967,2	8,101	3,047	3,237
			$p=0,2$		
100	719,03	156,8	2,185	—	—
110	708,99	174,3	2,357	—	—
120	698,77	193,5	2,524	—	—
130	688,57	213,5	2,684	1,417	2,018
140	678,40	233,7	2,834	1,408	2,031
150	668,22	254,1	2,974	1,392	2,034
160	658,01	274,4	3,105	1,381	2,039
170	647,72	294,9	3,229	1,377	2,050
180	637,33	315,4	3,347	1,379	2,067
190	626,79	336,2	3,459	1,388	2,090
200	616,06	357,2	3,567	1,400	2,118
210	605,11	378,6	3,671	1,417	2,152
220	593,90	400,3	3,772	1,437	2.190
230	582,39	422,4	3,871	1,461	2,234
240	570,52	445,0	3,967	1,488	2,284
250	4,5104	873,4	5,697	1,305	1,555
260	4,3013	889,1	5,759	1,347	1.588
270	4,1138	905,2	5,819	1,388	1,622
280	3,9449	921,6	5,879	1,429	1,656
290	3,7911	938,3	5,938	1,470	1,692
300	3,6500	955,4	5,996	1,512	1,729
310	3,5206	972,9	6,053	1,555	1,768
320	3,4008	990,8	6,110	1,598	1,808

Table II.7 (*Continued*)

T	p	h	s	c_v	c_p

$p=0,2$

330	3,2897	1009,1	6,166	1,641	1,849
340	3,1861	1027,8	6,222	1,685	1,891
350	3,0894	1046,9	6,277	1,729	1,933
360	2,9987	1066,4	6,332	1,773	1,975
370	2,9132	1086,4	6,387	1,817	2,018
380	2,8330	1106,8	6,441	1,860	2,061
390	2,7570	1127,6	6,495	1,904	2,104
400	2,6854	1148,8	6,549	1,948	2,146
410	2,6175	1170,5	6,603	1,991	2,189
420	2,5529	1192,6	6,656	2,034	2,231
430	2,4917	1215,1	6,709	2,076	2,273
440	2,4334	1238,1	6,762	2,118	2,315
450	2,3776	1261,4	6,814	2,160	2,356
460	2,3245	1285,2	6,866	2,202	2,398
470	2,2739	1309,4	6,918	2,243	2,439
480	2,2254	1334,0	6,970	2,284	2,479
490	2,1789	1359,0	7,022	2,325	2,519
500	2,1343	1384,4	7,073	2,365	2,559
520	2,0505	1436,3	7,175	2,444	2,637
540	1,9732	1489,8	7,276	2,520	2,713
560	1,9016	1544,8	7,376	2,593	2,786
580	1,8351	1601,3	7,475	2,663	2,856
600	1,7730	1659,1	7,573	2,730	2,922
620	1,7151	1718,1	7,670	2,793	2,985
640	1,6609	1778,5	7,765	2,855	3,047
660	1,6101	1840,0	7,860	2,916	3,108
680	1,5621	1902,8	7,954	2,979	3,171
700	1,5172	1966,9	8,047	3,047	3,238

$p=0,3$

100	719,07	156,9	2,184	—	—
110	709,04	174,5	2,357	—	—
120	698,81	193,6	2,524	—	—
130	688,62	213,6	2,683	1,417	2,018
140	678,45	233,9	2,834	1,408	2,031
150	668,27	254,2	2,974	1,392	2,034
160	658,07	274,5	3,105	1,380	2,039
170	647,78	295,0	3,229	1,376	2,050
180	637,40	315,5	3,347	1,379	2,067
190	626,86	336,3	3,459	1,387	2,089
200	616,15	357,4	3,567	1,400	2,118
210	605,21	378,7	3,671	1,417	2,151
220	594,01	400,4	3,772	1,437	2,190
230	582,51	422,5	3,870	1,461	2,234
240	570,61	445,0	3,966	1,489	2,283
250	558,40	468,2	4,061	1,518	2,339
260	6,6496	884,5	5,669	1,366	1,643
270	6,3324	901,0	5,732	1,406	1,669
280	6,0510	917,8	5,793	1,445	1,696
290	5,7985	934,9	5,853	1,485	1,726
300	5,5699	952,4	5,912	1,524	1,759
310	5,3621	970,1	5,970	1,565	1,793
320	5,1713	988,2	6,028	1,607	1,830

Table II.7 (*Continued*)

T	ρ	h	s	c_v	c_p
			$p=0,3$		
330	4,9953	1006,7	6,085	1,649	1,868
340	4,8319	1025,6	6,141	1,691	1,907
350	4,6803	1044,9	6,197	1,734	1,947
360	4,5391	1064,5	6,252	1,777	1,988
370	4,4065	1084,6	6,307	1,820	2,029
380	4,2817	1105,1	6,362	1,863	2,070
390	4,1647	1126,0	6,416	1,907	2,112
400	4,0540	1147,4	6,470	1,950	2,154
410	3,9494	1169,1	6,524	1,992	2,195
420	3,8503	1191,3	6,577	2,035	2,237
430	3,7566	1213,8	6,630	2,077	2,278
440	3,6671	1236,8	6,683	2,119	2,320
450	3,5818	1260,2	6,736	2,161	2,361
460	3,5009	1284,1	6,788	2,203	2,402
470	3,4237	1308,3	6,840	2,244	2,442
480	3,3496	1332,9	6,892	2,285	2,483
490	3,2789	1357,9	6,944	2,325	2,523
500	3,2114	1383,4	6,995	2,365	2,562
520	3,0842	1435,4	7,097	2,444	2,640
540	2,9667	1488,9	7,198	2,520	2,716
560	2,8581	1544,0	7,298	2,593	2,788
580	2,7573	1600,5	7,397	2,663	2,858
600	2,6636	1658,3	7,495	2,729	2,924
620	2,5758	1717,4	7,592	2,793	2,987
640	2,4938	1777,8	7,688	2,854	3,048
660	2,4170	1839,3	7,783	2,916	3,109
680	2,3450	1902,1	7,877	2,979	3,172
700	2,2769	1966,2	7,969	3,046	3,240
			$p=0,4$		
100	719,12	157,0	2,184	—	—
110	709,08	174,6	2,357	—	—
120	698,86	193,7	2,523	—	—
130	688,66	213,7	2,683	1,417	2,018
140	678,50	234,0	2,833	1,408	2,031
150	668,33	254,3	2,974	1,392	2,034
160	658,12	274,6	3,105	1,380	2,039
170	647,85	295,1	3,229	1,376	2,049
180	637,47	315,7	3,346	1,379	2,066
190	626,94	336,4	3,459	1,387	2,089
200	616,23	357,5	3,567	1,400	2,117
210	605,31	378,8	3,671	1,417	2,151
220	594,12	400,5	3,772	1,437	2,190
230	582,63	422,6	3,870	1,461	2,233
240	570,79	445,2	3,966	1,488	2,283
250	558,55	468,3	4,060	1,518	2,338
260	545,84	492,0	4,153	1,552	2,400
270	8,6850	896,6	5,665	1,425	1,723
280	8,2642	913,9	5,728	1,462	1,742
290	7,8932	931,4	5,790	1,499	1,765
300	7,5623	949,2	5,850	1,537	1,791
310	7,2640	967,3	5,909	1,576	1,821
320	6,9932	985,6	5,968	1,616	1,853

Table II.7 (*Continued*)

T	ρ	h	s	c_v	c_p

$$p = 0,4$$

330	6,7450	1004,3	6,025	1,656	1,888
340	6,5166	1023,4	6,082	1,698	1,924
350	6,3047	1042,8	6,139	1,740	1,962
360	6,1086	1062,6	6,194	1,782	2,001
370	5,9251	1082,8	6,250	1,824	2,040
380	5,7537	1103,4	6,305	1,867	2,080
390	5,5923	1124,5	6,359	1,909	2,121
400	5,4408	1145,9	6,413	1,952	2,162
410	5,2977	1167,7	6,467	1,994	2,202
420	5,1626	1189,9	6,521	2,036	2,243
430	5,0343	1212,5	6.574	2,078	2,284
440	4,9129	1235,6	6 627	2,120	2,325
450	4,7974	1259,0	6,680	2,162	2,366
460	4,6873	1282,9	6,732	2,203	2,406
470	4,5820	1307,2	6,784	2,244	2,446
480	4,4820	1331,8	6,836	2,285	2,486
490	4,3865	1356,9	6,888	2,325	2,526
500	4,2949	1382,3	6,939	2,365	2,565
520	4,1230	1434,4	7,042	2,444	2,643
540	3,9647	1488,0	7,143	2,520	2,718
560	3,8181	1543,1	7,243	2,593	2,791
580	3,6827	1599,6	7,342	2,663	2,860
600	3,5564	1657,5	7,440	2,729	2,926
620	3,4386	1716,7	7,537	2,793	2,989
640	3,3288	1777,1	7,633	2,854	3,050
660	3,2256	1838,7	7,728	2,915	3,111
680	3,1289	1901,5	7,821	2,979	3,174
700	3,0377	1965,6	7,914	3,046	3,241

$$p = 0,5$$

100	719,16	157,1	2,184	—	—
110	709,12	174,7	2,357	—	—
120	698,90	193,9	2,523	—	—
130	688,71	213,8	2.683	1,417	2,018
140	678,55	234,1	2,833	1,408	2,031
150	668,38	254,4	2,973	1,392	2,033
160	658,18	274,8	3,105	1,380	2,038
170	647,91	295,2	3,229	1,376	2,049
180	637,45	315,7	3,345	1,381	2,066
190	626,88	336,3	3,459	1,387	2,089
200	616,32	357,6	3,566	1,400	2,117
210	605,40	378,9	3,670	1,417	2,150
220	594,22	400,6	3,771	1,437	2,189
230	582,75	422,7	3,870	1,460	2,233
240	570,92	445,3	3,966	1,488	2,282
250	558,70	468,4	4,060	1,518	2,337
260	546,01	492,0	4,153	1,552	2,399
270	532,74	516,4	4,244	1,589	2,469
280	10,603	909,7	5,675	1,480	1,794
290	10,089	927,7	5,739	1,515	1,808
300	9,6364	945,9	5,800	1,551	1,827
310	9,2339	964,3	5,861	1,587	1,851
320	8,8716	982,9	5,920	1,625	1,879

Table II.7 (*Continued*)

T	ρ	h	s	c_v	c_p
			p=0,5		
330	8,5424	1001,9	5,978	1,665	1,909
340	8,2421	1021,1	6,035	1,704	1,943
350	7,9645	1040,7	6,092	1,745	1,978
360	7,7093	1060,7	6,149	1,786	2,014
370	7,4712	1081,0	6,204	1,828	2,052
380	7,2489	1101,7	6,259	1,870	2,091
390	7,0411	1122,8	6,314	1,912	2,130
400	6,8466	1144,3	6,369	1,954	2,170
410	6,6624	1166,2	6,423	1,996	2,210
420	6,4898	1188,5	6,476	2,038	2,250
430	6,3258	1211,2	6,530	2,080	2,290
440	6,1707	1234,3	6,583	2,121	2,330
450	6,0231	1257,8	6,636	2,163	2,370
460	5,8831	1281,7	6,688	2,204	2,410
470	5,7497	1306,0	6,741	2,245	2,450
480	5,6225	1330,7	6,793	2,285	2,490
490	5,5010	1355,8	6,844	2,325	2,529
500	5,3850	1381,3	6,896	2,365	2,569
520	5,1674	1433,5	6,998	2,444	2,646
540	4,9669	1487,1	7,099	2,520	2,721
560	4,7823	1542,3	7,200	2,593	2,793
580	4,6109	1598,8	7,299	2,663	2,862
600	4,4518	1656,7	7,397	2,729	2,928
620	4,3038	1715,9	7,494	2,792	2,991
640	4,1651	1776,4	7,590	2,854	3,052
660	4,0357	1838,0	7,685	2,915	3,112
680	3,9138	1900,9	7,779	2,979	3,175
700	3,7993	1965,0	7,872	3,046	3,242
			p=0,6		
100	719,20	157,3	2,184	—	—
110	709,17	174,8	2,356	—	—
120	698,95	194,0	2,523	—	—
130	688,76	213,9	2,683	1,416	2,018
140	678,60	234,2	2,833	1,408	2,031
150	668,44	254,5	2,973	1,392	2,033
160	658,24	274,9	3,104	1,380	2,038
170	647,98	295,3	3,228	1,376	2,049
180	637,61	315,9	3,346	1,379	2,066
190	627,10	336,6	3,458	1,387	2,089
200	616,41	357,7	3,566	1,400	2,117
210	605,50	379,0	3,670	1,417	2,150
220	594,33	400,7	3,771	1,437	2,189
230	582,86	422,8	3,869	1,460	2,232
240	571,06	445,4	3,965	1,488	2,281
250	558,86	468,4	4,059	1,518	2,336
260	546,18	492,1	4,152	1,552	2,398
270	532,94	516,4	4,244	1,588	2,468
280	518,97	541,5	4,335	1,628	2,549
290	12,400	923,8	5,694	1,532	1,857
300	11,802	942,5	5,758	1,565	1,867
310	11,278	961,2	5,819	1,599	1,884
320	10,811	980,2	5,879	1,635	1,906

Table II.7 (*Continued*)

T	ρ	h	s	c_v	c_p
		$p = 0,6$			
330	10,392	999,3	5,938	1,673	1,932
340	10,011	1018,8	5,996	1,712	1,962
350	9,6621	1038,6	6,054	1,751	1,995
360	9,3420	1058,7	6,110	1,791	2,029
370	9,0446	1079,2	6,166	1,832	2,065
380	8,7688	1100,0	6,222	1,873	2,102
390	8,5119	1121,2	6,277	1,915	2,140
400	8,2706	1142,8	6,332	1,956	2,178
410	8,0446	1164,8	6,386	1,998	2,217
420	7,8320	1187,1	6,440	2,039	2,257
430	7,6309	1209,9	6,493	2,081	2,296
440	7,4406	1233,1	6,547	2,122	2,336
450	7,2603	1256,6	6,600	2,163	2,375
460	7,0887	1280,6	6,652	2,204	2,415
470	6,9250	1304,9	6,705	2,245	2,454
480	6,7711	1329,7	6,757	2,285	2,494
490	6,6228	1354,8	6,808	2,326	2,533
500	6,4817	1380,3	6,860	2,365	2,572
520	6,2172	1432,5	6,962	2,444	2,649
540	5,9742	1486,2	7,064	2,519	2,723
560	5,7498	1541,4	7,164	2,593	2,795
580	5,5425	1598,0	7,263	2,662	2,864
600	5,3502	1656,0	7,362	2,729	2,930
620	5,1708	1715,2	7,459	2,792	2,992
640	5,0032	1775,7	7,555	2,854	3,053
660	4,8467	1837,3	7,650	2,915	3,114
680	4,7000	1900,2	7,743	2,978	3,177
700	4,5620	1964,4	7,836	3,046	3,244
		$p = 0,7$			
100	719,25	157,4	2,184	—	—
110	709,21	174,9	2,356	—	—
120	698,99	194,1	2,523	—	—
130	688,80	214,1	2,683	1,416	2,018
140	678,65	234,3	2,833	1,407	2,031
150	668,49	254,6	2,973	1,391	2,033
160	658,30	275,0	3,104	1,380	2,038
170	648,04	295,4	3,228	1,376	2,049
180	637,68	316,0	3,346	1,379	2,066
190	627,17	336,8	3,458	1,387	2,088
200	616,49	357,8	3,566	1,400	2,116
210	605,59	379,1	3,670	1,416	2,150
220	594,43	400,8	3,771	1,437	2,188
230	582,98	422,9	3,869	1,460	2,232
240	571,19	445,4	3,965	1,487	2,281
250	559,01	468,5	4,059	1,518	2,335
260	546,35	492,2	4,152	1,551	2,397
270	533,14	516,5	4,244	1,588	2,467
280	519,20	541,5	4,335	1,627	2,547
290	14,845	919,7	5,655	1,549	1,914
300	14,072	938,8	5,720	1,580	1,912
310	13,406	958,0	5,782	1,612	1,920
320	12,820	977,3	5,844	1,646	1,936

Table II.7 (*Continued*)

T	ρ	h	s	c_v	c_p
			$p=0,7$		
330	12,298	996,7	5,903	1,682	1,957
340	11,828	1016,4	5,962	1,719	1,983
350	11,400	1036,4	6,020	1,757	2,012
360	11,008	1056,7	6,077	1,796	2,044
370	10,648	1077,3	6,134	1,836	2,078
380	10,316	1098,2	6,190	1,877	2,113
390	10,005	1119,6	6,245	1,918	2,150
400	9,7157	1141,2	6,300	1,959	2,187
410	9,4441	1163,3	6,354	2,000	2,225
420	9,1896	1185,7	6,408	2,041	2,264
430	8,9498	1208,6	6,462	2,082	2,302
440	8,7229	1231,8	6,516	2,123	2,341
450	8,5079	1255,4	6,569	2,164	2,380
460	8,3050	1279,4	6,621	2,205	2,420
470	8,1116	1303,8	6,674	2,246	2,459
480	7,9279	1328,6	6,726	2,286	2,498
490	7,7526	1353,7	6,778	2,326	2,537
500	7,5852	1379,3	6,829	2,366	2,575
520	7,2726	1431,6	6,932	2,444	2,652
540	6,9854	1485,3	7,033	2,519	2,726
560	6,7217	1540,6	7,134	2,592	2,798
580	6,4774	1597,2	7,233	2,662	2,866
600	6,2507	1655,2	7,332	2,729	2,932
620	6,0399	1714,5	7,429	2,792	2,994
640	5,8435	1775,0	7,525	2,854	3,055
660	5,6596	1836,7	7,620	2,915	3,115
680	5,4871	1899,6	7,714	2,978	3,178
700	5,3255	1963,8	7,807	3,046	3,245
			$p=0,8$		
100	719,29	157,5	2,183	—	—
110	709,25	175,1	2,356	—	—
120	699,04	194,2	2,523	—	—
130	688,85	214,2	2,682	1,416	2,018
140	678,70	234,4	2,832	1,407	2,030
150	668,55	254,8	2,973	1,391	2,033
160	658,36	275,1	3,104	1,380	2,038
170	648,10	295,5	3,228	1,376	2,049
180	637,75	316,1	3,345	1,379	2,066
190	627,25	336,9	3,458	1,387	2,088
200	616,58	357,9	3,565	1,400	2,116
210	605,69	379,2	3,669	1,416	2,149
220	594,54	400,9	3,770	1,437	2,188
230	583,10	423,0	3,868	1,460	2,231
240	571,32	445,5	3,964	1,487	2,280
250	559,16	468,6	4,059	1,518	2,335
260	546,53	492,2	4,151	1,551	2,396
270	533,33	516,5	4,243	1,588	2,465
280	519,43	541,6	4,334	1,627	2,545
290	504,62	567,5	4,425	1,669	2,639
300	16,465	935,0	5,685	1,595	1,964
310	15,628	954,6	5,749	1,625	1,961
320	14,903	974,3	5,812	1,657	1,969

Table II.7 (*Continued*)

T	v	h	s	c_v	c_p
			$p=0,8$		
330	14,264	994,0	5,872	1,691	1.984
340	13,694	1014,0	5,932	1,726	2.006
350	13,180	1034,1	5,990	1,763	2.031
360	12,713	1054,6	6,048	1,802	2.060
370	12,283	1075,4	6,105	1,841	2,092
380	11,888	1096,4	6,161	1,880	2.125
390	11,522	1117,9	6,217	1,921	2,160
400	11,181	1139,7	6,272	1,961	2,196
410	10,862	1161,8	6,327	2,002	2,233
420	10,564	1184,3	6,381	2,043	2,271
430	10,283	1207,2	6,435	2,083	2,309
440	10,019	1230,5	6,488	2,124	2,347
450	9,7677	1254,2	6,541	2,165	2,386
460	9,5306	1278,2	6,594	2,206	2,424
470	9,3058	1302,6	6,647	2,246	2,463
480	9,0922	1327,5	6,699	2,286	2,502
490	8,8893	1352,7	6,751	2,326	2,540
500	8,6954	1378,3	6,803	2,366	2,579
520	8,3337	1430,6	6,905	2,444	2,655
540	8,0013	1484,4	7,007	2,519	2,728
560	7,6967	1539,7	7,108	2,592	2,800
580	7,4148	1596,4	7,207	2,662	2,868
600	7,1538	1654,4	7,305	2,729	2,934
620	6,9112	1713,7	7,403	2,792	2,996
640	6,6851	1774,3	7,499	2,854	3,057
660	6,4733	1836,0	7,594	2,915	3,117
680	6,2754	1899,0	7,688	2,978	3,180
700	6,0893	1963,2	7,781	3,046	3,246
			$p=0,9$		
100	719,33	157,6	2,183	—	—
110	709,30	175,2	2,356	—	—
120	699,08	194,3	2,522	—	—
130	688,90	214,3	2,682	1,416	2,018
140	678,75	234,6	2,832	1,407	2,030
150	668,60	254,9	2,972	1,391	2,033
160	658,42	275,2	3,104	1,380	2,038
170	648,17	295,6	3,228	1,376	2,049
180	637,82	316,2	3,345	1,379	2,065
190	627,33	337,0	3,457	1,387	2,088
200	616,66	358,0	3,565	1,400	2,116
210	605,78	379,3	3,669	1,416	2,149
220	594,65	401,0	3,770	1,436	2,187
230	583,22	423,1	3,868	1,460	2,231
240	571,46	445,6	3,964	1,487	2,279
250	559,31	468,7	4,058	1,517	2,334
260	546,70	492,3	4,151	1,551	2,395
270	533,53	516,6	4,243	1,587	2,464
280	519,67	541,6	4,334	1,627	2,543
290	504,90	567,5	4,424	1,669	2,637
300	18,996	930,9	5,653	1,612	2,024
310	17,957	951,1	5,719	1,639	2,007
320	17,069	971,1	5,782	1,668	2,005

Table II.7 (*Continued*)

T	ρ	h	s	c_v	c_p
		$p=0,9$			
330	16,298	991,2	5,844	1,700	2,014
340	15,616	1011,4	5,904	1,734	2,030
350	15,005	1031,8	5,964	1,770	2,052
360	14,453	1052,5	6,022	1,807	2,077
370	13,951	1073,4	6,079	1,845	2,106
380	13,489	1094,6	6,136	1,884	2,138
390	13,062	1116,2	6,192	1,924	2,171
400	12,667	1138,0	6,247	1,964	2,206
410	12,299	1160,3	6,302	2,004	2,242
420	11,954	1182,9	6,356	2,044	2,278
430	11,630	1205,9	6,410	2,085	2,315
440	11,326	1229,2	6,464	2,125	2,353
450	11,039	1252,9	6,517	2,166	2,391
460	10,768	1277,0	6,570	2,206	2,429
470	10,510	1301,5	6,623	2,247	2,468
480	10,266	1326,4	6,675	2,287	2,506
490	10,033	1351,6	6,727	2,326	2,544
500	9,8125	1377,3	6,779	2,366	2,582
520	9,3994	1429,7	6,882	2,444	2,658
540	9,0226	1483,5	6,984	2,519	2,731
560	8,6756	1538,9	7,084	2,592	2,802
580	8,3560	1595,6	7,184	2,662	2,870
600	8,0595	1653,7	7,282	2,728	2,936
620	7.7845	1713,0	7,379	2,792	2,998
640	7,5283	1773,6	7,476	2,853	3,058
660	7 2891	1835,3	7,571	2,915	3,119
680	7,0648	1898,3	7,665	2,978	3,181
700	6,8546	1962,6	7,758	3,046	3,248
		$p=1,0$			
100	719,38	157,7	2,183	—	—
110	709,34	175,3	2,356	—	—
120	699,13	194,4	2,522	—	—
130	688,95	214,4	2,682	1,415	2,017
140	678,80	234,7	2,832	1,407	2,030
150	668,65	255,0	2,972	1,391	2,033
160	658,48	275,3	3,103	1,380	2,038
170	648,23	295,8	3,227	1,376	2,048
180	637,89	316,3	3,345	1,379	2,065
190	627,41	337,1	3,457	1,387	2,088
200	616,75	358,1	3,565	1,400	2,116
210	605,87	379,4	3,669	1,416	2,149
220	594.75	401,1	3,770	1,436	2,187
230	583,34	423,2	3,868	1,460	2.230
240	571,59	445,7	3,964	1,487	2,279
250	559 46	468,8	4,058	1,517	2 333
260	546.87	492,4	4,151	1,551	2.394
270	533.73	516,7	4,242	1,587	2.463
280	519,90	541,7	4,333	1,626	2.541
290	505,17	567,5	4,424	1,668	2 634
300	489,24	594.4	4,515	1,714	2.748
310	20,406	947,4	5,690	1,653	2 059
320	19,328	967,9	5,755	1,680	2,046

Table II.7 (*Continued*)

T	ρ	h	s	c_v	c_p
		$p=1,0$			
330	18,403	988,3	5,818	1,710	2,046
340	17,596	1008,8	5,879	1,742	2,056
350	16,880	1029,5	5,939	1,776	2,073
360	16,236	1050,3	5,998	1,812	2,096
370	15,652	1071,4	6,055	1,850	2,122
380	15,120	1092,8	6,112	1,888	2,151
390	14,629	1114,4	6,169	1,927	2,183
400	14,175	1136,4	6,224	1,966	2,216
410	13,754	1158,8	6,279	2,006	2,250
420	13,361	1181,4	6,334	2,046	2,286
430	12,993	1204,5	6,388	2,086	2,322
440	12,647	1227,9	6,442	2,127	2,359
450	12,322	1251,7	6,495	2,167	2,397
460	12,014	1275,8	6,549	2,207	2,434
470	11,724	1300,4	6,601	2,247	2,472
480	11,448	1325,3	6,654	2,287	2,510
490	11,185	1350,6	6,706	2,327	2,548
500	10,937	1376,2	6,758	2,366	2,586
520	10,472	1428,7	6,861	2,444	2,661
540	10,048	1482,6	6,962	2,519	2,734
560	9,6582	1538,0	7,063	2,592	2,805
580	9,2995	1594,8	7,163	2,662	2,873
600	8,9677	1652,9	7,261	2,728	2,938
620	8,6599	1712,3	7,359	2,792	3,000
640	8,3732	1772,9	7,455	2,853	3,060
660	8,1056	1834,7	7,550	2,915	3,120
680	7,8553	1897,7	7,644	2,978	3,183
700	7,6209	1962,0	7,737	3,046	3,249
		$p=1,2$			
100	719,46	158,0	2,183	—	—
110	709,42	175,5	2,355	—	—
120	699,21	194,7	2,522	—	—
130	689,04	214,6	2,681	1,415	2,017
140	678,90	234,9	2,832	1,407	2,030
150	668,76	255,2	2,972	1,391	2,033
160	658,59	275,6	3,103	1,380	2,037
170	648,36	296,0	3,227	1,376	2,048
180	638,03	316,5	3,344	1,379	2,065
190	627,56	337,3	3,457	1,387	2,087
200	616,92	358,3	3,564	1,400	2,115
210	606,06	379,6	3,668	1,416	2,148
220	594,96	401,3	3,769	1,436	2,186
230	583,57	423,3	3,867	1,460	2,229
240	571,85	445,9	3,963	1,487	2,277
250	559,75	468,9	4,057	1,517	2,331
260	547,21	492,5	4,150	1,550	2,392
270	534,12	516,8	4,241	1,587	2,460
280	520,35	541,8	4,332	1,626	2,538
290	505,72	567,6	4,423	1,668	2,630
300	489,91	594,4	4,514	1,712	2,741
310	25,763	939,2	5,635	1,686	2,192

Table II.7 (*Continued*)

T	ρ	h	s	c_v	c_p
			$p=1,2$		
320	24,178	960,9	5,704	1,706	2,143
330	22,870	982,2	5,770	1,731	2,121
340	21,758	1003,4	5,833	1,759	2,116
350	20,789	1024,5	5,894	1,790	2,122
360	19,933	1045,8	5,954	1,824	2,136
370	19,168	1067,3	6,013	1,859	2,156
380	18,474	1088,9	6,071	1,896	2,180
390	17,843	1110,9	6,128	1,933	2,207
400	17,261	1133,1	6,184	1,971	2,237
410	16,725	1155,6	6,240	2,010	2,269
420	16,228	1178,5	6,295	2,050	2,303
430	15,764	1201,7	6,349	2,089	2,337
440	15,331	1225,2	6,403	2,129	2,372
450	14,924	1249,1	6,457	2,169	2,408
460	14,541	1273,4	6,510	2,209	2,445
470	14,179	1298,0	6,563	2,248	2,482
480	13,836	1323,0	6,616	2,288	2,519
490	13,513	1348,4	6,668	2,328	2,556
500	13,204	1374,2	6,720	2,367	2,593
520	12,632	1426,8	6,824	2,444	2,667
540	12,113	1480,8	6,926	2,520	2,739
560	11,635	1536,3	7,027	2,592	2,810
580	11,197	1593,2	7,126	2,662	2,877
600	10,792	1651,4	7,225	2,728	2,942
620	10,417	1710,9	7,322	2,792	3,003
640	10,068	1771,5	7,419	2,853	3,063
660	9,7430	1833,4	7,514	2,914	3,123
680	9,4396	1896,5	7,608	2,978	3,186
700	9,1552	1960,8	7,701	3,045	3,252
			$p=1,4$		
100	719,55	158,2	2,182	—	—
110	709,51	175,8	2,355	—	—
120	699,30	194,9	2,521	—	—
130	689,13	214,9	2,681	1,415	2,017
140	679,00	235,1	2,831	1,406	2,030
150	668,87	255,4	2,971	1,391	2,032
160	658,71	275,8	3,103	1,380	2,037
170	648,49	296,2	3,226	1,376	2,048
180	638,17	316,8	3,344	1,379	2,064
190	627,71	337,5	3,456	1,387	2,087
200	617,09	358,5	3,564	1,400	2,114
210	606,25	379,8	3,668	1,416	2,147
220	595,17	401,5	3,768	1,436	2,185
230	583,80	423,5	3,866	1,460	2,228
240	572,11	446,1	3,962	1,487	2,276
250	560,05	469,1	4,056	1,517	2,330
260	547,54	492,7	4,149	1,550	2,390
270	534,50	516,9	4,240	1,586	2,457
280	520,81	541,9	4,331	1,625	2,535
290	506,25	567,6	4,421	1,667	2,625
300	490,57	594,4	4,512	1,711	2,734

Table II.7 (*Continued*)

T	ρ	h	s	c_v	c_p
			$p=1,4$		
310	473,29	622,4	4,604	1,759	2,874
320	29,589	953,1	5,656	1,735	2,273
330	27,744	975,5	5,725	1,754	2,215
340	26,229	997,5	5,791	1,778	2,188
350	24,944	1019,3	5,854	1,805	2,179
360	23,828	1041,1	5,916	1,836	2,182
370	22,843	1063,0	5,975	1,869	2,194
380	21,965	1085,0	6,034	1,904	2,212
390	21,168	1107,2	6,092	1,940	2,234
400	20,446	1129,7	6,149	1,977	2,261
410	19,782	1152,4	6,205	2,015	2,289
420	19,168	1175,5	6,261	2,053	2,320
430	18,599	1198,8	6,316	2,092	2,353
440	18,070	1222,5	6,370	2,131	2,386
450	17,576	1246,6	6,424	2,171	2,421
460	17,110	1271,0	6,478	2,210	2,456
470	16,673	1295,7	6,531	2,250	2,492
480	16,261	1320,8	6,584	2,289	2,528
490	15,871	1346,3	6,636	2,328	2,565
500	15,501	1372,1	6,688	2,367	2,601
520	14,817	1424,8	6,792	2,445	2,674
540	14,195	1479,0	6,894	2,520	2,745
560	13,627	1534,6	6,995	2,592	2,815
580	13,106	1591,6	7,095	2,662	2,882
600	12,626	1649,9	7,194	2,728	2,946
620	12,182	1709,4	7,291	2,792	3,007
640	11,770	1770,2	7,388	2,853	3,067
660	11,387	1832,1	7,483	2,914	3,127
680	11,028	1895,2	7,577	2,978	3,189
700	10,693	1959,7	7,671	3,045	3,255
			$p=1,6$		
100	719,63	158,5	2,182	—	—
110	709,60	176,0	2,354	—	—
120	699,39	195,2	2,521	—	—
130	689,23	215,1	2,681	1,414	2,017
140	679,10	235,4	2,831	1,406	2,030
150	668,98	255,7	2,971	1,390	2,032
160	658,83	276,0	3,102	1,379	2,037
170	648,61	296,4	3,226	1,376	2,047
180	638,31	317,0	3,343	1,379	2,064
190	627,87	337,7	3,455	1,387	2,086
200	617,26	358,7	3,563	1,400	2,114
210	606,44	380,0	3,667	1,416	2,146
220	595,38	401,7	3,768	1,436	2,184
230	584,04	423,7	3,866	1,459	2,227
240	572,38	446,2	3,962	1,486	2,275
250	560,34	469,2	4,055	1,516	2,328
260	547,88	492,8	4,148	1,550	2,388
270	534,89	517,0	4,239	1,586	2,455
280	521,25	541,9	4,330	1,625	2,531
290	506,79	567,7	4,420	1,666	2,621
300	491,22	594,4	4,511	1,710	2,728

Table II.7 (*Continued*)

T	ρ	h	s	c_v	c_p
		$p=1,6$			
310	474,11	622,3	4,602	1,758	2,863
320	454,74	651,8	4,696	1,810	3,049
330	33,146	968,1	5,683	1,780	2,338
340	31,081	991,1	5,752	1,798	2,276
350	29,384	1013,7	5,817	1,822	2,246
360	27,949	1036,1	5,880	1,849	2,235
370	26,702	1058,4	5,941	1,880	2,237
380	25,603	1080,9	6,001	1,912	2,247
390	24,619	1103,4	6,060	1,947	2,264
400	23,735	1126,2	6,117	1,983	2,286
410	22,926	1149,1	6,174	2,020	2,311
420	22,186	1172,4	6,230	2,057	2,339
430	21,502	1195,9	6,286	2,096	2,369
440	20,868	1219,8	6,340	2,134	2,401
450	20,279	1244,0	6,395	2,173	2,434
460	19,726	1268,5	6,449	2,212	2,468
470	19,207	1293,3	6,502	2,251	2,503
480	18,720	1318,5	6,555	2,290	2,538
490	18,260	1344,1	6,608	2,329	2,573
500	17,826	1370,0	6,660	2,368	2,609
520	17,022	1422,9	6,764	2,445	2,681
540	16,296	1477,2	6,866	2,520	2,751
560	15,633	1532,9	6,968	2,592	2,820
580	15,028	1590,0	7,068	2,662	2,886
600	14,470	1548,4	7,167	2,728	2,950
620	13,956	1708,0	7,264	2,791	3,011
640	13,479	1768,8	7,361	2,853	3,070
660	13,035	1830,8	7,456	2,914	3,130
680	12,622	1894,0	7,551	2,978	3,192
700	12,234	1958,5	7,644	3,045	3,258
		$p=1,8$			
100	719,72	158,7	2,182	—	—
110	709,68	176,2	2,354	—	—
120	699,48	195,4	2,520	—	—
130	689,32	215,4	2,680	1,414	2,017
140	679,20	235,6	2,830	1,406	2,030
150	669,08	255,9	2,970	1,390	2,032
160	658,94	276,2	3,102	1,379	2,077
170	648,74	296,7	3,225	1,376	2,047
180	638,44	317,2	3,343	1,379	2,063
190	628,02	337,9	3,455	1,387	2,086
200	617,42	358,9	3,563	1,399	2,113
210	606,62	380,2	3,666	1,416	2,146
220	595,59	401,9	3,767	1,436	2,183
230	584,27	423,9	3,865	1,459	2,226
240	572,64	446,4	3,961	1,486	2,274
250	560,64	469,4	4,055	1,516	2,327
260	548,21	493,0	4,147	1,549	2,386
270	535,27	517,0	4,238	1,585	2,453
280	521,70	542,0	4,329	1,624	2,528
290	507,31	567,8	4,419	1,665	2,616
300	491,86	594,4	4,510	1,709	2,721

Table II.7 (*Continued*)

T	ρ	h	s	c_v	c_p

$p=1,8$

T	ρ	h	s	c_v	c_p
310	474,92	622,3	4,601	1,756	2,854
320	455,82	651 6	4,694	1,808	3,032
330	39,263	959,7	5,641	1,809	2,509
340	36,413	984,2	5,714	1,820	2,389
350	34,172	1007,7	5,782	1,839	2,328
360	32,332	1030,8	5,847	1,863	2,297
370	30,767	1053,7	5,910	1,891	2,286
380	29,407	1076,6	5,971	1,922	2,287
390	28,207	1099,5	6,030	1,955	2,297
400	27,136	1122,5	6,089	1,989	2,314
410	26,168	1145,8	6,146	2,025	2,335
420	25,283	1169,2	6,203	2,062	2,360
430	24,475	1193,0	6,258	2,099	2,387
440	23,728	1217,0	6,314	2,137	2,417
450	23,034	1241,3	6,368	2,175	2,448
460	22,388	1266,0	6,422	2,214	2,480
470	21,785	1290,9	6,476	2,253	2,514
480	21,217	1316,2	6,529	2,292	2,548
490	20,682	1341,9	6,582	2,330	2,583
500	20,179	1367,9	6,635	2,369	2,618
520	19,252	1420,9	6,739	2,446	2,688
540	18,415	1475,4	6,842	2,520	2,757
560	17,655	1531,2	6,943	2,593	2,825
580	16,962	1588,4	7,043	2,662	2,891
600	16,324	1646,9	7,143	2,728	2,954
620	15,738	1706,6	7,240	2,791	3,015
640	15,193	1767,4	7,337	2,853	3,074
660	14,689	1829,5	7,433	2,914	3,133
680	14,218	1892,8	7,527	2,978	3,195
700	13,779	1957,3	7,620	3,045	3,261

$p=2,0$

T	ρ	h	s	c_v	c_p
100	719,80	158,9	2,181	—	—
110	709,76	176,5	2,354	—	—
120	699,57	195,6	2,520	—	—
130	689,41	215,6	2,680	1,413	2,017
140	679,30	235,8	2,830	1,405	2,029
150	669,19	256,1	2,970	1,390	2,032
160	659,06	276,5	3,101	1,379	2,036
170	648,87	296,9	3,225	1,375	2,047
180	638,58	317,4	3,342	1,378	2,063
190	628,17	338,2	3,454	1,387	2,085
200	617,59	359,1	3,562	1,399	2,112
210	606,81	380,4	3,666	1,416	2,145
220	595,79	402,1	3,767	1,436	2,182
230	584,50	424,1	3,864	1,459	2,225
240	572,89	446,6	3,960	1,486	2,272
250	560,93	469,6	4,054	1,516	2,325
260	548,54	493,1	4,146	1,549	2,384
270	535,65	517,3	4,237	1,585	2,450
280	522,14	542,1	4,328	1,623	2,525
290	507,83	567,8	4,418	1,665	2,612
300	492,49	594,4	4,508	1,708	2,715

Table II.7 (*Continued*)

T	ρ	h	s	c_v	c_p
		$p=2,0$			
310	475,71	622,2	4,599	1,755	2,844
320	456,86	651,5	4,692	1,806	3,016
330	434,71	682,8	4,789	1,864	3,275
340	42,377	976,4	5,676	1,846	2,541
350	39,390	1001,2	5,784	1,859	2,429
360	37,025	1025,2	5,815	1,878	2,371
370	35,064	1048,7	5,880	1,903	2,342
380	33,398	1072,1	5,942	1,931	2,332
390	31,945	1095,4	6,003	1,962	2,334
400	30,661	1118,8	6,062	1,996	2,344
410	29,510	1142,3	6,120	2,030	2,361
420	28,470	1166,0	6,177	2,066	2,382
430	27,522	1190,0	6,233	2,103	2,406
440	26,650	1214,2	6,289	2,140	2,433
450	25,843	1238,6	6,344	2,178	2,463
460	25,097	1263,4	6,399	2,216	2,493
470	24,401	1288,5	6,453	2,254	2,526
480	23,749	1313,9	6,506	2,293	2,558
490	23,138	1339,7	6,559	2,332	2,592
500	22,561	1365,8	6,612	2,370	2,626
520	21,504	1419,0	6,716	2,446	2,695
540	20,554	1473,6	6,819	2,521	2,764
560	19,692	1529,5	6,921	2,593	2,831
580	18,907	1586,8	7,021	2,662	2,896
600	18,188	1645,4	7,121	2,728	2,959
620	17,527	1705,1	7,219	2,791	3,019
640	16,916	1766,1	7,315	2,853	3,078
660	16,348	1828,2	7,411	2,914	3,136
680	15,819	1891,6	7,506	2,978	3,198
700	15,326	1956,2	7,599	3,045	3,263
		$p=2,2$			
100	719,89	159,2	2,181	—	—
110	709,85	176,7	2,353	—	—
120	699,66	195,9	2,520	—	—
130	689,51	215,8	2,679	1,413	2,017
140	679,40	236,1	2,829	1,405	2,029
150	669,30	256,4	2,969	1,390	2,031
160	659,17	276,7	3,101	1,379	2,036
170	648,99	297,1	3,224	1,375	2,046
180	638,72	317,7	3,342	1,378	2,063
190	628,32	338,4	3,454	1,387	2,085
200	617,76	359,4	3,561	1,399	2,112
210	607,00	380,6	3,665	1,416	2,144
220	596,00	402,3	3,766	1,436	2,182
230	584,73	424,3	3,864	1,459	2,224
240	573,15	446,8	3,959	1,486	2,271
250	561,22	469,7	4,053	1,516	2,324
260	548,87	493,3	4,145	1,549	2,382
270	536,03	517,4	4,237	1,584	2,448
280	522,58	542,2	4,327	1,623	2,522
290	508,35	567,9	4,417	1,664	2,608
300	493,11	594,4	4,507	1,707	2,709

Table II.7 (*Continued*)

T	ρ	h	s	c_v	c_p
		$p=2,2$			
310	476,49	622,1	4,598	1,754	2,835
320	457,89	651,3	4,690	1,804	3,002
330	436,17	682,4	4,786	1,861	3,246
340	49,222	967,6	5,637	1,876	2,758
350	45,148	994,1	5,714	1,880	2,560
360	42,093	1019,1	5,784	1,895	2,461
370	39,637	1043,5	5,851	1,916	2,409
380	37,597	1067,4	5,915	1,942	2,383
390	35,847	1091,2	5,977	1,971	2,375
400	34,318	1114,9	6,037	2,002	2,378
410	32,965	1138,8	6,096	2,036	2,389
420	31,746	1162,7	6,153	2,071	2,405
430	30,644	1186,9	6,210	2,106	2,427
440	29,639	1211,3	6,266	2,143	2,451
450	28,711	1235,9	6,322	2,180	2,478
460	27,856	1260,8	6,377	2,218	2,507
470	27,061	1286,1	6,431	2,256	2,538
480	26,319	1311,6	6,485	2,294	2,570
490	25,627	1337,5	6,538	2,333	2,602
500	24,974	1363,7	6,591	2,371	2,635
520	23,780	1417,0	6,695	2,447	2,703
540	22,709	1471,8	6,799	2,521	2,770
560	21,745	1527,8	6,901	2,593	2,837
580	20,864	1585,2	7,001	2,662	2,901
600	20,063	1643,9	7,101	2,728	2,963
620	19,324	1703,7	7,199	2,791	3,023
640	18,643	1764,8	7,296	2,853	3,081
660	18,012	1827,0	7,391	2,914	3,140
680	17,426	1890,4	7,486	2,978	3,201
700	16,878	1955,0	7,580	3,045	3,266
		$p=2,4$			
100	719,97	159,4	2,180	—	—
110	709,93	177,0	2,353	—	—
120	699,74	196,1	2,519	—	—
130	689,60	216,1	2,679	1,412	2,016
140	679,50	236,3	2,829	1,405	2,029
150	669,40	256,6	2,969	1,390	2,031
160	659,29	276,9	3,100	1,379	2,036
170	649,12	297,3	3,224	1,375	2,046
180	638,86	317,9	3,341	1,378	2,062
190	628,47	338,6	3,453	1,387	2,084
200	617,93	359,6	3,561	1,399	2,111
210	607,18	380,8	3,665	1,416	2,144
220	596,20	402,5	3,765	1,436	2,181
230	584,96	424,5	3,863	1,459	2,223
240	573,41	446,9	3,959	1,486	2,270
250	561,51	469,9	4,052	1,515	2,322
260	549,20	493,4	4,145	1,548	2,381
270	536,40	517,5	4,236	1,584	2,445
280	523,01	542,3	4,326	1,622	2,519
290	508,86	567,9	4,416	1,663	2,603
300	493,73	594,5	4,506	1,706	2,704

Table II.7 (*Continued*)

T	ρ	h	s	c_v	c_p
		$p=2,4$			
310	477,26	622,1	4,596	1,752	2,827
320	458,88	651,1	4,688	1,802	2,987
330	437,58	682,1	4,784	1,858	3,220
340	57,420	957,2	5,595	1,912	3,106
350	51,630	986,2	5,679	1,905	2,738
360	47,613	1012,6	5,754	1,913	2,573
370	44,533	1037,9	5,823	1,930	2,487
380	42,034	1062.5	5,889	1,953	2,442
390	39,929	1086,8	5,952	1,980	2,421
400	38,124	1111,0	6,013	2,010	2,415
410	36,533	1135,1	6,073	2,042	2,419
420	35,122	1159,4	6,131	2,075	2,431
430	33,850	1183,8	6,189	2,110	2,448
440	32,696	1208,3	6,245	2,146	2,470
450	31,640	1233,2	6,301	2,183	2,495
460	30,667	1258,2	6,356	2,220	2,522
470	29,768	1283,6	6,411	2,258	2,551
480	28,931	1309,3	6,465	2,296	2,581
490	28,147	1335,2	6,518	2,334	2,612
500	27,415	1361,5	6,571	2,372	2,645
520	26,080	1415,1	6,676	2,448	2,711
540	24,886	1469,9	6,780	2,522	2,777
560	23,811	1526,1	6,882	2,594	2,842
580	22,834	1583,6	6,983	2,663	2,906
600	21,945	1642,4	7,082	2,728	2,968
620	21,128	1702,3	7,181	2,792	3,027
640	20,378	1763,4	7,278	2,853	3,085
660	19,682	1825,7	7,373	2,914	3,143
680	19,035	1889,2	7,468	2,978	3,204
700	18,433	1953,9	7,562	3,045	3,269
		$p=2,6$			
100	720,05	159,6	2,180	—	—
110	710,02	177,2	2,352	—	—
120	699,83	196,3	2,519	—	—
130	689,69	216,3	2,678	1,412	2,016
140	679,59	236,5	2,828	1,405	2,029
150	669,51	256,8	2,969	1,389	2,031
160	659,40	277,2	3,100	1,379	2,036
170	649,24	297,6	3,223	1,375	2,046
180	638,99	318,1	3,341	1,378	2,062
190	628,62	338,8	3,453	1,387	2,084
200	618,09	359,8	3,560	1,399	2,111
210	607,36	381,0	3,664	1,416	2,143
220	596,39	402,6	3,765	1,438	2,180
230	585,19	424,7	3,862	1,459	2,222
240	573,66	447,1	3,958	1,485	2,269
250	561,79	470,1	4,052	1,515	2,321
260	549,52	493,5	4,144	1,548	2,379
270	536,77	517,7	4,235	1,584	2,443
280	523,44	542,4	4,325	1,622	2,516
290	509,36	568,0	4,415	1,662	2,599
300	494,33	594,5	4,504	1,705	2,698

Table II.7 (*Continued*)

T	ρ	h	s	c_v	c_p

$p=2,6$

T	ρ	h	s	c_v	c_p
310	478,01	622,0	4,595	1,751	2,818
320	459,86	651,0	4,686	1,800	2,974
330	438,93	681,7	4,781	1,855	3,195
340	413,16	715,4	4,881	1,920	3,572
350	59,130	977,1	5,643	1,934	2,997
360	53,723	1005,5	5,723	1,933	2,717
370	49,805	1031,9	5,795	1,945	2,582
380	46,739	1057,3	5,863	1,965	2,510
390	44,219	1082,2	5,928	1,989	2,472
400	42,082	1106,8	5,990	2,017	2,455
410	40,231	1131,4	6,051	2,048	2,452
420	38,599	1155,9	6,110	2,080	2,459
430	37,142	1180,6	6,168	2,115	2,472
440	35,825	1205,4	6,225	2,150	2,490
450	34,625	1230,4	6,281	2,186	2,512
460	33,528	1255,6	6,337	2,223	2,537
470	32,516	1281,1	6,392	2,260	2,564
480	31,580	1306,9	6,446	2,298	2,593
490	30,706	1333,0	6,500	2,335	2,623
500	29,889	1359,4	6,553	2,373	2,654
520	28,401	1413,1	6,658	2,448	2,719
540	27,080	1468,1	6,762	2,522	2,784
560	25,890	1524,4	6,865	2,594	2,848
580	24,817	1582,0	6,966	2,663	2,911
600	23,838	1640,9	7,065	2,729	2,972
620	22,941	1700,9	7,164	2,792	3,031
640	22,117	1762,1	7,261	2,853	3,089
660	21,354	1824,4	7,357	2,914	3,147
680	20,648	1888,0	7,452	2,978	3,207
700	19,989	1952,8	7,545	3,045	3,272

$p=2,8$

T	ρ	h	s	c_v	c_p
100	720,14	159,9	2,180	—	—
110	710,10	177,4	2,352	—	—
120	699,92	196,6	2,518	—	—
130	689,78	216,5	2,678	1,412	2,016
140	679,69	236,8	2,828	1,404	2,029
150	669,62	257,1	2,968	1,389	2,031
160	659,52	277,4	3,099	1,379	2,035
170	649,37	297,8	3,223	1,375	2,045
180	639,13	318,3	3,340	1,378	2,061
190	628,77	339,0	3,452	1,387	2,083
200	618,26	360,0	3,560	1,399	2,110
210	607,55	381,3	3,664	1,416	2,142
220	596,61	402,9	3,764	1,435	2,179
230	585,41	424,9	3,862	1,459	2,221
240	573,92	447,3	3,957	1,485	2,268
250	562,08	470,2	4,051	1,515	2,319
260	549,84	493,7	4,143	1,548	2,377
270	537,14	517,8	4,234	1,583	2,441
280	523,86	542,5	4,324	1,621	2,513
290	509,86	568,1	4,413	1,662	2,596
300	494,93	594,5	4,503	1,705	2,692

Table II.7 (*Continued*)

T	ρ	h	s	c_v	c_p
			$p=2,8$		
310	478,75	622,0	4,593	1,750	2,810
320	460,81	650,8	4,685	1,799	2,961
330	440,24	681,4	4,779	1,852	3,172
340	415,21	714,7	4,878	1,915	3,519
350	68,214	966,3	5,603	1,969	3,425
360	60,592	997,5	5,691	1,956	2,912
370	55,551	1025,5	5,768	1,962	2,699
380	51,761	1051,8	5,838	1,978	2,590
390	48,736	1077,4	5,905	1,999	2,531
400	46,216	1102,6	5,968	2,025	2,501
410	44,063	1127,5	6,030	2,054	2,489
420	42,188	1152,4	6,090	2,086	2,488
430	40,520	1177,3	6,148	2,119	2,497
440	39,027	1202,3	6,206	2,153	2,511
450	37,676	1227,5	6,263	2,189	2,530
460	36,445	1252,9	6,318	2,225	2,553
470	35,315	1278,6	6,374	2,262	2,578
480	34,269	1304,5	6,428	2,299	2,605
490	33,297	1330,7	6,482	2,337	2,634
500	32,392	1357,2	6,536	2,374	2,664
520	30,747	1411,1	6,641	2,449	2,727
540	29,290	1466,3	6,746	2,523	2,791
560	27,988	1522,7	6,848	2,594	2,854
580	26,810	1580,5	6,949	2,663	2,917
600	25,741	1639,4	7,049	2,729	2,977
620	24,762	1699,5	7,148	2,792	3,035
640	23,865	1760,8	7,245	2,853	3,092
660	23,035	1823,2	7,341	2,914	3,150
680	22,264	1886,8	7,436	2,978	3,210
700	21,551	1951,6	7,530	3,045	3,275
			$p=3,0$		
100	720,22	160,1	2,179	—	—
110	710,18	177,7	2,352	—	—
120	700,00	196,8	2,518	—	—
130	689,88	216,8	2,678	1,411	2,016
140	679,79	237,0	2,828	1,404	2,028
150	669,72	257,3	2,968	1,389	2,031
160	659,63	277,6	3,099	1,378	2,035
170	649,49	298,0	3,222	1,375	2,045
180	639,27	318,5	3,340	1,378	2,061
190	628,92	339,3	3,452	1,387	2,083
200	618,42	360,2	3,559	1,399	2,110
210	607,73	381,5	3,663	1,416	2,141
220	596,81	403,1	3,763	1,435	2,178
230	585,64	425,0	3,861	1,459	2,220
240	574,17	447,5	3,957	1,485	2,266
250	562,36	470,4	4,050	1,515	2,318
260	550,17	493,8	4,142	1,547	2,375
270	537,51	517,9	4,233	1,583	2,439
280	524,29	542,6	4,323	1,621	2,510
290	510,36	568,1	4,412	1,661	2,592
300	495,52	594,5	4,502	1,704	2,687

Table II.7 (*Continued*)

T	ρ	h	s	c_v	c_p
			$p=3,0$		
310	479,48	622,0	4,592	1,749	2,802
320	461,74	650,7	4,683	1,797	2,949
330	441,51	681,1	4,777	1,850	3,150
340	417,15	714,1	4,875	1,911	3,471
350	383,74	751,9	4,985	1,993	4,223
360	68,544	988,5	5,658	1,983	3,194
370	61,874	1018,4	5,740	1,981	2,848
380	57,162	1046,0	5,813	1,991	2,685
390	53,512	1072,4	5,882	2,010	2,599
400	50,543	1098,1	5,947	2,034	2,552
410	48,047	1123,5	6,010	2,061	2,528
420	45,891	1148,7	6,070	2,091	2,521
430	43,998	1174,0	6,130	2,123	2,523
440	42,314	1199,2	6,188	2,157	2,534
450	40,796	1224,6	6,245	2,192	2,550
460	39,418	1250,2	6,301	2,228	2,570
470	38,157	1276,0	6,357	2,264	2,593
480	36,995	1302,1	6,411	2,301	2,618
490	35,924	1328,4	6,466	2,338	2,646
500	34,923	1355,0	6,520	2,376	2,675
520	33,118	1409,1	6,626	2,450	2,736
540	31,523	1464,5	6,730	2,524	2,798
560	30,098	1521,0	6,833	2,595	2,860
580	28,816	1578,9	6,934	2,663	2,922
600	27,652	1637,9	7,034	2,729	2,981
620	26,591	1698,1	7,133	2,792	3,039
640	25,617	1759,5	7,231	2,853	3,096
660	24,718	1822,0	7,327	2,915	3,153
680	23,886	1885,6	7,422	2,978	3,213
700	23,114	1950,5	7,516	3,045	3,278
			$p=3,2$		
100	720,30	160,4	2,179	——	—
110	710,27	177,9	2,351	—	—
120	700,09	197,0	2,518	—	—
130	689,97	217,0	2,677	1,411	2,016
140	679,89	237,2	2,827	1,404	2,028
150	669,83	257,5	2,967	1,389	2,030
160	659,75	277,8	3,098	1,378	2,035
170	649,62	298,2	3,222	1,375	2,045
180	639,40	318,8	3,339	1,378	2,061
190	629,07	339,5	3,451	1,386	2,082
200	618,59	360,4	3,559	1,399	2,109
210	607,91	381,7	3,662	1,415	2,141
220	597,02	403,3	3,763	1,435	2,177
230	585,86	425,2	3,860	1,458	2,219
240	574,42	447,6	3,956	1,485	2,265
250	562,64	470,6	4,049	1,515	2,317
260	550,48	494,0	4,141	1,547	2,373
270	537,87	518,0	4,232	1,583	2,437
280	524,70	542,8	4,322	1,620	2,507
290	510,85	568,2	4,411	1,661	2,588
300	496,11	594,6	4,501	1,703	2,682

Table II.7 (*Continued*)

T	ρ	h	s	c_v	c_p
			p=3,2		
310	480,19	621,9	4,590	1,748	2,795
320	462,65	650,6	4,681	1,796	2,937
330	442,74	680,8	4,774	1,848	3,130
340	418,99	713,5	4,872	1,907	3,429
350	387,53	750,4	4,979	1,984	4,060
360	78,130	977,8	5,620	2,015	3,650
370	68,946	1010,7	5,710	2,001	3,045
380	63,002	1039,8	5,788	2,006	2,801
390	58,585	1067,1	5,859	2,021	2,677
400	55,089	1093,5	5,926	2,042	2,609
410	52,187	1119,4	5,990	2,068	2,572
420	49,722	1145,0	6,052	2,097	2,556
430	47,576	1170,5	6,112	2,128	2,552
440	45,680	1196,1	6,170	2,161	2,558
450	43,982	1221,7	6,228	2,191	2,570
460	42,447	1247,5	6,285	2,231	2,587
470	41,048	1273,5	6,340	2,267	2,608
480	39,768	1299,7	6,396	2,303	2,632
490	38,588	1326,1	6,450	2,340	2,658
500	37,490	1352,8	6,504	2,377	2,686
520	35,511	1407,1	6,611	2,451	2,744
540	33,773	1462,6	6,715	2,524	2,805
560	32,224	1519,3	6,818	2,595	2,867
580	30,834	1577,3	6,920	2,664	2,927
600	29,573	1036,4	7,020	2,729	2,986
620	28,425	1696,7	7,119	2,792	3,044
640	27,375	1758,2	7,217	2,854	3,100
660	26,407	1820,7	7,313	2,915	3,157
680	25,512	1884,5	7,408	2,978	3,216
700	24,680	1949,4	7,502	3,045	3,281
			p=3,4		
100	720,38	160,6	2,178	—	—
110	710,35	178,1	2,351	—	—
120	700,18	197,3	2,517	—	—
130	690,06	217,2	2,677	1,410	2,016
140	679,99	237,5	2,827	1,404	2,028
150	669,93	257,8	2,967	1,389	2,030
160	659,86	278,1	3,098	1,378	2,034
170	649,74	298,5	3,221	1,375	2,044
180	639,54	319,0	3,339	1,378	2,060
190	629,22	339,7	3,451	1,386	2,082
200	618,75	360,6	3,558	1,399	2,108
210	608,10	381,9	3,662	1,415	2,140
220	597,22	403,5	3,762	1,435	2,177
230	586,09	425,4	3,860	1,458	2,218
240	574,67	447,8	3,955	1,485	2,264
250	562,92	470.7	4,049	1,514	2.315
260	550,80	494,2	4,140	1,547	2,372
270	538,23	518,2	4,231	1,582	2,434
280	525,12	542,9	4,321	1,620	2,505
290	511,33	568,3	4,410	1,660	2,584
300	496,69	594,6	4,499	1,702	2,677

Table II.7 (*Continued*)

T	ρ	h	s	c_v	c_p
			$p=3,4$		
310	480,89	621,9	4,589	1,747	2,788
320	463,55	650,4	4,679	1,794	2,926
330	443,94	680,6	4,772	1,845	3,111
340	420,74	713,0	4,869	1,904	3,391
350	390,70	749,2	4,974	1,976	3,940
360	90,694	964,2	5,576	2,056	4,564
370	77,654	1002,0	5,680	2,025	3,320
380	69,387	1033,1	5,762	2,022	2,945
390	64,005	1061,5	5,836	2,033	2,768
400	59,869	1088,7	5,905	2,052	2,673
410	56,510	1115,1	5,970	2,075	2,621
420	53,688	1141,2	6,033	2,103	2,594
430	51,260	1167,1	6,094	2,133	2,583
440	49,128	1192,9	6,153	2,165	2,583
450	47,238	1218,8	6,212	2,199	2,591
460	45,535	1244,7	6,269	2,233	2,606
470	43,994	1270,9	6,325	2,269	2,624
480	42,584	1297,2	6,380	2,305	2,646
490	41,284	1323,8	6,435	2,342	2,670
500	40,085	1350,6	6,489	2,379	2,697
520	37,928	1405,1	6,596	2,452	2,753
540	36,039	1460,8	6,701	2,525	2,813
560	34,362	1517,7	6,805	2,596	2,873
580	32,862	1575,7	6,907	2,664	2,933
600	31,503	1635,0	7,007	2,730	2,991
620	30,268	1695,3	7,106	2,793	3,048
640	29,137	1756,9	7,204	2,854	3,104
660	28,100	1819,5	7,300	2,915	3,160
680	27,140	1883,3	7,395	2,978	3,220
700	26,249	1948,3	7,489	3,046	3,284
			$p=3,6$		
100	720,47	160,8	2,178	—	—
110	710,43	178,4	2,350	—	—
120	700,26	197,5	2,517	—	—
130	690,15	217,5	2,676	1,410	2,016
140	680,09	237,7	2,826	1,403	2,028
150	670,04	258,0	2,966	1,389	2,030
160	659,97	278,3	3,097	1,378	2,034
170	649,86	298,7	3,221	1,375	2,044
180	639,67	319,2	3,338	1,378	2,060
190	629,37	339,9	3,450	1,386	2,081
200	618,92	360,8	3,558	1,399	2,108
210	608,28	382,1	3,661	1,415	2,139
220	597,42	403,7	3,762	1,435	2,176
230	586,31	425,6	3,859	1,458	2,217
240	574,92	448,0	3,954	1,485	2,263
250	563,20	470,9	4,048	1,514	2,314
260	551,12	494,3	4,140	1,547	2,370
270	538,59	518,3	4,230	1,582	2,432
280	525,53	543,0	4,320	1,620	2,502
290	511,81	568,4	4,409	1,659	2,581
300	497,26	594,6	4,498	1,702	2,672

Table II.7 (*Continued*)

T	ρ	h	s	c_v	c_p
		$p=3,6$			
310	481,59	621,9	4,587	1,746	2,781
320	464,42	650,3	4,678	1,793	2,916
330	445,10	680,3	4,770	1,843	3,094
340	422,42	712,5	4,866	1,900	3,356
350	393,60	748,1	4,969	1,970	3,842
360	346,77	793,1	5,096	2,080	5,816
370	—	—	—	—	—
380	76,453	1025,8	5,736	2,040	3,128
390	69,829	1055,6	5,814	2,046	2,876
400	64,916	1083,7	5,885	2,062	2,746
410	61,018	1110,7	5,951	2,083	2,674
420	57,800	1137,3	6,015	2,109	2,635
430	55,056	1163,5	6,077	2,138	2,616
440	52,668	1189,6	6,137	2,169	2,610
450	50,565	1215,7	6,196	2,202	2,614
460	48,683	1241,9	6,253	2,236	2,625
470	46,986	1268,3	6,310	2,271	2,641
480	45,439	1294,8	6,366	2,307	2,661
490	44,019	1321,5	6,421	2,344	2,683
500	42,714	1348,4	6,475	2,380	2,708
520	40,371	1403,1	6,583	2,453	2,763
540	38,328	1459,0	6,688	2,526	2,820
560	36,520	1516,0	6,752	2,597	2,880
580	34,902	1574,1	6,894	2,665	2,938
600	33,444	1633,5	6,994	2,730	2,996
620	32,119	1694,0	7,093	2,793	3,052
640	30,909	1755,6	7,191	2,854	3,108
660	29,796	1818,3	7,288	2,915	3,164
680	28,772	1882,1	7,383	2,978	3,223
700	27,821	1947,2	7,477	3,046	3,286
		$p=3,8$			
100	720,55	161,1	2,178	—	—
110	710,51	178,6	2,350	—	—
120	700,35	197,7	2,516	—	—
130	690,24	217,7	2,676	1,410	2,015
140	680,18	237,9	2,826	1,403	2,028
150	670,14	258,2	2,966	1,388	2,030
160	660,09	278,5	3,097	1,378	2,034
170	649,99	298,9	3,221	1,375	2,044
180	639,81	319,4	3,338	1,378	2,059
190	629,52	340,1	3,450	1 386	2,081
200	619,08	361,1	3,557	1,399	2,107
210	608,46	382,3	3,661	1,415	2,139
220	597,62	403,9	3,761	1,435	2,175
230	586,53	425,8	3,858	1,458	2,216
240	575,17	448,2	3,954	1,484	2,262
250	563,48	471,1	4,047	1,514	2,312
260	551,43	494,5	4,139	1,546	2,368
270	538,95	518,4	4,229	1,582	2,430
280	525,94	543,1	4,319	1,619	2,499
290	512,29	568,5	4,408	1,659	2,577
300	497,82	594,7	4,497	1,701	2,667

Table II.7 (*Continued*)

T	ρ	h	s	c_v	c_p
		$p=3,8$			
310	482,28	621,9	4,586	1,745	2,774
320	465,28	650,2	4,676	1,791	2,905
330	446,23	680,1	4,768	1,841	3,077
340	424,03	712,0	4,863	1,897	3,324
350	396,27	747,2	4,965	1,964	3,759
360	354,29	789,8	5,085	2,062	5,139
370	98,660	979,7	5,607	2,086	4,467
380	84,413	1017,7	5,708	2,060	3,372
390	76,129	1049,3	5,790	2,060	3,007
400	70,272	1078,4	5,864	2,072	2,830
410	65,743	1106,2	5,933	2,091	2,734
420	62,061	1133,2	5,998	2,116	2,680
430	58,970	1159,9	6,060	2,143	2,651
440	56,305	1186,3	6,121	2,174	2,639
450	53,973	1212,7	6,180	2,206	2,638
460	51,899	1239,1	6,239	2,239	2,645
470	50,030	1265,6	6,296	2,274	2,658
480	48,339	1292,3	6,352	2,309	2,676
490	46,792	1319,1	6,407	2,345	2,697
500	45,371	1346,2	6,462	2,382	2,720
520	42,840	1401,1	6,569	2,455	2,772
540	40,634	1457,1	6,675	2,527	2,828
560	38,686	1514,3	6,779	2,597	2,886
580	36,950	1572,6	6,881	2,665	2,944
600	35,391	1632,0	6,982	2,731	3,001
620	33,974	1692,6	7,081	2,793	3,056
640	32,683	1754,3	7,179	2,854	3,111
660	31,500	1817,1	7,276	2,915	3,167
680	30,407	1881,0	7,371	2,979	3,226
700	29,397	1946,1	7,466	3,046	3,289
		$p=4,0$			
100	720,63	161,3	2,177	—	—
110	710,60	178,9	2,349	—	—
120	700,44	198,0	2,516	—	—
130	690,33	217,9	2,676	1,409	2,015
140	680,28	238,2	2,825	1,403	2,028
150	670,25	258,5	2,965	1,388	2,030
160	660,20	278,8	3,097	1,378	2,034
170	650,11	299,1	3,220	1,375	2,043
180	639,94	319,7	3,337	1,378	2,059
190	629,67	340,3	3,449	1,386	2,080
200	619,24	361,3	3,557	1,399	2,107
210	608,64	382,5	3,660	1,415	2,138
220	597,82	404,1	3,760	1,435	2,174
230	586,76	426,0	3,858	1,458	2,215
240	575,42	448,4	3,953	1,484	2,261
250	563,76	471,2	4,046	1.514	2,311
260	551,74	494,6	4,138	1,546	2,367
270	539,30	518,6	4,229	1,581	2,428
280	526,35	543,2	4,318	1,619	2,496
290	512,76	568,5	4,407	1,658	2,574
300	498,38	594,7	4,496	1,700	2,662

Table II.7 (*Continued*)

T	ρ	h	s	c_v	c_p
		$p=4,0$			
310	482,95	621,8	4,585	1,744	2,767
320	466,12	650,1	4,674	1,790	2,895
330	447,33	679,9	4,766	1,840	3,061
340	425,57	711,6	4,861	1,894	3,295
350	398,75	746,3	4,961	1,958	3,689
360	360,23	787,3	5,076	2,048	4,472
370	115,49	963,4	5,558	2,130	6,142
380	93,577	1008,6	5,678	2,082	3,712
390	83,010	1042,6	5,766	2,075	3,167
400	75,973	1072,9	5,843	2,083	2,928
410	70,697	1101,5	5,914	2,100	2,801
420	66,497	1129,1	5,980	2,122	2,729
430	63,009	1156,2	6,044	2,149	2,689
440	60,044	1182,9	6,106	2,178	2,670
450	57,457	1209,6	6,166	2,209	2,663
460	55,173	1236,2	6,224	2,242	2,666
470	53,128	1262,9	6,282	2,277	2,676
480	51,284	1289,8	6,338	2,312	2,691
490	49,605	1316,8	6,394	2,347	2,710
500	48,068	1344,0	6,449	2,383	2,732
520	45,330	1399,1	6,557	2,456	2,782
540	42,955	1455,3	6,663	2,528	2,836
560	40,871	1512,6	6,767	2,598	2,893
580	39,015	1571,0	6,870	2,666	2,950
600	37,348	1630,6	6,970	2,731	3,006
620	35,841	1691,2	7,070	2,794	3,061
640	34,464	1753,0	7,168	2,855	3,115
660	33,206	1815,9	7,265	2,916	3,171
680	32,047	1879,9	7,360	2,979	3,229
700	30,973	1945,1	7,455	3,046	3,292
		$p=4,2$			
100	720,71	161,6	2,177	—	—
110	710,68	179,1	2,349	—	—
120	700,52	198,2	2,515	—	—
130	690,43	218,2	2,675	1,409	2,015
140	680,38	238,4	2,825	1,403	2,027
150	670,35	258,7	2,965	1,388	2,029
160	660,32	279,0	3,096	1,378	2,033
170	650,23	299,4	3,220	1,375	2,043
180	640,08	319,9	3,337	1,378	2,059
190	629,81	340,6	3,449	1,386	2,080
200	619,41	361,5	3,556	1,399	2,106
210	608,82	382,7	3,659	1,415	2,137
220	598,02	404,3	3,760	1,435	2,173
230	586,98	426,2	3,857	1,458	2,214
240	575,66	448,6	3,952	1,484	2,260
250	564,03	471,4	4,046	1,514	2,310
260	552,05	494,8	4,137	1,546	2,365
270	539,65	518,7	4,228	1,581	2,426
280	526,75	543,3	4,317	1,618	2,494
290	513,23	568,6	4,406	1,658	2,570
300	498,93	594,7	4,494	1,700	2,658

Table II.7 (*Continued*)

T	ρ	h	s	c_v	c_p

$p=4,2$

310	483,61	621,8	4,583	1,743	2,761
320	466,95	650,0	4,673	1,789	2,886
330	448,40	679,7	4,764	1,838	3,046
340	427,06	711,2	4,858	1,891	3,268
350	401,08	745,5	4,957	1,954	3,628
360	365,20	785,2	5,069	2,036	4,475
370	149,64	933,1	5,472	2,206	16,52
380	104,48	998,1	5,645	2,108	4,222
390	90,618	1035,2	5,742	2,091	3,368
400	82,076	1067,1	5,822	2,095	3,041
410	75,923	1096,6	5,895	2,109	2,875
420	71,114	1124,8	5,963	2,129	2,783
430	67,190	1152,4	6,028	2,154	2,730
440	63,881	1179,5	6,091	2,183	2,702
450	61,022	1206,4	6,151	2,213	2,690
460	58,519	1233,3	6,210	2,246	2,688
470	56,288	1260,2	6,268	2,279	2,695
480	54,277	1287,3	6,325	2,314	2,708
490	52,458	1314,4	6,381	2,349	2,724
500	50,793	1341,8	6,436	2.385	2,745
520	47,841	1397,1	6,545	2,457	2,792
540	45,301	1453,5	6,651	2,529	2,845
560	43,068	1510,9	6,755	2,599	2,900
580	41,089	1569,5	6,858	2,667	2,956
600	39,311	1629,1	6,959	2,732	3,011
620	37,712	1689,9	7,059	2,794	3,065
640	36,252	1751,7	7,157	2,855	3,119
660	34,917	1814,7	7,254	2,916	3,174
680	33,688	1878,7	7,350	2,979	3,232
700	32,554	1944,0	7,444	3,046	3,295

$p=4,24746$

100	720,73	161,6	2,177	—	—
110	710,70	179,1	2,349	—	—
120	700,54	198,3	2,515	—	—
130	690,45	218,2	2,675	1,409	2,015
140	680,40	238,5	2,825	1,403	2,027
150	670,38	258,7	2,965	1,388	2,029
160	660,34	279,0	3,096	1,378	2,033
170	650,26	299,4	3,219	1,375	2,043
180	640,11	319,9	3,377	1,378	2,059
190	629,85	340,6	3,449	1,386	2,080
200	619,44	361,5	3,556	1,399	2,106
210	608,86	382,8	3,659	1,415	2,137
220	598,07	404,3	3,760	1,435	2,173
230	587,03	426,2	3,857	1,458	2,214
240	575,72	448,6	3,952	1,484	2,259
250	564,10	471,4	4,045	1,514	2,309
260	552,12	494,8	4,137	1,546	2,365
270	539,73	518,7	4,227	1,581	2,426
280	526,84	543,3	4,317	1,618	2,493
290	513,34	568,6	4,406	1,658	2,569
300	499,06	594,8	4,494	1,699	2,657

Table II.7 (*Continued*)

T	ρ	h	s	c_v	c_p
		$p=4,24746$			
310	483,77	621,8	4,583	1,743	2,759
320	467,14	650,0	4,672	1,789	2,884
330	448,65	679,6	4,763	1,837	3,043
340	427,40	711,1	4,857	1,891	3,262
350	401,61	745,3	4,956	1,953	3,614
360	366,27	784,7	5,068	2,034	4,424
370	176,77	912,0	5,414	2,248	5,401
380	107,41	995,3	5,637	2,115	4,382
390	92,537	1033,4	5,736	2,095	3,423
400	83,595	1065,6	5,817	2,098	3,071
410	77,201	1095,4	5,891	2,111	2,895
420	72,237	1123,8	5,959	2,131	2,797
430	68,205	1151,4	6,024	2,156	2,741
440	64,808	1178,7	6,087	2,184	2,710
450	61,887	1205,7	6,148	2,214	2,696
460	59,317	1232,6	6,207	2,246	2,694
470	57,039	1259,6	6,265	2,280	2,700
480	54,993	1286,6	6,322	2,314	2,712
490	53,136	1313,8	6,378	2,350	2,728
500	51,443	1341,2	6,433	2.385	2,748
520	48,442	1396,6	6,542	2,457	2,794
540	45,858	1453,0	6,648	2,529	2,846
560	43,592	1510,5	6,753	2,599	2,901
580	41,580	1569,1	6,856	2,667	2,957
600	39,781	1628,8	6,957	2,732	3,012
620	38,154	1689,6	7,056	2,794	3,066
640	36,677	1751,4	7,155	2,855	3,120
660	35,322	1814,4	7,252	2,916	3,175
680	34,079	1878,5	7,347	2,979	3,233
700	32,929	1943,7	7,442	3,047	3,296
		$p=4,4$			
100	720,79	161,5	2,177	—	—
110	710,76	179,3	2,349	—	—
120	700,61	198,5	2,515	—	—
130	690,52	218,4	2,675	1,409	2,015
140	680,48	238,6	2,825	1,402	2,027
150	670,46	258,9	2,965	1,388	2,029
160	660,43	279,2	3,096	1,378	2,033
170	650,36	299,6	3,219	1,375	2,043
180	640,21	320,1	3,336	1,378	2,058
190	629,96	340,8	3,448	1,386	2,079
200	619,57	361,7	3,555	1,399	2,106
210	609,00	382,9	3,659	1,415	2,137
220	598,22	404,5	3,759	1,435	2,173
230	587,20	426,4	3,857	1,458	2,213
240	575,91	448,7	3,952	1,484	2,258
250	564,31	471,6	4,045	1,513	2,308
260	552,36	494,9	4,136	1,546	2,363
270	540,00	518,8	4,227	1,581	2,424
280	527,15	543,4	4,316	1,618	2,491
290	513,69	568,7	4,405	1,657	2,567
300	499,47	594,8	4,493	1,699	2,653

Table II.7 (*Continued*)

T	ρ	h	s	c_v	c_p

$$p=4,4$$

310	484,27	621,8	4,582	1,742	2,755
320	467,76	650,0	4,671	1,788	2,877
330	449,45	679,5	4,762	1,836	3,032
340	428,49	710,8	4,855	1,889	3,243
350	403,27	744,7	4,954	1,949	3,574
360	369,50	783,4	5,063	2,026	4,280
370	298,23	841,3	5,221	2,176	10,91
380	118,07	985,5	5,607	2,138	5,062
390	99,099	1027,2	5,716	2,109	3,626
400	88,649	1060,9	5,801	2,107	3,174
410	81,427	1091,4	5,877	2,118	2,960
420	75,936	1120,4	5,946	2,137	2,842
430	71,511	1148,5	6,012	2,160	2,775
440	67,830	1176,0	6,076	2,187	2,737
450	64,677	1203,3	6,137	2,217	2,718
460	61,927	1230,4	6,197	2,249	2,712
470	59,498	1257,5	6,255	2,282	2,715
480	57,318	1284,7	6,312	2,316	2,724
490	55,345	1312,0	6,368	2,351	2,739
500	53,548	1339,5	6,424	2,387	2,757
520	50,384	1395,1	6,533	2,458	2,802
540	47,657	1451,6	6,640	2,530	2,853
560	45,280	1509,2	6,744	2,600	2,907
580	43,170	1567,9	6,847	2,667	2,962
600	41,288	1627,7	6,949	2,732	3,016
620	39,587	1688,5	7,048	2,795	3,070
640	38,041	1750,5	7,147	2,855	3,123
660	36,633	1813,5	7,244	2,916	3,178
680	35,335	1877,6	7,339	2,979	3,236
700	34,136	1942,9	7,434	3,047	3,298

$$p=4,6$$

100	720,87	162,0	2,176	—	—
110	710,84	179,6	2,348	—	—
120	700,70	198,7	2,515	—	—
130	690,61	218,6	2,674	1,408	2,015
140	680,57	238,9	2,824	1,402	2,027
150	670,56	259,1	2,964	1,388	2,029
160	660,54	279,5	3,095	1,378	2,033
170	650,48	299,8	3,219	1,375	2,042
180	640,35	320,3	3,336	1,378	2,058
190	630,11	341,0	3,448	1,386	2,079
200	619,73	361,9	3,555	1,399	2,105
210	609,17	383,1	3,658	1,415	2,136
220	598,41	404,7	3,758	1,435	2,172
230	587,42	425,6	3,856	1,458	2,212
240	576,15	448,9	3,951	1,484	2,257
250	564,58	471,7	4,044	1,513	2,307
260	552,67	495,1	4,136	1,545	2,362
270	540,35	519,0	4,226	1,580	2,422
280	527,54	543,5	4,315	1,618	2,489
290	514,15	568,8	4,404	1,657	2,564
300	500,01	594,8	4,492	1,698	2,649

Table II.7 (*Continued*)

T	ρ	h	s	c_v	c_p
			$p=4,6$		
310	484,92	621,8	4,581	1,742	2,748
320	468,56	649,9	4,670	1,787	2,868
330	450,47	679,3	4,760	1,835	3,018
340	429,88	710,4	4,853	1,887	3,220
350	405,34	744,0	4,950	1,945	3,526
360	373,32	781,9	5,057	2,018	4,129
370	317,13	832,3	5,195	2,138	7,049
380	136,41	969,4	5,561	2,175	6,657
390	108,76	1018,3	5,688	2,128	3,967
400	95,760	1054,3	5,779	2,120	3,333
410	87,250	1086,1	5,858	2,128	3,055
420	80,958	1115,9	5,929	2,144	2,907
430	75,983	1144,5	5,997	2,166	2,822
440	71,888	1172,4	6,061	2,192	2,774
450	68,419	1200,0	6,123	2,221	2,747
460	65,413	1227,4	6,183	2,252	2,736
470	62,767	1254,8	6,242	2,285	2,735
480	60,400	1282,1	6,300	2,319	2,742
490	58,274	1309,6	6,356	2,353	2,754
500	56,344	1337,2	6,412	2,388	2,771
520	52,948	1393,1	6,522	2,460	2,812
540	50,038	1449,8	6,629	2,531	2,861
560	47,504	1507,5	6,734	2,600	2,914
580	45,269	1566,3	6,837	2,668	2,968
600	43,271	1626,2	6,938	2,733	3,021
620	41,473	1687,2	7,038	2,795	3,074
640	39,841	1749,2	7,137	2,856	3,127
660	38,349	1812,3	7,234	2,917	3,182
680	36,984	1876,5	7,330	2,980	3,239
700	35,721	1941,9	7,424	3,047	3,301
			$p=4,8$		
100	720,95	162,3	2,176	—	—
110	710,92	179,8	2,348	—	—
120	700,78	198,9	2,514	—	—
130	690,70	218,9	2,674	1,408	2,015
140	680,67	239,1	2,824	1,402	2,027
150	670,67	259,4	2,964	1,388	2,029
160	660,65	279,7	3,095	1,378	2,033
170	650,60	300,1	3,218	1,375	2,042
180	640,48	320,5	3,335	1,378	2,058
190	630,26	341,2	3,447	1,386	2,078
200	619,89	362,1	3,554	1,399	2,104
210	609,35	383,3	3,658	1,415	2,135
220	598,61	404,9	3,758	1,435	2,171
230	587,63	426,8	3,855	1,458	2,211
240	576,39	449,1	3,950	1,484	2,256
250	564,85	471,9	4,043	1,513	2,306
260	552,97	495,2	4,135	1,545	2,360
270	540,69	519,1	4,225	1,580	2,420
280	527,94	543,7	4,314	1,617	2,486
290	514,60	568,9	4,403	1,657	2,560
300	500,54	594,9	4,491	1,698	2,645

Table II.7 (*Continued*)

T	ρ	h	s	c_v	c_p
			$p=4,8$		
310	485,56	621,8	4,579	1,741	2,743
320	469,34	649,8	4,668	1,786	2,860
330	451,47	679,1	4,758	1,833	3,005
340	431,22	710,1	4,851	1,884	3,198
350	407,31	743,4	4,947	1,942	3,483
360	376,76	780,5	5,052	2,011	4,008
370	328,30	827,1	5,179	2,114	5,844
380	164,23	947,1	5,499	2,218	10,07
390	119,96	1008,4	5,658	2,149	4,429
400	103,52	1047,3	5,757	2,134	3,523
410	93,441	1080,5	5,839	2,138	3,163
420	86,221	1111,2	5,913	2,152	2,979
430	80,626	1140,4	5,981	2,172	2,874
440	76,079	1168,8	6,047	2,197	2,813
450	72,251	1196,7	6,109	2,225	2,778
460	68,963	1224,4	6,170	2,256	2,761
470	66,087	1252,0	6,230	2,288	2,756
480	63,532	1279,6	6,288	2,321	2,760
490	61,245	1307,2	6,345	2,355	2,770
500	59,169	1335,0	6,401	2,390	2,784
520	55,532	1391,0	6,511	2,461	2,823
540	52,434	1447,9	6,618	2,532	2,870
560	49,743	1505,8	6,723	2,601	2,921
580	47,374	1564,8	6,827	2,669	2,974
600	45,258	1624,8	6,928	2,733	3,026
620	43,363	1685,9	7,029	2,795	3,079
640	41,643	1748,0	7,127	2,856	3,131
660	40,075	1811,1	7,224	2,917	3,185
680	38,635	1875,4	7,320	2,980	3,242
700	37,308	1940,8	7,415	3,047	3,304
			$p=5,0$		
100	721,03	162,5	2,175	—	—
110	711,01	180,0	2,347	—	—
120	700,87	199,2	2,514	—	—
130	690,79	219,1	2,673	—	—
140	680,77	239,3	2,823	1,402	2,027
150	670,77	259,6	2,963	1,388	2,029
160	660,77	279,9	3,094	1,378	2,032
170	650,72	300,3	3,218	1,374	2,042
180	640,61	320,8	3,335	1,378	2,057
190	630,40	341,4	3,447	1,386	2,078
200	620,05	362,3	3,554	1,399	2,104
210	609,53	383,5	3,657	1,415	2,135
220	598,81	405,1	3,757	1,435	2,170
230	587,85	427,0	3,855	1,458	2,210
240	576,63	449,3	3,950	1,484	2,255
250	565,12	472,1	4,043	1,513	2,305
260	553,27	495,4	4,134	1,545	2,359
270	541,03	519,3	4,224	1,580	2,418
280	528,33	543,8	4,313	1,617	2,484
290	515,05	569,0	4,402	1,656	2,557
300	501,07	594,9	4,490	1,697	2,641

Table II.7 (*Continued*)

T	ρ	h	s	c_v	c_p
		$p=5,0$			
310	486,19	621,8	4,578	1,740	2,737
320	470,11	649,7	4,667	1,785	2,851
330	452,45	678,9	4,756	1,832	2,993
340	432,52	709,7	4,848	1,882	3,178
350	409,18	742,8	4,944	1,938	3,444
360	379,91	779,3	5,047	2,005	3,909
370	336,48	823,4	5,168	2,097	5,226
380	207,64	917,0	5,417	2,248	14,02
390	133,21	997,1	5,625	2,172	6,067
400	112,04	1039,8	5,733	2,148	3,753
410	100,03	1074,7	5,820	2,148	3,286
420	91,732	1106,3	5,896	2,160	3,057
430	85,440	1136,2	5,966	2,178	2,929
440	80,380	1165,1	6,032	2,202	2,855
450	76,186	1193,4	6,096	2,229	2,811
460	72,596	1221,4	6,158	2,259	2,788
470	69,477	1249,2	6,217	2,291	2,778
480	66,715	1277,0	6,276	2,324	2,778
490	64,251	1304,8	6,333	2,357	2,786
500	62,031	1332,7	6,390	2,392	2,798
520	58,142	1389,0	6,500	2,462	2,834
540	54,851	1446,1	6,608	2,533	2,879
560	51,997	1504,2	6,713	2,602	2,928
580	49,487	1563,2	6,817	2,669	2,980
600	47,260	1623,4	6,919	2,734	3,032
620	45,256	1684,5	7,019	2,796	3,084
640	43,446	1746,7	7,118	2,857	3,135
660	41,801	1809,9	7,215	2,917	3,189
680	40,291	1874,3	7,311	2,980	3,245
700	38,899	1939,8	7,406	3,048	3,307
		$p=5,5$			
100	721,22	163,1	2,174	—	—
110	711,21	180,6	2,346	—	—
120	701,08	199,8	2,513	—	—
130	691,02	219,7	2,672	—	—
140	681,01	239,9	2,822	1,402	2,026
150	671,03	260,2	2,962	1,387	2,028
160	661,05	280,5	3,093	1,377	2,032
170	651,03	300,8	3,216	1,374	2,041
180	640,95	321,3	3,334	1,378	2,056
190	630,77	342,0	3,445	1,386	2,077
200	620,45	362,9	3,552	1,399	2,103
210	609,97	384,1	3,656	1,415	2,133
220	599,29	405,6	3,756	1,435	2,168
230	588,39	427,4	3,853	1,458	2,208
240	577,24	449,7	3,948	1,484	2,253
250	565,79	472,5	4,041	1,513	2,301
260	554,02	495,8	4,132	1,545	2,355
270	541,88	519,6	4,222	1,579	2,413
280	529,29	544,1	4,311	1,616	2,478
290	516,16	569,2	4,399	1,655	2,550
300	502,37	595,1	4,487	1,696	2,631

Table II.7 (*Continued*)

T	ρ	h	s	c_v	c_p
			$p=5,5$		
310	487,73	621,9	4,575	1,738	2,723
320	471,99	649,6	4,663	1,782	2,832
330	454,80	678,6	4,752	1,829	2,964
340	435,61	709,0	4,843	1,877	3,132
350	413,52	741,4	4,937	1,931	3,361
360	386,79	776,7	5,036	1,991	3,720
370	351,01	817,0	5,146	2,067	4,466
380	287,53	872,5	5,294	2,182	7,500
390	179,45	961,7	5,526	2,227	7,437
400	137,63	1018,5	5,670	2,187	4,547
410	118,60	1058,9	5,770	2,175	3,671
420	106,76	1093,4	5,853	2,180	3,290
430	98,287	1125,2	5,928	2,195	3,088
440	91,746	1155,5	5,997	2,215	2,971
450	86,442	1184,8	6,063	2,240	2,901
460	82,014	1213,6	6,127	2,268	2,859
470	78,206	1242,1	6,188	2,298	2,837
480	74,887	1270,4	6,247	2,330	2,828
490	71,958	1298,6	6,306	2,363	2,828
500	69,321	1326,9	6,363	2,397	2,834
520	64,784	1383,9	6,475	2,466	2,862
540	60,962	1441,5	6,583	2,536	2,901
560	57,684	1500,0	6,690	2,604	2,947
580	54,822	1559,4	6,794	2,671	2,995
600	52,287	1619,8	6,896	2,735	3,045
620	50,025	1681,2	6,997	2,797	3,095
640	47,987	1743,6	7,096	2,858	3,146
660	46,135	1807,0	7,193	2,918	3,198
680	44,439	1871,5	7,290	2,981	3,253
700	42,885	1937,2	7,385	3,048	3,314
			$p=6,0$		
100	721,42	163,7	2,174	—	—
110	711,41	181,2	2,345	—	—
120	701,29	200,3	2,512	—	—
130	691,24	220,3	2,671	—	—
140	681,25	240,5	2,821	1,401	2,026
150	671,29	260,8	2,961	1,387	2,028
160	661,33	281,1	3,092	1,377	2,031
170	651,33	301,4	3,215	1,374	2,040
180	641,28	321,9	3,332	1,378	2,055
190	631,13	342,5	3,444	1,386	2,076
200	620,85	363,4	3,551	1,399	2,101
210	610,41	384,6	3,654	1,415	2,132
220	599,78	406,1	3,754	1,435	2,167
230	588,93	427,9	3,851	1,458	2,206
240	577,83	450,2	3,946	1,483	2,250
250	566,45	472,9	4,039	1,512	2,298
260	554,77	496,2	4,130	1,544	2,351
270	542,72	520,0	4,220	1,579	2,409
280	530,24	544,4	4,309	1,615	2,472
290	517,25	569,4	4,397	1,654	2,542
300	503,63	595,3	4,484	1,695	2,621

Table II.7 (*Continued*)

T	ρ	h	s	c_v	c_p
		$p=6{,}0$			
310	489,22	621,9	4,572	1,737	2,710
320	473,79	649,5	4,659	1,780	2,814
330	457,04	678,3	4,748	1,826	2,938
340	438,50	708,4	4,838	1,873	3,092
350	417,46	740,2	4,930	1,924	3,294
360	392,64	774,5	5,026	1,981	3,585
370	361,37	812,6	5,131	2,048	4,092
380	316,01	858,7	5,254	2,134	5,341
390	236,54	925,7	5,427	2,229	7,760
400	170,79	993,5	5,599	2,221	5,567
410	140,76	1041,1	5,717	2,203	4,174
420	123,82	1079,5	5,809	2,201	3,577
430	112,44	1113,6	5,890	2,211	3,275
440	104,01	1145,4	5,963	2,228	3,104
450	97,364	1175,9	6,031	2,251	3,001
460	91,931	1205,6	6,097	2,277	2,938
470	87,340	1234,8	6,159	2,306	2,901
480	83,380	1263,7	6,220	2,336	2,881
490	79,910	1292,4	6,280	2,369	2,873
500	76,837	1321,2	6,338	2,402	2,873
520	71,566	1378,8	6,451	2,470	2,892
540	67,187	1436,9	6,560	2,539	2,925
560	63,456	1495,8	6,667	2,607	2,966
580	60,219	1555,6	6,772	2,673	3,011
600	57,368	1616,3	6,875	2,737	3,059
620	54,829	1677,9	6,976	2,799	3,107
640	52,549	1740,6	7,076	2,859	3,156
660	50,487	1804,2	7,173	2,920	3,207
680	48,605	1868,9	7,270	2,982	3,262
700	46,880	1934,7	7,365	3,049	3,322
		$p=6{,}5$			
100	721,62	164,3	2,173	—	—
110	711,61	181,8	2,344	—	—
120	701,50	200,9	2,511	—	—
130	691,46	220,9	2,670	—	—
140	681,49	241,1	2,820	1,401	2,026
150	671,54	261,3	2,960	1,387	2,027
160	661,60	281,6	3,091	1,377	2,031
170	651,63	302,0	3,214	1,374	2,040
180	641,61	322,4	3,331	1,378	2,054
190	631,49	343,1	3,443	1,386	2,075
200	621,24	364,0	3,550	1,399	2,100
210	610,84	385,1	3,653	1,415	2,130
220	600,26	406,6	3,753	1,435	2,165
230	589,46	428,4	3,850	1,457	2,204
240	578,42	450,7	3,945	1,483	2,247
250	567,11	473,4	4,037	1,512	2,295
260	555,50	496,6	4,128	1,544	2,348
270	543,54	520,3	4,218	1,578	2,405
280	531,17	544,7	4,306	1,615	2,467
290	518,32	569,7	4,394	1,653	2,536
300	504,87	595,4	4,481	1,694	2,612

Table II.7 (*Continued*)

T	ρ	h	s	c_v	c_p
			$p=6,5$		
310	490,67	622,0	4,568	1,735	2,698
320	375,53	649,4	4,656	1,779	2,798
330	459,18	678,0	4,744	1,823	2,915
340	441,22	707,8	4,833	1,870	3,057
350	421,07	739,2	4,924	1,919	3,237
360	397,78	772,8	5,018	1,973	3,483
370	369,57	809,3	5,118	2,033	3,862
380	332,44	851,1	5,230	2,105	4,584
390	276,76	903,9	5,367	2,192	6,116
400	209,02	968,1	5,529	2,235	6,125
410	166,71	1021,9	5,662	2,226	4,717
420	143,11	1064,6	5,765	2,221	3,909
430	128,01	1101,4	5,852	2,227	3,490
440	117,23	1135,0	5,929	2,242	3,254
450	108,99	1166,8	6,000	2,262	3,112
460	102,36	1197,4	6,068	2,286	3,024
470	96,868	1227,4	6,132	2,314	2,970
480	92,184	1256,9	6,194	2,343	2,937
490	88,131	1286,2	6,255	2,374	2,920
500	84,557	1315,3	6,313	2,407	2,914
520	78,494	1373,6	6,428	2,474	2,922
540	73,515	1432,3	6,539	2,542	2,949
560	69,301	1491,7	6,646	2,609	2,986
580	65,670	1551,8	6,752	2,675	3,028
600	62,486	1612,8	6,855	2,739	3,073
620	59,666	1674,7	6,957	2,800	3,119
640	57,139	1737,5	7,057	2,861	3,166
660	54,861	1801,4	7,155	2,921	3,216
680	52,782	1866,2	7,252	2,984	3,270
700	50,887	1932,2	7,347	3,051	3,329
			$p=7,0$		
100	721,81	164,9	2,172	—	—
110	711,81	182,4	2,343	—	—
120	701,71	201,5	2,510	—	—
130	691,69	221,5	2,669	—	—
140	681,73	241,7	2,819	1,401	2,025
150	671,80	261,9	2,959	1,387	2,027
160	661,88	282,2	3,090	1,377	2,030
170	651,93	302,5	3,213	1,374	2,039
180	641,93	323,0	3,330	1,378	2,054
190	631,84	343,6	3,441	1,387	2,074
200	621,64	364,5	3,548	1,399	2,099
210	611,28	385,6	3,651	1,415	2,128
220	600,73	407,1	3,751	1,435	2,163
230	589,98	428,9	3,848	1,457	2,202
240	579,00	451,1	3,943	1,483	2,245
250	567,76	473,8	4,035	1,512	2,292
260	556,22	497,0	4,126	1,544	2,344
270	544,35	520,7	4,216	1,578	2,400
280	532,09	545,0	4,304	1,614	2,462
290	519,36	570,0	4,392	1,653	2,529
300	506,07	595,6	4,479	1,693	2,603

Table II.7 (*Continued*)

T	ρ	h	s	c_v	c_p
\multicolumn{6}{c}{$p=7,0$}					
310	492,08	622,1	4,565	1,734	2,687
320	477,21	649,4	4,652	1,777	2,783
330	461,23	677,8	4,740	1,821	2,893
340	443,79	707,3	4,828	1,867	3,025
350	424,41	738,4	4,918	1,915	3,189
360	402,37	771,3	5,010	1,966	3,401
370	376,42	806,7	5,107	2,022	3,702
380	344,18	845,9	5,212	2,086	4,192
390	300,84	891,9	5,331	2,159	5,081
400	244,02	947,5	5,472	2,222	5,829
410	195,05	1002,8	5,609	2,238	5,088
420	164,44	1049,1	5,720	2,237	4,236
430	144,96	1088,7	5,813	2,242	3,718
440	131,42	1124,2	5,895	2,254	3,415
450	121,29	1157,4	5,970	2,272	3,231
460	113,31	1189,1	6,039	2,295	3,116
470	106,79	1219,8	6,106	2,321	3,043
480	101,30	1250,0	6,169	2,350	2,997
490	96,592	1279,8	6,231	2,380	2,970
500	92,467	1309,4	6,290	2,412	2,956
520	85,565	1368,5	6,406	2,478	2,954
540	79,943	1427,7	6,518	2,545	2,974
560	75,225	1487,5	6,627	2,612	3,006
580	71,177	1548,0	6,733	2,678	3,044
600	67,646	1609,3	6,837	2,741	3,087
620	64,533	1671,5	6,939	2,802	3,131
640	61,751	1734,6	7,039	2,862	3,177
660	59,249	1798,6	7,137	2,922	3,226
680	56,978	1863,6	7,234	2,985	3,278
700	54,898	1929,8	7,330	3,052	3,337
\multicolumn{6}{c}{$p=7,5$}					
100	722,0	165,5	2,171	—	→
110	712,01	183,0	2,342	—	—
120	701,92	202,1	2,509	—	—
130	691,91	222,0	2,668	—	—
140	681,97	242,2	2,818	1,400	2,025
150	672,06	262,5	2,958	1,387	2,026
160	662,15	282,8	3,089	1,377	2,029
170	652,23	303,1	3,212	1,374	2,038
180	642,26	323,6	3,329	1,378	2,053
190	632,20	344,2	3,440	1,387	2,073
200	622,03	365,0	3,547	1,399	2,097
210	611,70	386,1	3,650	1,416	2,127
220	601,21	407,6	3,750	1,435	2,161
230	590,51	429,4	3,847	1,457	2,200
240	579,58	451,6	3,941	1,483	2,243
250	568,40	474,2	4,034	1,512	2,289
260	556,93	497,4	4,124	1,543	2,341
270	545,15	521,1	4,214	1,577	2,396
280	532,99	545,3	4,302	1,614	2,456
290	520,39	570,2	4,389	1,652	2,523
300	507,25	595,8	4,476	1,692	2,595

Table II.7 (*Continued*)

T	ρ	h	s	c_v	c_p

$$p = 7,5$$

310	493,45	622,2	4,563	1,733	2,676
320	478,84	649,4	4,649	1,775	2,768
330	463,19	677,6	4,736	1,819	2,874
340	446,22	706,9	4,823	1,864	2,997
350	427,53	737,6	4,912	1,911	3,147
360	406,53	770,0	5,003	1,960	3,334
370	382,34	804,5	5,098	2,013	3,584
380	353,41	842,0	5,198	2,071	3,948
390	317,07	884,2	5,307	2,136	4,525
400	270,75	933,1	5,431	2,200	5,220
410	222,71	985,8	5,561	2,237	5,141
420	186,94	1033,9	5,677	2,247	4,481
430	163,05	1075,8	5,776	2,254	3,932
440	146,47	1113,2	5,862	2,265	3,578
450	134,24	1147,8	5,940	2,282	3,354
460	124,75	1180,6	6,012	2,304	3,211
470	117,09	1212,2	6,080	2,329	3,119
480	110,72	1243,1	6,145	2,356	3,059
490	105,29	1273,5	6,207	2,386	3,021
500	100,59	1303,5	6,268	2,417	2,999
520	92,766	1363,3	6,386	2,482	2,987
540	86,458	1423,2	6,498	2,548	2,999
560	81,206	1483,4	6,608	2,615	3,026
580	76,731	1541,3	6,715	2,680	3,061
600	72,844	1605,9	6,819	2,743	3,101
620	69,427	1668,3	6,922	2,804	3,143
640	66,381	1731,6	7,022	2,864	3,188
660	63,649	1795,9	7,121	2,924	3,235
680	61,176	1861,1	7,218	2,986	3,286
700	58,922	1927,4	7,314	3,053	3,344

$$p = 8,0$$

100	722,19	166,1	2,170	—	—
110	712,20	183,6	2,341	—	—
120	702,13	202,7	2,508	—	—
130	692,13	222,6	2,667	—	—
140	682,20	242,8	2,817	1,400	2,024
150	672,31	263,1	2,957	1,387	2,026
160	662,43	283,4	3,087	1,377	2,029
170	652,53	303,7	3,211	1,374	2,037
180	642,58	324,1	3,327	1,378	2,052
190	632,55	344,7	3,439	1,387	2,072
200	622,41	365,6	3,546	1,400	2,096
210	612,13	386,7	3,649	1,416	2,125
220	601,67	408,1	3,748	1,435	2,159
230	591,02	429,9	3,845	1,457	2,198
240	580,15	452,1	3,940	1,483	2,240
250	569,03	474,7	4,032	1,512	2,287
260	557,64	497,8	4,123	1,543	2,337
270	545,94	521,5	4,212	1,577	2,392
280	533,88	545,7	4,300	1,613	2,452
290	521,39	570,5	4,387	1,651	2,516
300	508,40	596,0	4,474	1,691	2,588

Table II.7 (*Continued*)

T	ρ	h	s	c_v	c_p
		$p=8,0$			
310	494,79	622,3	4,560	1,732	2,666
320	480,42	649,4	4,646	1,774	2,755
330	465,08	677,4	4,732	1,817	2,856
340	448,54	706,5	4,819	1,861	2,972
350	430,45	736,9	4,907	1,907	3,110
360	410,35	768,8	4,997	1,955	3,277
370	387,57	802,6	5,089	2,006	3,491
380	361,07	838,9	5,186	2,060	3,779
390	329,22	878,6	5,289	2,119	4,190
400	290,29	923,1	5,402	2,179	4,715
410	246,95	971,9	5,522	2,226	4,943
420	209,23	1019,9	5,638	2,250	4,587
430	181,72	1063,3	5,740	2,262	4,100
440	162,15	1102,3	5,830	2,274	3,727
450	147,73	1138,2	5,911	2,291	3,475
460	136,61	1172,1	5,985	2,311	3,307
470	127,72	1204,6	6,055	2,335	3,196
480	120,39	1236,1	6,121	2,362	3,122
490	114,20	1267,1	6,185	2,391	3,074
500	108,87	1297,7	6,247	2,422	3,044
520	100,09	1358,2	6,366	2,486	3,020
540	93,063	1418,6	6,480	2,551	3,025
560	87,248	1479,3	6,590	2,617	3,047
580	82,325	1540,6	6,698	2,682	3,078
600	78,070	1602,5	6,802	2,745	3,115
620	74,337	1665,2	6,905	2,806	3,155
640	71,034	1728,7	7,006	2,865	3,198
660	68,064	1793,2	7,105	2,925	3,244
680	65,384	1858,5	7,203	2,987	3,295
700	62,949	1925,0	7,299	3,054	3,351
		$p=8,5$			
100	722,38	166,7	2,169	—	—
110	712,40	184,2	2,340	—	—
120	702,34	203,3	2,507	—	—
130	692,35	223,2	2,666	—	—
140	682,44	243,4	2,816	1,400	2,024
150	672,56	263,7	2,955	1,387	2,025
160	662,70	283,9	3,086	1,377	2,028
170	652,82	304,2	3,209	1,375	2,037
180	642,90	324,7	3,326	1,378	2,051
190	632,90	345,3	3,438	1,387	2,071
200	622,80	366,1	3,544	1,400	2,095
210	612,55	387,2	3,647	1,416	2,124
220	602,14	408,6	3,747	1,435	2,158
230	591,54	430,4	3,844	1,458	2,196
240	580,72	452,5	3,938	1,483	2,238
250	569,66	475,1	4,030	1,512	2,284
260	558,34	498,2	4,121	1,543	2,334
270	546,72	521,8	4,210	1,577	2,388
280	534,75	546,0	4,298	1,613	2,447
290	522,38	570,8	4,385	1,651	2,510
300	509,53	596,2	4,471	1,690	2,580

Table II.7 (*Continued*)

T	ρ	h	s	c_v	c_p
			$p=8,5$		
310	496,10	622,4	4,557	1,731	2,657
320	481,95	649,4	4,642	1,773	2,742
330	466,90	677,3	4,728	1,816	2,839
340	450,75	706,2	4,815	1,859	2,949
350	433,21	736,3	4,902	1,904	3,077
360	413,89	767,8	4,991	1,951	3,229
370	392,28	801,0	5,082	1,999	3,415
380	367,64	836,3	5,176	2,051	3,653
390	338,95	874,3	5,274	2,106	3,967
400	305,09	915,9	5,380	2,162	4,360
410	266,83	961,2	5,492	2,212	4,656
420	229,95	1007,7	5,603	2,245	4,557
430	200,26	1051,5	5,707	2,265	4,199
440	178,12	1091,7	5,799	2,280	3,848
450	161,56	1128,8	5,882	2,298	3,584
460	148,80	1163,6	5,959	2,318	3,398
470	138,63	1196,9	6,031	2,342	3,271
480	130,30	1229,2	6,099	2,368	3,185
490	123,30	1260,7	6,164	2,396	3,126
500	117,31	1291,8	6,226	2,426	3,089
520	107,51	1353,1	6,347	2,489	3,053
540	99,729	1414,1	6,462	2,555	3,051
560	93,346	1475,3	6,573	2,620	3,068
580	87,955	1536,9	6,681	2,684	3,096
600	83,326	1599,2	6,787	2,747	3,130
620	79,275	1662,1	6,890	2,807	3,168
640	75,692	1725,9	6,991	2,867	3,209
660	72,487	1790,5	7,090	2,927	3,254
680	69,602	1856,1	7,188	2,989	3,303
700	66,980	1922,7	7,285	3,055	3,359
			$p=9,0$		
100	722,56	167,3	2,167	—	—
110	712,59	184,8	2,340	—	—
120	702,54	203,9	2,506	—	—
130	692,57	223,8	2,665	—	—
140	682,67	244,0	2,815	1,400	2,024
150	672,82	264,2	2,954	1,387	2,025
160	662,97	284,5	3,085	1,377	2,028
170	653,12	304,8	3,208	1,375	2,036
180	643,22	325,2	3,325	1,378	2,050
190	633,25	345,8	3,436	1,387	2,070
200	623,18	366,7	3,543	1,400	2,094
210	612,97	387,7	3,646	1,416	2,123
220	602,60	409,1	3,745	1,435	2,156
230	592,04	430,9	3,842	1,458	2,194
240	581,28	453,0	3,936	1,483	2,236
250	570,28	475,6	4,029	1,512	2,281
260	559,03	498,6	4,119	1,543	2,331
270	547,48	522,2	4,208	1,577	2,384
280	535,61	546,3	4,296	1,612	2,442
290	523,36	571,1	4,382	1,650	2,505
300	510,64	596,5	4,468	1,690	2,573

Table II.7 (*Continued*)

T	ρ	h	s	c_v	c_p
			$p=9{,}0$		
310	497,37	622,6	4,554	1,730	2,648
320	483,43	649,4	4,639	1,772	2,730
330	468,66	677,2	4,725	1,814	2,823
340	452,87	705,9	4,811	1,857	2,928
350	435,82	735,8	4,897	1,902	3,048
360	417,18	767,0	4,985	1,947	3,187
370	396,56	799,6	5,074	1,994	3,353
380	373,43	834,1	5,166	2,044	3,554
390	347,10	870,9	5,262	2,095	3,806
400	316,84	910,4	5,362	2,148	4,110
410	282,87	953,0	5,467	2,198	4,388
420	248,27	997,4	5,574	2,237	4,438
430	217,94	1040,9	5,676	2,264	4,224
440	193,95	1081,6	5,770	2,284	3,930
450	175,51	1119,6	5,855	2,303	3,674
460	161,16	1155,3	5,934	2,324	3,481
470	149,72	1189,4	6,007	2,347	3,342
480	140,36	1222,3	6,077	2,373	3,245
490	132,54	1254,4	6,143	2,401	3,178
500	125,87	1286,0	6,206	2,431	3,133
520	115,01	1348,1	6,328	2,493	3,086
540	106,46	1409,7	6,444	2,558	3,077
560	99,483	1471,3	6,557	2,623	3,089
580	93,616	1533,3	6,665	2,687	3,113
600	88,591	1595,8	6,771	2,749	3,144
620	84,223	1659,1	6,875	2,809	3,180
640	80,361	1723,1	6,977	2,869	3,220
660	76,914	1787,9	7,076	2,928	3,263
680	73,820	1853,6	7,174	2,990	3,311
700	71,009	1920,4	7,271	3,057	3,366
			$p=9{,}5$		
100	722,75	167,9	2,167	—	—
110	712,79	185,4	2,339	—	—
120	702,75	204,5	2,505	—	—
130	692,79	224,4	2,664	—	—
140	682,91	244,6	2,814	1,400	2,023
150	673,07	264,8	2,953	1,387	2,024
160	663,24	285,1	3,084	1,377	2,027
170	653,41	305,4	3,207	1,375	2,035
180	643,54	325,8	3,324	1,379	2,049
190	633,60	346,4	3,435	1,387	2,069
200	623,56	367,2	3,542	1,400	2,093
210	613,39	388,3	3,645	1,416	2,121
220	603,06	409,6	3,744	1,435	2,154
230	592,55	431,4	3,841	1,458	2,192
240	581,83	453,5	3,935	1,483	2,233
250	570,89	476,0	4,027	1,511	2,279
260	559,71	499,1	4,117	1,543	2,328
270	548,24	522,6	4,206	1,576	2,381
280	536,46	546,7	4,294	1,612	2,438
290	524,31	571,4	4,380	1,650	2,499
300	511,73	596,7	4,466	1,689	2,566

Table II.7 (*Continued*)

T	ρ	h	s	c_v	c_p

$$p=9,5$$

T	ρ	h	s	c_v	c_p
310	498,62	622,7	4,551	1,729	2,639
320	484,88	649,5	4,636	1,771	2,719
330	470,37	677,1	4,721	1,813	2,808
340	454,91	705,7	4,807	1,856	2,908
350	438,30	735,3	4,893	1,899	3,021
360	420,27	766,2	4,979	1,944	3,150
370	400,51	798,4	5,068	1,990	3,299
380	378,62	832,3	5,158	2,037	3,475
390	354,12	868,0	5,251	2,087	3,684
400	326,55	906,1	5,347	2,137	3,928
410	295,96	946,6	5,447	2,185	4,168
420	264,03	989,0	5,549	2,227	4,283
430	234,24	1031,5	5,649	2,260	4,189
440	209,24	1072,4	5,743	2,284	3,970
450	189,32	1110,9	5,830	2,306	3,741
460	173,55	1147,3	5,910	2,328	3,551
470	160,89	1182,1	5,985	2,352	3,407
480	150,53	1215,6	6,055	2,377	3,302
490	141,87	1248,2	6,123	2,405	3,227
500	134,51	1280,2	6,187	2,435	3,176
520	122,58	1343,1	6,311	2,497	3,119
540	113,24	1405,2	6,428	2,561	3,103
560	105,65	1467,3	6,541	2,625	3,109
580	99,303	1529,7	6,650	2,689	3,130
600	93,879	1592,6	6,757	2,751	3,159
620	89,176	1656,1	6,861	2,811	3,192
640	85,036	1720,3	6,963	2,870	3,230
660	81,343	1785,3	7,063	2,930	3,272
680	78,036	1851,2	7,161	2,992	3,319
700	75,032	1918,1	7,258	3,058	3,374

$$p=10,0$$

T	ρ	h	s	c_v	c_p
100	722,93	168,5	2,166	—	—
110	712,98	186,0	2,338	—	—
120	702,95	205,1	2,504	—	—
130	693,01	225,0	2,663	—	—
140	683,14	245,2	2,813	1,400	2,023
150	673,32	265,4	2,952	1,387	2,024
160	663,51	285,7	3,083	1,377	2,027
170	653,70	306,0	3,206	1,375	2,035
180	643,86	326,4	3,323	1,379	2,049
190	633,95	346,9	3,434	1,388	2,068
200	623,94	367,7	3,541	1,400	2,091
210	613,80	388,8	3,643	1,416	2,120
220	603,51	410,1	3,743	1,436	2,153
230	593,05	431,9	3,839	1,458	2,190
240	582,38	454,0	3,933	1,483	2,231
250	571,50	476,5	4,025	1,511	2,276
260	560,38	499,5	4,115	1,543	2,325
270	548,99	523,0	4,204	1,576	2,377
280	537,30	547,0	4,291	1,612	2,433
290	525,25	571,7	4.378	1,649	2,494
300	512,79	596,9	4,464	1,689	2,560

Table II.7 (*Continued*)

T	ρ	h	s	c_v	c_p
		$p=10,0$			
310	499,84	622,9	4,549	1,729	2,631
320	486,29	649,6	4,633	1,770	2,709
330	472,01	677,1	4,718	1,812	2,795
340	456,87	705,5	4,803	1,854	2,890
350	440,67	734,9	4,888	1,898	2,997
360	423,19	765,5	4,974	1,941	3,117
370	404,17	797,3	5,061	1,986	3,253
380	383,32	830,6	5,150	2,032	3,409
390	360,30	865,6	5,241	2,080	3,588
400	334,82	902,5	5,334	2,128	3,790
410	306,88	941,4	5,431	2,175	3,993
420	277,46	982,1	5,529	2,218	4,128
430	248,89	1023,5	5,626	2,254	4,116
440	223,65	1064,0	5,719	2,282	3,972
450	202,73	1102,8	5,806	2,307	3,781
460	185,78	1139,7	5,887	2,330	3,605
470	172,02	1175,0	5,963	2,355	3,462
480	160,71	1209,0	6,035	2,381	3,353
490	151,24	1242,1	6,103	2,409	3,273
500	143,19	1274,6	6,169	2,438	3,216
520	130,19	1338,1	6,293	2,500	3,151
540	120,05	1400,9	6,412	2,564	3,128
560	111,83	1463,4	6,526	2,628	3,130
580	105,00	1526,2	6,636	2,691	3,147
600	99,175	1589,4	6,743	2,753	3,173
620	94,138	1653,1	6,847	2,813	3,205
640	89,705	1717,6	6,950	2,872	3,241
660	85,771	1782,8	7,050	2,931	3,281
680	82,247	1848,9	7,149	2,993	3,328
700	79,055	1915,9	7,246	3,059	3,381
		$p=11$			
100	723,29	169,7	2,164	—	—
110	713,36	187,2	2,336	—	—
120	703,36	206,2	2,502	—	—
130	693,45	226,2	2,661	—	—
140	683,61	246,3	2,811	1,399	2,022
150	673,81	266,6	2,950	1,387	2,023
160	664,05	286,8	3,081	1,378	2,026
170	654,28	307,1	3,204	1,375	2,033
180	644,49	327,5	3,320	1,379	2,047
190	634,63	348,1	3,431	1,388	2,066
200	624,69	368,8	3,538	1,401	2,089
210	614,62	389,8	3,641	1,417	2,117
220	604,41	411,2	3,740	1,436	2,150
230	594,04	432,9	3,836	1,458	2,186
240	583,47	454,9	3,930	1,483	2,227
250	572,70	477,4	4,022	1,512	2,271
260	561,71	500,3	4,112	1,542	2,319
270	550,46	523,8	4,200	1,576	2,370
280	538,93	547,8	4,287	1,611	2,425
290	527,09	572,3	4,374	1,649	2,484
300	514,86	597,5	4,459	1,688	2,547

Table II.7 (*Continued*)

T	ρ	h	s	c_v	c_p

$$p=11$$

310	502,20	623,3	4,543	1,728	2,616
320	489,00	649,8	4,628	1,769	2,689
330	475,17	677,1	4,712	1,810	2,770
340	460,58	705,2	4,796	1,852	2,858
350	445,10	734,3	4,880	1,894	2,955
360	428,56	764,3	4,965	1,937	3,061
370	410,80	795,5	5,050	1,980	3,177
380	391,62	827,9	5,136	2,024	3,305
390	370,87	861,7	5,224	2,069	3,445
400	348,42	896,9	5,313	2,114	3,594
410	324,30	933,6	5,404	2,159	3,745
420	298,93	971,7	5,496	2,201	3,870
430	273,36	1010,7	5,587	2,240	3,929
440	249,15	1049,9	5,678	2,275	3,896
450	227,60	1088,4	5,764	2,304	3,793
460	209,20	1125,7	5,846	2,332	3,663
470	193,74	1161,7	5,923	2 359	3,539
480	180,79	1196,6	5,997	2,386	3,435
490	169,85	1230,5	6,067	2,415	3,352
500	160,51	1263,7	6,134	2,444	3,289
520	145,41	1328,6	6,261	2,506	3,211
540	133,67	1392,4	6,382	2,569	3,176
560	124,22	1455,8	6,497	2,633	3,170
580	116,39	1519,3	6,608	2,696	3,180
600	109,76	1583,1	6,716	2,757	3,201
620	104,04	1647,4	6,822	2,817	3,229
640	99,040	1712,3	6,925	2,875	3,261
660	94,610	1777,9	7,026	2,934	3,300
680	90,651	1844,3	7,125	2,996	3,344
700	87,079	1911,7	7,222	3,062	3,395

$$p=12$$

100	723,65	170,9	2,162	—	—
110	713,74	188,3	2,334	—	—
120	703,77	207,4	2,500	—	—
130	693,88	227,3	2,659	—	—
140	684,07	247,5	2,808	1,400	2,022
150	674,31	267,7	2,948	1,387	2,022
160	664,58	288,0	3,078	1,378	2,025
170	654,86	308,2	3,201	1,376	2,032
180	645,11	328,6	3,318	1,380	2,045
190	635,31	349,2	3,429	1,389	2,064
200	625,43	369,9	3,535	1,401	2,087
210	615,43	390,9	3,638	1,417	2,114
220	605,30	412,2	3,737	1,437	2,146
230	595,01	433,9	3,833	1,459	2,183
240	584,54	455,9	3,927	1,484	2,223
250	573,88	478,3	4,018	1,512	2,266
260	563,00	501,2	4,108	1,542	2,313
270	551,90	524,6	4,197	1,576	2,364
280	540,53	548,5	4,283	1,611	2,418
290	528,87	573,0	4,369	1,648	2,475
300	516,86	598,0	4,454	1,687	2,536

Table II.7 (*Continued*)

T	ρ	h	s	c_v	c_p
			$p=12$		
310	504,46	623,7	4,538	1,727	2,601
320	491,58	650,1	4,622	1,768	2,672
330	478,15	677,2	4,705	1,809	2,748
340	464,06	705,0	4,789	1,850	2,830
350	449,20	733,8	4,872	1,892	2,919
360	433,45	763,4	4,956	1,934	3,014
370	416,69	794,1	5,040	1,976	3,117
380	398,82	825,8	5,124	2,019	3,226
390	379,74	858,6	5,209	2,062	3,341
400	359,41	892,6	5,295	2,105	3,460
410	337,89	927,8	5,382	2,147	3,578
420	315,42	964,2	5,470	2,189	3,684
430	292,55	1001,4	5,558	2,229	3,757
440	270,16	1039,1	5,644	2,265	3,777
450	249,23	1076,8	5,729	2,299	3,741
460	230,47	1113,8	5,810	2,330	3,664
470	214,10	1150,0	5,888	2,360	3,573
480	200,01	1185,3	5,962	2,389	3,485
490	187,91	1219,7	6,033	2,418	3,408
500	177,47	1253,5	6,102	2,448	3,346
52˙	160,46	1319,5	6,231	2,510	3,262
540	147,20	1384,3	6,353	2,574	3,220
560	136,54	1448,5	6,470	2,637	3,207
580	127,73	1512,7	6,583	2,700	3,211
600	120,29	1577,0	6,692	2,761	3,228
620	113,90	1641,8	6,798	2,820	3,252
640	108,32	1707,1	6,902	2,879	3,282
660	103,40	1773,1	7,003	2,937	3,317
680	99,000	1839,9	7,103	2,999	3,359
700	95,055	1907,5	7,201	3,065	3,409
			$p=13$		
100	724,00	172,1	2,160	—	—
110	714,12	189,5	2,332	—	—
120	704,17	208,6	2,498	—	—
130	694,31	228,5	2,657	—	—
140	684,53	248,7	2,806	1,400	2,021
150	674,80	268,9	2,946	1,387	2,021
160	665,11	289,1	3,076	1,378	2,024
170	655,43	309,4	3,199	1,376	2,031
180	645,73	329,7	3,316	1,380	2,044
190	635,98	350,3	3,427	1,389	2,062
200	626,16	371,0	3,533	1,402	2,085
210	616,23	392,0	3,635	1,418	2,112
220	606,18	413,3	3,734	1,437	2,143
230	595,97	434,9	3,830	1,459	2,179
240	585,59	456,9	3,924	1,484	2,219
250	575,03	479,3	4,015	1,512	2,262
260	564,28	502,1	4,105	1,543	2,308
270	553,30	525,4	4,193	1,576	2,358
280	542,08	549,3	4,280	1,611	2,410
290	530,60	573,6	4,365	1,648	2,466
300	518,80	598,6	4,450	1,687	2,525

Table II.7 (*Continued*)

T	ρ	h	s	c_v	c_p

$$p = 13$$

T	ρ	h	s	c_v	c_p
310	506,64	624,2	4,534	1,726	2,589
320	494,06	650,4	4,617	1,767	2,656
330	480,98	677,3	4,700	1,808	2,728
340	467,33	705,0	4,782	1,849	2,805
350	453,01	733,4	4,865	1,890	2,887
360	437,94	762,7	4,947	1,932	2,975
370	422,02	792,9	5,030	1,973	3,067
380	405,19	824,1	5,113	2,015	3,164
390	387,41	856,2	5,196	2,056	3,263
400	368,67	889,3	5,280	2,098	3,362
410	349,04	923,5	5,365	2,139	3,460
420	328,70	958,5	5,449	2,180	3,549
430	307,98	994,4	5,533	2,219	3,620
440	287,40	1030,8	5,617	2,257	3,660
450	267,61	1067,4	5,699	2,293	3,662
460	249,22	1103,9	5,780	2,326	3,628
470	232,62	1139,9	5,857	2,358	3,571
480	217,92	1175,3	5,932	2,389	3,505
490	205,03	1210,0	6,003	2,420	3,442
500	193,75	1244,2	6,072	2,451	3,386
520	175,13	1311,0	6,203	2,514	3,304
540	160,50	1376,6	6,327	2,577	3,259
560	148,69	1441,5	6,445	2,641	3,241
580	138,94	1506,3	6,559	2,704	3,241
600	130,71	1571,2	6,669	2,764	3,253
620	123,66	1636,5	6,776	2,824	3,274
640	117,52	1702,2	6,880	2,882	3,301
660	112,10	1768,6	6,982	2,941	3,334
680	107,28	1835,6	7,082	3,002	3,375
700	102,95	1903,6	7,181	3,068	3,423

$$p = 14$$

T	ρ	h	s	c_v	c_p
100	724,35	173,3	2,158	—	—
110	714,49	190,7	2,330	—	—
120	704,57	209,8	2,496	—	—
130	694,73	229,7	2,655	—	—
140	684,98	249,8	2,804	1,400	2,020
150	675,29	270,1	2,944	1,388	2,020
160	665,63	290,3	3,074	1,379	2,023
170	655,99	310,5	3,197	1,377	2,030
180	646,34	330,9	3,313	1,381	2,043
190	636,64	351,4	3,424	1,390	2,060
200	626,88	372,1	3,530	1,403	2,083
210	617,02	393,1	3,633	1,419	2,109
220	607,04	414,3	3,731	1,438	2,141
230	596,91	435,9	3,827	1,460	2,176
240	586,63	457,8	3,921	1,485	2,215
250	576,17	480,2	4,012	1,512	2,257
260	565,52	503,0	4,101	1,543	2,303
270	554,67	526,3	4,189	1,576	2,352
280	543,60	550,0	4,276	1,611	2,403
290	532,28	574,3	4,361	1,648	2,458
300	520,67	599,2	4,445	1,686	2,516

Table II.7 (*Continued*)

T	ρ	h	s	c_v	c_p
			$p=14$		
310	508,74	624,7	4,529	1,726	2,577
320	496,43	650,7	4,612	1,766	2,641
330	483,68	677,5	4,694	1,807	2,710
340	470,42	705,0	4,776	1,848	2,783
350	456,58	733,2	4,858	1,889	2,860
360	442,09	762,2	4,939	1,930	2,941
370	426,89	792,0	5,021	1,971	3,026
380	410,93	822,7	5,103	2,012	3,112
390	394,19	854,3	5,185	2,052	3,200
400	376,70	886,7	5,267	2,093	3,287
410	358,52	920,0	5,349	2,133	3,370
420	339,79	954,1	5,431	2,173	3,447
430	320,75	988,9	5,513	2,212	3,512
440	301,73	1024,3	5,595	2,250	3,559
450	283,17	1060,0	5,675	2,287	3,580
460	265,52	1095,8	5,754	2,322	3,573
470	249,16	1131,4	5,830	2,355	3,545
480	234,29	1166,6	5,904	2,388	3,502
490	220,98	1201,4	5,976	2,420	3,455
500	209,13	1235,7	6,045	2,452	3,409
520	189,25	1303,1	6,178	2,516	3,336
540	173,42	1369,3	6,303	2,581	3,291
560	160,58	1434,9	6,422	2,644	3,270
580	149,95	1500,3	6,536	2,707	3,267
600	140,98	1565,7	6,647	2,768	3,276
620	133,29	1631,4	6,755	2,827	3,294
640	126,59	1697,5	6,860	2,885	3,319
660	120,71	1764,2	6,963	2,943	3,351
680	115,47	1831,6	7,063	3,005	3,389
700	110,77	1899,8	7,162	3,070	3,436
			$p=15$		
100	724,70	174,5	2,156	—	—
110	714,86	191,9	2,328	—	—
120	704,97	211,0	2,494	—	—
130	695,16	230,9	2,653	—	—
140	685,43	251,0	2,802	1,400	2,020
150	675,77	271,2	2,942	1,388	2,020
160	666,15	291,4	3,072	1,379	2,022
170	656,55	311,7	3,195	1,377	2,029
180	646,94	332,0	3,311	1,382	2,041
190	637,30	352,5	3,422	1,391	2,059
200	627,60	373,2	3,528	1,404	2,081
210	617,80	394,1	3,630	1,419	2,107
220	607,89	415,4	3,729	1,438	2,138
230	597,85	436,9	3,825	1,460	2,173
240	587,65	458,8	3,918	1,485	2,211
250	577,29	481,1	4,009	1,513	2,253
260	566,75	503,9	4,098	1,543	2,298
270	556,02	527,1	4,186	1,576	2,346
280	545,08	550,8	4,272	1,611	2,397
290	533,91	575,1	4,357	1,648	2,450
300	522,49	599,8	4,441	1,686	2,506

Table II.7 (*Continued*)

T	ρ	h	s	c_v	c_p
		$p=15$			
310	510,77	625,2	4,524	1,726	2,566
320	498,71	651,2	4,607	1,766	2,628
330	486,26	677,8	4,688	1,806	2,694
340	473,36	705,0	4,770	1,847	2,763
350	459,94	733,0	4,851	1,888	2,836
360	445,97	761,8	4,932	1,929	2,912
370	431,38	791,3	5,013	1,969	2,990
380	416,16	821,6	5,094	2,010	3,069
390	400,29	852,7	5,174	2,050	3,148
400	383,80	884,5	5,255	2,090	3,226
410	366,78	917,2	5,336	2,129	3,300
420	349,33	950,5	5,416	2,169	3,369
430	331,62	984,5	5,496	2,207	3,428
440	313,91	1019,0	5,575	2,245	3,475
450	296,48	1054,0	5,654	2,282	3,505
460	279,67	1089,1	5,731	2,318	3,515
470	263,79	1124,2	5,807	2,352	3,506
480	249,07	1159,2	5,880	2,386	3,484
490	235,63	1193,8	5,952	2,420	3,452
500	223,47	1228,2	6,021	2,453	3,419
520	202,70	1295,9	6,154	2,518	3,358
540	185,89	1362,6	6,280	2,583	3,316
560	172,14	1428,7	6,400	2,647	3,295
580	160,70	1494,6	6,515	2,710	3,291
600	151,04	1560,4	6,627	2,771	3,298
620	142,74	1626,5	6,735	2,830	3,314
640	135,53	1693,0	6,841	2,888	3,336
660	129,18	1760,0	6,944	2,946	3,366
680	123,54	1827,7	7,045	3,007	3,403
700	118,48	1896,2	7,144	3,073	3,448
		$p=16$			
100	725,04	175,7	2,155	—	—
110	715,23	193,1	2,326	—	—
120	705,36	212,1	2,492	—	—
130	695,58	232,0	2,651	—	—
140	685,88	252,2	2,800	1,401	2,019
150	676,25	272,4	2,939	1,389	2,019
160	666,67	292,6	3,070	1,380	2,021
170	657,11	312,8	3,192	1,378	2,028
180	647,54	333,1	3,309	1,382	2,040
190	637,95	353,6	3,419	1,392	2,057
200	628,31	374,3	3,525	1,404	2,079
210	618,58	395,2	3,627	1,420	2,105
220	608,73	416,4	3,726	1,439	2,135
230	598,77	437,9	3,822	1,461	2,169
240	588,65	459,8	3,915	1,486	2,208
250	578,38	482,1	4,006	1,513	2,249
260	567,95	504,8	4,095	1,543	2,293
270	557,33	528,0	4,182	1,576	2,341
280	546,53	551,6	4,268	1,611	2,391
290	535,51	575,8	4,353	1,648	2,443
300	524,26	600,5	4,437	1,686	2,498

Table II.7 (*Continued*)

T	ρ	h	s	c_v	c_p
			$p=16$		
310	512,74	625,7	4,520	1,726	2,555
320	500,91	651,6	4,602	1,766	2,616
330	488,73	678,1	4,683	1,806	2,679
340	476,15	705,2	4,764	1,847	2,745
350	463,13	733,0	4,845	1,887	2,814
360	449,61	761,5	4,925	1,928	2,886
370	435,56	790,7	5,005	1,968	2,959
380	420,97	820,7	5,085	2,008	3,032
390	405,83	851,4	5,165	2,048	3,105
400	390,19	882,8	5,244	2,087	3,176
410	374,11	914,9	5,323	2,126	3,244
420	357,69	947,6	5,402	2,165	3,306
430	341,08	981,0	5,481	2,204	3,361
440	324,44	1014,8	5,559	2,241	3,406
450	308,01	1049,0	5,636	2,278	3,440
460	292,03	1083,5	5,711	2,315	3,459
470	276,74	1118,2	5,786	2,350	3,464
480	262,34	1152,8	5,859	2,385	3,456
490	248,98	1187,3	5,930	2,419	3,439
500	236,72	1221,5	5,999	2,453	3,417
520	215,38	1289,4	6,132	2,519	3,371
540	197,82	1356,4	6,259	2,585	3,336
560	183,30	1422,9	6,380	2,650	3,316
580	171,15	1489,2	6,496	2,713	3,311
600	160,85	1555,5	6,608	2,773	3,317
620	151,99	1621,9	6,717	2,833	3,331
640	144,28	1688,7	6,823	2,890	3,352
660	137,50	1756,1	6,927	2,949	3,380
680	131,47	1824,0	7,028	3,010	3,416
700	126,07	1892,8	7,128	3,075	3,460
			$p=17$		
100	725,38	176,8	2,153	—	—
110	715,59	194,3	2,324	—	—
120	705,75	213,3	2,490	—	—
130	696,00	233,2	2,649	—	—
140	686,33	253,4	2,798	1,402	2,018
150	676,73	273,5	2,937	1,389	2,018
160	667,18	293,7	3,068	1,381	2,020
170	657,66	314,0	3,190	1,379	2,027
180	648,14	334,3	3,306	1,383	2,038
190	638,60	354,7	3,417	1,392	2,055
200	629,01	375,4	3,523	1,405	2,077
210	619,34	396,3	3,625	1,421	2,102
220	609,57	417,5	3,723	1,440	2,132
230	599,67	439,0	3,819	1,462	2,166
240	589,64	460,8	3,912	1,486	2,204
250	579,46	483,0	4,003	1,514	2,245
260	569,13	505,7	4,092	1,544	2,289
270	558,62	528,8	4,179	1,577	2,336
280	547,94	552,4	4,265	1,612	2,385
290	537,07	576,5	4,349	1,648	2,436
300	525,98	601,2	4,433	1,686	2,490

Table II.7 (*Continued*)

T	ρ	h	s	c_v	c_p
			$p=17$		
310	514,64	626,3	4,515	1,726	2,546
320	503,03	652,1	4,597	1,766	2,604
330	491,10	678,4	4,678	1,806	2,666
340	478,82	705,4	4,759	1,847	2,729
350	466,15	733,0	4,839	1,887	2,795
360	453,04	761,3	4,918	1,927	2,863
370	439,47	790,3	4,998	1,968	2,931
380	425,43	819,9	5,077	2,007	3,000
390	410,93	850,3	5,156	2,047	3,068
400	396,01	881,3	5,234	2,086	3,134
410	380,72	913,0	5,312	2,125	3,196
420	365,16	945,2	5,390	2,163	3,254
430	349,44	978,0	5,467	2,201	3,305
440	333,72	1011,3	5,544	2,239	3,349
450	318,15	1045,0	5,619	2,276	3,384
460	302,93	1078,9	5,694	2,312	3,409
470	288,24	1113,1	5,768	2,348	3,422
480	274,26	1147,3	5,840	2,383	3,425
490	261,12	1181,6	5,910	2,418	3,419
500	248,90	1215,7	5,979	2,452	3,407
520	227,29	1283,5	6,112	2,520	3,376
540	209,18	1350,8	6,239	2,587	3,349
560	194,03	1417,6	6,361	2,652	3,333
580	181,26	1484,2	6,477	2,715	3,328
600	170,39	1550,8	6,590	2,776	3,334
620	161,01	1617,6	6,700	2,835	3,347
640	152,84	1684,7	6,806	2,893	3,367
660	145,65	1752,3	6,910	2,951	3,394
680	139,25	1820,5	7,012	3,012	3,428
700	133,52	1889,5	7,112	3,078	3,471
			$p=18$		
100	725,71	178,0	2,151	—	—
110	715,96	195,4	2,322	—	—
120	706,14	214,5	2,488	—	—
130	696,42	234,4	2,647	—	—
140	686,78	254,5	2,796	1,402	2,018
150	677,21	274,7	2,935	1,390	2,017
160	667,69	294,9	3,065	1,381	2,019
170	658,21	315,1	3,188	1,380	2,025
180	648,73	335,4	3,304	1,384	2,037
190	639,24	355,9	3,415	1,393	2,054
200	629,70	376,5	3,521	1,406	2,075
210	620,09	397,4	3,622	1,422	2,100
220	610,39	418,5	3,721	1,441	2,130
230	600,57	440,0	3,816	1,462	2,163
240	590,62	461,8	3,909	1,487	2,201
250	580,52	484,0	4,000	1,514	2,241
260	570,28	506,6	4,088	1,544	2,285
270	559,89	529,7	4,175	1,577	2,331
280	549,33	553,3	4,261	1,612	2,379
290	538,59	577,3	4,345	1,648	2,430
300	527,65	601,9	4,429	1,687	2,482

Table II.7 (*Continued*)

T	ρ	h	s	c_v	c_p
			$p=18$		
310	516,49	627,0	4,511	1,726	2,537
320	505,08	652,6	4,592	1,766	2,594
330	493,39	678,8	4,673	1,806	2,653
340	481,38	705,7	4,753	1,847	2,714
350	469,03	733,1	4,833	1,887	2,777
360	456,29	761,2	4,912	1,927	2,842
370	443,15	790,0	4,991	1,967	2,907
380	429,60	819,4	5,069	2,007	2,972
390	415,65	849,4	5,147	2,046	3,036
400	401,35	880,1	5,225	2,085	3,098
410	386,74	911,4	5,302	2,123	3,156
420	371,90	943,2	5,379	2,162	3,210
430	356,95	975,6	5,455	2,199	3,259
440	342,00	1008,4	5,530	2,237	3,301
450	327,18	1041,6	5,605	2,274	3,337
460	312,64	1075,1	5,679	2,310	3,364
470	298,53	1108,8	5,751	2,347	3,383
480	284,99	1142,7	5,822	2,382	3,393
490	272,15	1176,7	5,893	2,417	3,396
500	260,08	1210,6	5,961	2,452	3,392
520	238,40	1278,3	6,094	2,521	3,375
540	219,94	1345,6	6,221	2,588	3,357
560	204,30	1412,6	6,343	2,654	3,345
580	191,01	1479,5	6,460	2,717	3,343
600	179,63	1546,4	6,573	2,778	3,348
620	169,79	1613,5	6,683	2,837	3,361
640	161,19	1680,9	6,790	2,895	3,380
660	153,61	1748,7	6,895	2,954	3,406
680	146,87	1817,2	6,997	3,015	3,440
700	140,82	1886,4	7,097	3,080	3,482
			$p=19$		
100	726,04	179,2	2,149	—	—
110	716,32	196,6	2,320	—	—
120	706,53	215,7	2,486	—	—
130	696,83	235,6	2,645	—	—
140	687,22	255,7	2,794	1,403	2,017
150	677,68	275,9	2,933	1,391	2,017
160	668,20	296,0	3,063	1,382	2,018
170	658,75	316,3	3,186	1,380	2,024
180	649,32	336,5	3,302	1,385	2,036
190	639,87	357,0	3,412	1,394	2,052
200	630,39	377,6	3,518	1,407	2,073
210	620,84	398,5	3,620	1,423	2,098
220	611,20	419,6	3,718	1,442	2,127
230	601,45	441,0	3,813	1,463	2,161
240	591,58	462,8	3,906	1,488	2,197
250	581,57	485,0	3,997	1,515	2,238
260	571,42	507,6	4,085	1,545	2,281
270	561,13	530,6	4,172	1,577	2,326
280	550,69	554,1	4,258	1,612	2,374
290	540,07	578,1	4,342	1,649	2,423
300	529,28	602,6	4,425	1,687	2,475

Table II.7 (*Continued*)

T	ρ	h	s	c_v	c_p
		$p = 19$			
310	518,29	627,6	4,507	1,726	2,529
320	507,07	653,2	4,588	1,766	2,584
330	495,60	679,3	4,668	1,806	2,642
340	483,84	706,0	4,748	1,847	2,701
350	471,78	733,3	4,827	1,887	2,761
360	459,37	761,2	4,906	1,927	2,823
370	446,62	789,8	4,984	1,967	2,886
380	433,51	818,9	5,062	2,007	2,948
390	420,05	848,7	5,139	2,046	3,008
400	406,29	879,1	5,216	2,085	3,067
410	392,27	910,0	5,293	2,123	3,122
420	378,07	941,5	5,368	2,161	3,173
430	363,77	973,5	5,444	2,198	3,220
440	349,48	1005,9	5,518	2,236	3,261
450	335,31	1038,7	5,592	2,273	3,296
460	321,38	1071,8	5,665	2,309	3,325
470	307,81	1105,2	5,736	2,346	3,347
480	294,71	1138,7	5,807	2,382	3,363
490	282,19	1172,4	5,876	2,417	3,371
500	270,33	1206,1	5,945	2,452	3,375
520	248,77	1273,6	6,077	2,522	3,370
540	230,10	1340,9	6,204	2,589	3,361
560	214,09	1408,1	6,326	2,655	3,355
580	200,38	1475,1	6,444	2,719	3,354
600	188,56	1542,3	6,558	2,780	3,361
620	178,30	1609,6	6,668	2,840	3,374
640	169,32	1677,3	6,775	2,898	3,393
660	161,38	1745,3	6,880	2,956	3,418
680	154,31	1814,0	6,983	3,017	3,451
700	147,97	1883,4	7,083	3,082	3,492
		$p = 20$			
100	726,37	180,4	2,147	—	—
110	716,67	197,8	2,318	—	—
120	706,92	216,9	2,484	—	—
130	697,25	236,7	2,643	—	—
140	687,66	256,9	2,792	1,404	2,016
150	678,15	277,0	2,931	1,392	2,016
160	668,70	297,2	3,061	1,383	2,017
170	659,29	317,4	3,184	1,381	2,023
180	649,90	337,7	3,300	1,386	2,035
190	640,50	358,1	3,410	1,395	2,051
200	631,07	378,7	3,516	1,408	2,071
210	621,58	399,6	3,617	1,424	2,096
220	612,00	420,7	3,716	1,443	2,125
230	602,32	442,1	3,811	1,464	2,158
240	592,52	463,8	3,903	1,488	2,194
250	582,60	486,0	3,994	1,516	2,234
260	572,55	508,5	4,082	1,546	2,277
270	562,35	531,5	4,169	1,578	2,321
280	552,02	554,9	4,254	1,613	2,369
290	541,53	578,9	4,338	1,649	2,418
300	530,87	603,3	4,421	1,687	2,468

Table II.7 (*Continued*)

T	ρ	h	s	c_v	c_p
			$p=20$		
310	520,04	628,2	4,503	1,726	2,521
320	509,00	653,7	4,584	1,766	2,575
330	497,73	679,8	4,664	1,807	2,631
340	486,21	706,3	4,743	1,847	2,688
350	474,41	733,5	4,822	1,887	2,747
360	462,32	761,3	4,900	1,928	2,806
370	449,91	789,6	4,978	1,967	2,866
380	437,19	818,6	5,055	2,007	2,925
390	424,18	848,1	5,132	2,046	2,983
400	410,90	878,3	5,208	2,085	3,039
410	397,40	908,9	5,284	2,123	3,092
420	383,75	940,1	5,359	2,161	3,141
430	370,02	971,7	5,433	2,198	3,186
440	356,31	1003,8	5,507	2,235	3,226
450	342,71	1036,2	5,580	2,272	3,261
460	329,32	1069,0	5,652	2,309	3,291
470	316,24	1102,0	5,723	2,345	3,315
480	303,57	1135,3	5,793	2,381	3,334
490	291,38	1168,7	5,862	2,417	3,347
500	279,77	1202,2	5,930	2,452	3,356
520	258,42	1269,4	6,061	2,522	3,362
540	239,68	1336,7	6,188	2,591	3,361
560	223,42	1403,9	6,310	2,657	3,360
580	209,36	1471,1	6,428	2,721	3,363
600	197,17	1538,4	6,543	2,782	3,371
620	186,55	1606,0	6,653	2,842	3,385
640	177,21	1673,9	6,761	2,900	3,403
660	168,95	1742,2	6,866	2,958	3,428
680	161,58	1811,0	6,969	3,019	3,461
700	154,96	1880,6	7,070	3,085	3,501
			$p=21$		
100	726,69	181,6	2,145	—	—
110	717,03	199,0	2,316	—	—
120	707,30	218,1	2,482	—	—
130	697,66	237,9	2,641	—	—
140	688,10	258,0	2,790	1,405	2,015
150	678,62	278,2	2,929	1,393	2,015
160	669,20	298,4	3,059	1,384	2,016
170	659,82	318,5	3,182	1,382	2,023
180	650,47	338,8	3,297	1,387	2,034
190	641,12	359,2	3,408	1,396	2,049
200	631,74	379,8	3,513	1,409	2,070
210	622,31	400,6	3,615	1,425	2,094
220	612,79	421,7	3,713	1,444	2,123
230	603,18	443,1	3,808	1,465	2,155
240	593,46	464,8	3,901	1,489	2,191
250	583,62	486,9	3,991	1,516	2,231
260	573,65	509,5	4,079	1,546	2,273
270	563,55	532,4	4,166	1,579	2,317
280	553,32	555,8	4,251	1,613	2,364
290	542,95	579,7	4,335	1,650	2,412
300	532,43	604,1	4,417	1,688	2,462

Table II.7 (*Continued*)

T	ρ	h	s	c_v	c_p
			$p=21$		
310	521,74	628,9	4,499	1,727	2,513
320	510,87	654,3	4,579	1,767	2,566
330	499,79	680,3	4,659	1,807	2,621
340	488,49	706,7	4,738	1,847	2,677
350	476,94	733,8	4,817	1,888	2,733
360	465,13	761,4	4,894	1,928	2,791
370	453,04	789,6	4,972	1,968	2,848
380	440,69	818,4	5,048	2,007	2,905
390	428,07	847,7	5,125	2,046	2,961
400	415,22	877,6	5,200	2,085	3,015
410	402,18	908,0	5,275	2,123	3,066
420	389,01	938,9	5,350	2,161	3,113
430	375,79	970,2	5,424	2,198	3,156
440	362,59	1002,0	5,497	2,235	3,195
450	349,50	1034,1	5,569	2,272	3,230
460	336,59	1066,6	5,640	2,309	3,261
470	323,96	1099,3	5,711	2,345	3,287
480	311,68	1132,3	5,780	2,381	3,308
490	299,83	1165,5	5,848	2,417	3,325
500	288,47	1198,8	5,916	2,453	3,337
520	267,41	1265,7	6,047	2,523	3,352
540	248,69	1332,8	6,173	2,592	3,359
560	232,28	1400,1	6,296	2,658	3,364
580	217,96	1467,4	6,414	2,723	3,370
600	205,47	1534,9	6,528	2,784	3,380
620	194,52	1602,6	6,639	2,844	3,394
640	184,87	1670,7	6,747	2,902	3,413
660	176,31	1739,2	6,853	2,960	3,438
680	168,66	1808,2	6,956	3,021	3,470
700	161,78	1878,0	7,057	3,087	3,510
			$p=22$		
100	727,02	182,8	2,144	—	—
110	717,38	200,2	2,314	—	—
120	707,69	219,2	2,480	—	—
130	698,07	239,1	2,639	—	—
140	688,54	259,2	2,788	1,406	2,015
150	679,09	279,4	2,927	1,393	2,015
160	669,70	299,5	3,057	1,385	2,016
170	660,36	319,7	3,179	1,383	2,022
180	651,05	340,0	3,295	1,388	2,032
190	641,74	360,4	3,406	1,397	2,048
200	632,41	380,9	3,511	1,410	2,068
210	623 03	401,7	3,613	1,426	2,092
220	613,58	422,8	3,710	1,445	2,120
230	604,03	444,2	3,805	1,466	2,152
240	594,38	465,9	3,898	1,490	2,188
250	584,62	487,9	3,988	1,517	2,227
260	574,73	510,4	4,076	1,547	2,269
270	564,73	533,3	4,163	1,579	2,313
280	554,60	556,7	4,247	1,614	2,359
290	544,34	580,5	4,331	1,650	2,407
300	533,95	604,8	4,413	1,688	2,456

Table II.7 (*Continued*)

T	ρ	h	s	c_v	c_p
		$p = 22$			
310	523,40	629,6	4,495	1,727	2,506
320	512,69	654,9	4,575	1,767	2,558
330	501,79	680,8	4,655	1,807	2,612
340	490,69	707,2	4,734	1,848	2,666
350	479,38	734,1	4,812	1,888	2,721
360	467,83	761,6	4,889	1,929	2,777
370	456,03	789,7	4,966	1,968	2,832
380	444,00	818,2	5,042	2,008	2,887
390	431,74	847,4	5,118	2,047	2,941
400	419,28	877,1	5,193	2,085	2,993
410	406,66	907,2	5,267	2,124	3,042
420	393,93	937,9	5,341	2,161	3,088
430	381,16	969,0	5,414	2,199	3,130
440	368,41	1000,5	5,487	2,236	3,169
450	355,77	1032,3	5,558	2,273	3,203
460	343,29	1064,5	5,629	2,309	3,234
470	331,07	1097,0	5,699	2,346	3,261
480	319,16	1129,7	5,768	2,382	3,284
490	307,62	1162,7	5,836	2,418	3,303
500	296,52	1195,8	5,903	2,454	3,319
520	275,80	1262,4	6,033	2,524	3,341
540	257,19	1329,4	6,160	2,593	3,355
560	240,70	1396,6	6,282	2,660	3,365
580	226,19	1464,0	6,400	2,724	3,375
600	213,45	1531,6	6,515	2,786	3,387
620	202,22	1599,5	6,626	2,846	3,402
640	192,30	1667,7	6,734	2,904	3,422
660	183,46	1736,4	6,840	2,962	3,446
680	175,56	1805,6	6,943	3,023	3,478
700	168,44	1875,5	7,045	3,089	3,518
		$p = 23$			
100	727,33	184,0	2,142	—	—
110	717,73	201,4	2,312	—	—
120	708,07	220,4	2,478	—	—
130	698,48	240,3	2,637	—	—
140	688,98	260,4	2,786	1,407	2,014
150	679,55	280,5	2,925	1,395	2,014
160	670,19	300,7	3,055	1,386	2,015
170	660,89	320,9	3,177	1,384	2,021
180	651,61	341,1	3,293	1,389	2,031
190	642,35	361,5	3,403	1,398	2,047
200	633,07	382,1	3,509	1,411	2,066
210	623,75	402,8	3,610	1,427	2,090
220	614,35	423,9	3,708	1,446	2,118
230	604,87	445,2	3,803	1,467	2,150
240	595,29	466,9	3,895	1,491	2,185
250	585,60	488,9	3,985	1,518	2,224
260	575,80	511,4	4,073	1,548	2,265
270	565,89	534,2	4,159	1,580	2,309
280	555,86	557,6	4,244	1,614	2,354
290	545,71	581,3	4,328	1,651	2,402
300	535,43	605,6	4,410	1,689	2,450

Table II.7 (*Continued*)

T	p	h	s	c_v	c_p
			$p=23$		
310	525,02	630,3	4,491	1,728	2,500
320	514,46	655,6	4,571	1,767	2,551
330	503,73	681,4	4,650	1,808	2,603
340	492,83	707,7	4,729	1,848	2,656
350	481,72	734,5	4,807	1,889	2,710
360	470,42	761,8	4,884	1,929	2,764
370	458,90	789,8	4,960	1,969	2,818
380	447,16	818,2	5,036	2,009	2,871
390	435,23	847,2	5,111	2,048	2,923
400	423,12	876,7	5,186	2,086	2,973
410	410,88	906,6	5,260	2,124	3,021
420	398,54	937,1	5,333	2,162	3,066
430	386,17	967,9	5,406	2,199	3,107
440	373,84	999,2	5,478	2,237	3,145
450	361,60	1030,8	5,549	2,273	3,179
460	349,52	1062,8	5,619	2,310	3,210
470	337,66	1095,0	5,688	2,346	3,238
480	326,09	1127,5	5,757	2,383	3,262
490	314,86	1160,2	5,824	2,419	3,283
500	304,01	1193,2	5,891	2,454	3,301
520	283,64	1259,5	6,021	2,525	3,329
540	265,19	1326,3	6,147	2,594	3,349
560	248,69	1393,4	6,269	2,661	3,364
580	234,05	1460,8	6,387	2,726	3,378
600	221,11	1528,5	6,502	2,788	3,392
620	209,66	1596,5	6,614	2,848	3,409
640	199,49	1664,9	6,722	2,906	3,429
660	190,41	1733,7	6,828	2,964	3,454
680	182,27	1803,1	6,932	3,025	3,486
700	174,93	1873,2	7,033	3,091	3,526
			$p=24$		
100	727,65	185,2	2,140	—	—
110	718,08	202,6	2,311	—	—
120	708,45	221,6	2,476	—	—
130	698,89	241,4	2,635	—	—
140	689,41	261,6	2,784	1,408	2,014
150	680,01	281,7	2,923	1,396	2,014
160	670,68	301,8	3,053	1,387	2,014
170	661,41	322,0	3,175	1,385	2,020
180	652,18	342,2	3,291	1,390	2,030
190	642,96	362,6	3,401	1,399	2,045
200	633,73	383,2	3,506	1,412	2,065
210	624,46	403,9	3,608	1,428	2,088
220	615,12	425,0	3,705	1,447	2,116
230	605,70	446,3	3,800	1,468	2,147
240	596,19	467,9	3,892	1,492	2,183
250	586,58	489,9	3,982	1,519	2,221
260	576,86	512,3	4,070	1,548	2,262
270	567,03	535,2	4,156	1,580	2,305
280	557,10	558,4	4,241	1,615	2,350
290	547,05	582,2	4,324	1,651	2,397

Table II.7 (*Continued*)

T	ρ	h	s	c_v	c_p
		$p=24$			
300	536,89	606,4	4,406	1,689	2,445
310	526,60	631,1	4,487	1,728	2,494
320	516,18	656,3	4,567	1,768	2,544
330	505,62	682,0	4,646	1,808	2,595
340	494,89	708,2	4,724	1,849	2,647
350	483,99	734,9	4,802	1,890	2,699
360	472,91	762,1	4,879	1,930	2,752
370	461,64	789,9	4,955	1,970	2,804
380	450,18	818,2	5,030	2,010	2,856
390	438,55	847,0	5,105	2,049	2,906
400	426,77	876,3	5,179	2,087	2,955
410	414,86	906,1	5,253	2,125	3,002
420	402,89	936,4	5,326	2,163	3,045
430	390,88	967,0	5,398	2,201	3,086
440	378,92	998,1	5,469	2,238	3,124
450	367,04	1029,5	5,540	2,274	3,158
460	355,32	1061,2	5,610	2,311	3,189
470	343,80	1093,3	5,678	2,348	3,217
480	332,54	1125,6	5,746	2,384	3,242
490	321,60	1158,1	5,814	2,420	3,265
500	311,00	1190,8	5,880	2,456	3,285
520	290,99	1256,9	6,009	2,526	3,317
540	272,73	1323,5	6,135	2,596	3,342
560	256,27	1390,5	6,257	2,663	3,362
580	241,56	1457,9	6,375	2,728	3,379
600	228,48	1525,7	6,490	2,790	3,396
620	216,83	1593,8	6,602	2,849	3,414
640	206,45	1662,3	6,710	2,908	3,435
660	197,16	1731,2	6,816	2,966	3,461
680	188,80	1800,8	6,920	3,027	3,493
700	181,26	1871,0	7,022	3,093	3,533
		$p=25$			
100	727,96	186,4	2,138	—	—
110	718,43	203,8	2,309	—	—
120	708,83	222,8	2,474	—	—
130	699,29	242,6	2,633	—	—
140	689,84	262,7	2,782	1,410	2,013
150	680,47	282,9	2,921	1,397	2,013
160	671,17	303,0	3,051	1,388	2,014
170	661,93	323,2	3,173	1,386	2,019
180	652,74	343,4	3,289	1,391	2,029
190	643,56	363,8	3,399	1,400	2,044
200	634,38	384,3	3,504	1,413	2,063
210	625,16	405,0	3,605	1,429	2,086
220	615,88	426,0	3,703	1,448	2,114
230	606,53	447,3	3,798	1,469	2,145
240	597,08	468,9	3,890	1,493	2,180
250	587,54	490,9	3,979	1,520	2,218
260	577,90	513,3	4,067	1,549	2,259
270	568,15	536,1	4,153	1,581	2,301
280	558,31	559,3	4,238	1,616	2,346

Table II.7 (*Continued*)

T	p	h	s	c_v	c_p
		$p = 25$			
290	548,36	583,0	4,321	1,652	2,392
300	538,31	607,2	4,403	1,690	2,439
310	528,15	631,8	4,484	1,729	2,488
320	517,87	657,0	4,563	1,769	2,537
330	507,45	682,6	4,642	1,809	2,587
340	496,90	708,7	4,720	1,850	2,638
350	486,19	735,3	4,797	1,890	2,689
360	475,31	762,5	4,874	1,931	2,740
370	464,28	790,1	4,950	1,971	2,792
380	453,08	818,3	5,025	2,010	2,842
390	441,72	847,0	5,099	2,050	2,891
400	430,23	876,1	5,173	2,088	2,939
410	418,65	905,7	5,246	2,127	2,984
420	406,99	935,8	5,319	2,164	3,027
430	395,33	966,3	5,390	2,202	3,067
440	383,70	997,1	5,461	2,239	3,104
450	372,16	1028,4	5,531	2,276	3,138
460	360,76	1059,9	5,601	2,312	3,170
470	349,55	1091,7	5,669	2,349	3,198
480	338,59	1123,9	5,737	2,385	3,224
490	327,90	1156,2	5,803	2,421	3,248
500	317,54	1188,8	5,869	2,457	3,269
520	297,90	1254,6	5,998	2,528	3,306
540	279,86	1321,0	6,124	2,597	3,335
560	263,48	1387,9	6,245	2,665	3,358
580	248,74	1455,3	6,364	2,729	3,379
600	235,55	1523,1	6,478	2,792	3,399
620	223,75	1591,3	6,590	2,851	3,419
640	213,18	1659,8	6,699	2,910	3,441
660	203,70	1728,9	6,805	2,968	3,467
680	195,16	1798,6	6,909	3,029	3,499
700	187,42	1868,9	7,011	3,094	3,539
		$p = 26$			
100	728,28	187,6	2,136	—	—
110	718,78	204,9	2,307	—	—
120	709,20	224,0	2,472	—	—
130	699,70	243,8	2,631	—	—
140	690,27	263,9	2,780	1,411	2,012
150	680,93	284,0	2,919	1,398	2,012
160	671,66	304,2	3,049	1,389	2,013
170	662,45	324,3	3,171	1,387	2,018
180	653,29	344,5	3,287	1,392	2,028
190	644,16	364,9	3,397	1,402	2,043
200	635,02	385,4	3,502	1,414	2,062
210	625,85	406,1	3,603	1,430	2,085
220	616,63	427,1	3,701	1,449	2,112
230	607,34	448,4	3,795	1,470	2,143
240	597,96	470,0	3,887	1,494	2,177
250	588,49	491,9	3,977	1,521	2,215
260	578,92	514,3	4,064	1,550	2,255

Table II.7 (*Continued*)

T	ρ	h	s	c_v	c_p
\multicolumn{6}{c}{$p=26$}					
270	569,26	537,1	4,150	1,582	2,298
280	559,50	560,3	4,235	1,616	2,342
290	549,65	583,9	4,318	1,652	2,388
300	539,71	608,0	4,399	1,690	2,434
310	529,66	632,6	4,480	1,729	2,482
320	519,51	657,7	4,559	1,769	2,531
330	509,24	683,2	4,638	1,810	2,580
340	498,84	709,3	4,716	1,850	2,630
350	488,31	735,8	4,793	1,891	2,680
360	477,64	762,9	4,869	1,932	2,730
370	466,82	790,4	4,944	1,972	2,780
380	455,85	818,5	5,019	2,011	2,829
390	444,75	847,0	5,093	2,051	2,877
400	433,54	876,0	5,167	2,089	2,924
410	422,24	905,5	5,240	2,128	2,969
420	410,89	935,4	5,312	2,166	3,011
430	399,53	965,7	5,383	2,203	3,050
440	388,21	996,4	5,453	2,240	3,087
450	376,98	1027,4	5,523	2,277	3,121
460	365,88	1058,8	5,592	2,314	3,152
470	354,96	1090,4	5,660	2,350	3,181
480	344,26	1122,4	5,728	2,387	3,208
490	333,83	1154,6	5,794	2,423	3,233
500	323,69	1187,0	5,859	2,458	3,255
520	304,41	1252,5	5,988	2,529	3,294
540	286,60	1318,8	6,113	2,599	3,327
560	270,33	1385,6	6,234	2,666	3,354
580	255,60	1452,9	6,353	2,731	3,378
600	242,33	1520,7	6,467	2,793	3,400
620	230,41	1588,9	6,579	2,853	3,422
640	219,70	1657,6	6,688	2,911	3,445
660	210,05	1726,7	6,795	2,970	3,472
680	201,33	1796,5	6,899	3,031	3,505
700	193,43	1867,0	7,001	3,096	3,545
\multicolumn{6}{c}{$p=28$}					
100	728,89	190,0	2,133	—	—
110	719,47	207,3	2,303	—	—
120	709,95	226,3	2,468	—	—
130	700,50	246,2	2,627	—	—
140	691,13	266,3	2,776	1,414	2,011
150	681,83	286,4	2,915	1,400	2,011
160	672,62	306,5	3,045	1,391	2,012
170	663,48	326,6	3,167	1,390	2,017
180	654,39	346,8	3,282	1,394	2,026
190	645,34	367,2	3,392	1,404	2,041
200	636,29	387,7	3,497	1,417	2,059
210	627,23	408,4	3,598	1,433	2,081
220	618,11	429,3	3,696	1,451	2,108
230	608,94	450,5	3,790	1,472	2,138
240	599,68	472,1	3,892	1,496	2,172
250	590,35	494,0	3,971	1,522	2,209

Table II.7 (*Continued*)

T	v	h	s	c_v	c_p

$p = 28$

260	580,93	516,3	4,059	1,552	2,249
270	571,42	539,0	4,144	1,583	2,291
280	561,83	562,1	4,228	1,618	2,334
290	552,17	585,7	4,311	1,654	2,379
300	542,42	609,7	4,393	1,691	2,425
310	532,59	634,2	4,473	1,731	2,472
320	522,68	659,1	4,552	1,770	2,519
330	512,68	684,5	4,630	1,811	2,567
340	502,58	710,5	4,708	1,852	2,615
350	492,37	736,9	4,784	1,893	2,663
360	482,06	763,7	4,860	1,933	2,712
370	471,63	791,1	4,935	1,974	2,759
380	461,09	818,9	5,009	2,014	2,807
390	450,46	847,2	5,082	2,053	2,853
400	439,74	876,0	5,155	2,092	2,898
410	428,96	905,2	5,227	2,131	2,941
420	418,14	934,8	5,299	2,169	2,982
430	407,33	964,8	5,369	2,206	3,020
440	396,56	995,2	5,439	2,243	3,056
450	385,87	1025,9	5,508	2,280	3,090
460	375,30	1057,0	5,577	2,317	3,122
470	364,89	1088,3	5,644	2,354	3,152
480	354,68	1120,0	5,711	2,390	3,179
490	344,70	1151,9	5,776	2,426	3,205
500	334,98	1184,1	5,841	2,462	3,229
520	316,39	1249,1	5,969	2,533	3,273
540	299,05	1315,0	6,093	2,602	3,311
560	283,06	1381,5	6,214	2,670	3,345
580	268,43	1448,7	6,332	2,735	3,374
600	255,11	1516,5	6,447	2,797	3,401
620	243,04	1584,7	6,559	2,857	3,426
640	232,10	1653,5	6,668	2,915	3,452
660	222,18	1722,8	6,775	2,974	3,481
680	213,17	1792,8	6,879	3,034	3,514
700	204,97	1863,5	6,982	3,100	3,555

$p = 30$

100	729,50	192,4	2,129	—	—
110	720,15	209,7	2,299	—	—
120	710,70	228,7	2,465	—	—
130	701,30	248,5	2,623	—	—
140	691,98	268,6	2,772	1,417	2,010
150	682,74	288,7	2,911	1,403	2,010
160	673,58	308,8	3,041	1,394	2,010
170	664,50	328,9	3,163	1,392	2,015
180	655,48	349,1	3,278	1,397	2,025
190	646,51	369,4	3,388	1,406	2,038
200	637,54	389,9	3,493	1,419	2,056
210	628,57	410,6	3,594	1,435	2,078
220	619,56	431,5	3,691	1,453	2,104
230	610,50	452,7	3,785	1,474	2,134

Table II.7 (*Continued*)

T	ρ	h	s	c_v	c_p

$p = 30$

T	ρ	h	s	c_v	c_p
240	601,37	474,2	3,877	1,498	2,167
250	592,16	496,0	3,966	1,524	2,204
260	582,88	518,3	4,053	1,553	2,243
270	573,52	540,9	4,139	1,585	2,284
280	564,09	564,0	4,222	1,619	2,327
290	554,60	587,4	4,305	1,655	2,372
300	545,03	611,4	4,386	1,693	2,417
310	535,41	635,8	4,466	1,732	2,462
320	525,72	660,6	4,545	1,772	2,509
330	515,96	686,0	4,623	1,812	2,555
340	506,12	711,7	4,700	1,853	2,602
350	496,21	738,0	4,776	1,894	2,649
360	486,21	764,7	4,851	1,935	2,695
370	476,13	791,9	4,926	1,976	2,742
380	465,98	819,6	4,999	2,016	2,787
390	455,74	847,6	5,072	2,056	2,832
400	445,45	876,2	5,145	2,095	2,875
410	435,12	905,1	5,216	2,133	2,917
420	424,77	934,5	5,287	2,172	2,957
430	414,43	964,3	5,357	2,209	2,995
440	404,13	994,4	5,426	2,247	3,031
450	393,92	1024,9	5,495	2,284	3,064
460	383,81	1055,7	5,562	2,321	3,096
470	373,85	1086,8	5,629	2,357	3,126
480	364,07	1118,2	5,695	2,394	3,155
490	354,48	1149,9	5,761	2,430	3,182
500	345,13	1181,8	5,825	2,466	3,207
520	327,18	1246,4	5,952	2,537	3,254
540	310,33	1312,0	6,076	2,606	3,296
560	294,66	1378,3	6,196	2,674	3,334
580	280,20	1445,3	6,314	2,739	3,368
600	266,92	1513,0	6,428	2,801	3,399
620	254,78	1581,2	6,540	2,860	3,427
640	243,70	1650,1	6,650	2,919	3,456
660	233,59	1719,5	6,756	2,977	3,487
680	224,36	1789,5	6,861	3,038	3,521
700	215,92	1860,4	6,964	3,103	3,563

$p = 32$

T	ρ	h	s	c_v	c_p
100	730,10	194,8	2,126	—	—
110	720,83	212,1	2,296	—	—
120	711,44	231,1	2,461	—	—
130	702,10	250,9	2,620	—	—
140	692,82	270,9	2,768	1,420	2,009
150	683,63	291,0	2,907	1,406	2,009
160	674,52	311,1	3,037	1,396	2,009
170	665,50	331,2	3,159	1,394	2,014
180	656,55	351,4	3,274	1,399	2,023
190	647,65	371,7	3,384	1,409	2,036
200	638,78	392,2	3,489	1,422	2,054

Table II.7 (*Continued*)

T	ρ	h	s	c_v	c_p

$$p=32$$

T	ρ	h	s	c_v	c_p
210	629,90	412,8	3,589	1,437	2,075
220	620,99	433,7	3,686	1,456	2,100
230	612,03	454,8	3,780	1,476	2,130
240	603,02	476,3	3,872	1,500	2,163
250	593,93	498,1	3,961	1,526	2,199
260	584,78	520,3	4,048	1,555	2,238
270	575,56	542,9	4,133	1,587	2,278
280	566,29	565,8	4,216	1,621	2,321
290	556,95	589,3	4,299	1,657	2,364
300	547,56	613,1	4,380	1,694	2,409
310	538,12	637,5	4,459	1,733	2,454
320	528,63	662,2	4,538	1,773	2,499
330	519,09	687,4	4,616	1,814	2,545
340	509,50	713,1	4,692	1,855	2,590
350	499,85	739,2	4,768	1,896	2,636
360	490,14	765,8	4,843	1,937	2,681
370	480,37	792,9	4,917	1,978	2,726
380	470,55	820,4	4,990	2,018	2,770
390	460,67	848,3	5,063	2,058	2,813
400	450,76	876,6	5,134	2,098	2,855
410	440,82	905,4	5,205	2,137	2,896
420	430,88	934,5	5,276	2,175	2,935
430	420,95	964,1	5,345	2,213	2,973
440	411,08	994,0	5,414	2,251	3,008
450	401,27	1024,2	5,482	2,288	3,042
460	391,57	1054,8	5,549	2,325	3,074
470	382,01	1085,7	5,616	2,361	3,105
480	372,60	1116,9	5,681	2,398	3,134
490	363,38	1148,4	5,746	2,434	3,161
500	354,35	1180,1	5,810	2,470	3,188
520	336,99	1244,4	5,936	2,541	3,237
540	320,61	1309,6	6,059	2,610	3,282
560	305,28	1375,5	6,179	2,678	3,323
580	291,03	1442,5	6,297	2,742	3,361
600	277,86	1510,1	6,411	2,805	3,395
620	265,72	1578,3	6,523	2,864	3,427
640	251,57	1617,1	6,632	2,922	3,458
660	244,33	1716,6	9,739	2,981	3,490
680	234,93	1786,8	6,844	3,041	3,527
700	226,30	1857,7	6,947	3,107	3,569

$$p=34$$

T	ρ	h	s	c_v	c_p
100	730,70	197,2	2,122	—	—
110	721,50	214,4	2,292	—	—
120	712,19	233,4	2,457	—	—
130	702,90	253,2	2,616	—	—
140	693,67	273,3	2,765	1,424	2,008
150	684,52	293,4	2,903	1,409	2,008
160	675,46	313,5	3,033	1,399	2,008
170	666,50	333,6	3,155	1,397	2,013
180	657,61	353,7	3,270	1,401	2,021
190	648,78	374,0	3,379	1,411	2,034

Table II.7 (*Continued*)

T	p	h	s	c_v	c_p
		$p=34$			
200	639,99	394,4	3,484	1,424	2,051
210	631,20	415,0	3,585	1,440	2,072
220	622,38	435,9	3,682	1,458	2,097
230	613,53	457,0	3,776	1,479	2,126
240	604,63	478,4	3,867	1,502	2,158
250	595,67	500,2	3,956	1,528	2,194
260	586,64	522,3	4,042	1,557	2,232
270	577,55	544,8	4,127	1,588	2,273
280	568,42	567,8	4,211	1,622	2,315
290	559,23	591,1	4,293	1,658	2,358
300	550,00	614,9	4,373	1,696	2,402
310	540,73	639,2	4,453	1,735	2,446
320	531,43	663,8	4,531	1,775	2,491
330	522,09	689,0	4,609	1,815	2,535
340	512,72	714,6	4,685	1,857	2,580
350	503,31	740,6	4,760	1,898	2,624
360	493,86	767,0	4,835	1,939	2,669
370	484,37	793,9	4,909	1,980	2,712
380	474,85	821,3	4,982	2,021	2,755
390	465,29	849,0	5,054	2,061	2,797
400	455,71	877,2	5,125	2,100	2,838
410	446,13	905,8	5,196	2,140	2,878
420	436,54	934,8	5,265	2,178	2,917
430	426,99	964,1	5,334	2,216	2,954
440	417,48	993,9	5,403	2,254	2,989
450	408,05	1023,9	5,470	2,292	3,023
460	398,71	1054,3	5,537	2,329	3,055
470	389,50	1085,0	5,603	2,366	3,086
480	380,43	1116,0	5,668	2,402	3,115
490	371,53	1147,3	5,733	2,438	3,144
500	362,81	1178,9	5,797	2,474	3,171
520	345,98	1242,8	5,922	2,545	3,222
540	330,05	1307,7	6,045	2,615	3,269
560	315,06	1373,6	6,164	2,682	3,313
580	301,05	1440,2	6,281	2,747	3,353
600	288,01	1507,7	6,396	2,809	3,390
620	275,93	1575,8	6,507	2,868	3,425
640	264,76	1644,7	6,617	2,926	3,458
660	254,45	1714,2	6,723	2,984	3,493
680	244,94	1784,4	6,828	3,045	3,531
700	236,16	1855,4	6,931	3,110	3,574
		$p=36$			
100	731,29	199,6	2,119	—	—
110	722,17	216,8	2,289	—	—
120	712,92	235,8	2,454	—	—
130	703,69	255,6	2,612	—	—
140	694,50	275,6	2,761	1,428	2,007
150	685,40	295,7	2,899	1,412	2,007
160	676,39	315,8	3,029	1,401	2,007
170	667,48	335,9	3,151	1,399	2,011
180	658,66	356,0	3,266	1,404	2,020

Table II.7 (*Continued*)

T	ρ	h	s	c_v	c_p
			$p=36$		
190	649,90	376,3	3,375	1,413	2,033
200	641,18	396,7	3,480	1,426	2,049
210	632,48	417,3	3,580	1,442	2,069
220	623,76	438,1	3,677	1,460	2,094
230	615,01	459,2	3,771	1,481	2,122
240	606,21	480,6	3,862	1,504	2,154
250	597,36	502,3	3,951	1,530	2,189
260	588,45	524,4	4,037	1,559	2,227
270	579,49	546,8	4,122	1,590	2,267
280	570,49	569,7	4,205	1,624	2,309
290	561,44	593,0	4,287	1,659	2,352
300	552,37	616,7	4,367	1,697	2,395
310	543,26	640,9	4,447	1,736	2,439
320	534,13	665,5	4,525	1,776	2,483
330	524,98	690,6	4,602	1,817	2,527
340	515,81	716,1	4,678	1,858	2,571
350	506,61	742,0	4,753	1,900	2,614
360	497,40	768,3	4,827	1,941	2,657
370	488,16	795,1	4,901	1,982	2,700
380	478,91	822,3	4,973	2,023	2,742
390	469,64	850,0	5,045	2,063	2,783
400	460,36	878,0	5,116	2,103	2,823
410	451,09	906,4	5,186	2,143	2,863
420	441,83	935,2	5,256	2,181	2,900
430	432,61	964,4	5,324	2,220	2,937
440	423,43	994,0	5,392	2,258	2,972
450	414,33	1023,9	5,459	2,295	3,006
460	405,32	1054,1	5,526	2,333	3,038
470	396,42	1084,6	5,591	2,370	3,069
480	387,66	1115,5	5,656	2,406	3,099
490	379,05	1146,6	5,721	2,443	3,128
500	370,61	1178,0	5,784	2,479	3,156
520	354,28	1241,7	5,909	2,550	3,208
540	338,77	1306,3	6,031	2,619	3,257
560	324,12	1372,0	6,150	2,686	3,303
580	310,35	1438,4	6,267	2,751	3,346
600	297,48	1505,8	6,381	2,813	3,385
620	285,49	1573,8	6,493	2,872	3,422
640	274,34	1642,6	6,602	2,930	3,457
660	264,00	1712,1	6,709	2,988	3,493
680	254,42	1782,4	6,814	3,048	3,533
700	245,53	1853,5	6,917	3,113	3,578
			$p=38$		
100	731,87	201,9	2,116	—	—
110	722,84	219,2	2,285	—	—
120	713,66	238,2	2,450	—	—
130	704,48	257,9	2,608	—	—
140	695,34	278,0	2,757	1,431	2,006
150	686,28	298,1	2,895	1,415	2,006
160	677,31	318,1	3,025	1,404	2,006
170	668,46	338,2	3,147	1,402	2,010
180	659,69	358,3	3,262	1,406	2,019

Table II.7 (*Continued*)

T	ρ	h	s	c_v	c_p
		$p=38$			
190	651,00	378,6	3,371	1,416	2,031
200	642,36	399,0	3,476	1,429	2,047
210	633,73	419,5	3,576	1,445	2,067
220	625,11	440,3	3,673	1,463	2,091
230	616,45	461,4	3,766	1,483	2,119
240	607,76	482,7	3,857	1,506	2,150
250	599,02	504,4	3,946	1,532	2,185
260	590,23	526,4	4,032	1,561	2,223
270	581,39	548,8	4,117	1,592	2,262
280	572,51	571,7	4,200	1,625	2,304
290	563,60	594,9	4,281	1,661	2,346
300	554,66	618,6	4,361	1,698	2,389
310	545,71	642,7	4,441	1,737	2,432
320	536,74	667,2	4,518	1,777	2,476
330	527,76	692,2	4,595	1,818	2,519
340	518,77	717,6	4,671	1,860	2,562
350	509,78	743,4	4,746	1,901	2,605
360	550,78	769,7	4,820	1,943	2,647
370	491,77	796,4	4,893	1,984	2,689
380	482,76	823,5	4,965	2,025	2,730
390	473,75	851,0	5,037	2,066	2,771
400	464,75	878,9	5,107	2,106	2,810
410	455,76	907,2	5,177	2,145	2,849
420	446,79	935,9	5,246	2,185	2,886
430	437,86	964,9	5,315	2,223	2,922
440	428,99	994,3	5,382	2,261	2,957
450	420,18	1024,1	5,449	2,299	2,991
460	411,47	1054,1	5,515	2,337	3,023
470	402,86	1084,5	5,581	2,374	3,055
480	394,38	1115,2	5,645	2,411	3,085
490	386,03	1146,2	5,709	2,447	3,114
500	377,85	1177,5	5,772	2,483	3,142
520	361,98	1240,9	5,897	2,554	3,196
540	346,86	1305,3	6,018	2,624	3,247
560	332,54	1370,7	6,137	2,691	3,294
580	319,03	1437,1	6,254	2,755	3,338
600	306,34	1504,2	6,367	2,817	3,379
620	294,46	1572,2	6,479	2,876	3,418
640	283,36	1641,0	6,588	2,934	3,455
660	273,03	1710,4	6,695	2,992	3,493
680	263,40	1780,7	6,800	3,052	3,534
700	254,45	1851,8	6,903	3,117	3,580
		$p=40$			
100	732,45	204,3	2,112	—	—
110	723,51	221,6	2,282	—	—
120	714,40	240,5	2,446	—	—
130	705,27	260,3	2,605	—	—
140	696,17	280,3	2,753	1,435	2,005
150	687,15	300,4	2,892	1,418	2,005
160	678,23	320,5	3,021	1,407	2,005
170	669,42	340,5	3,143	1,404	2,009

Table II.7 (*Continued*)

T	ρ	h	s	c_v	c_p

$p=40$

T	ρ	h	s	c_v	c_p
180	660,71	360,7	3,258	1,409	2,017
190	652,09	380,9	3,367	1,418	2,029
200	643,52	401,3	3,472	1,431	2,045
210	634,97	421,8	3,572	1,447	2,064
220	626,43	442,6	3,668	1,465	2,088
230	617,87	463,6	3,762	1,485	2,115
240	609,28	484,9	3,852	1,508	2,146
250	600,64	506,5	3,941	1,534	2,181
260	591,96	528,5	4,027	1,562	2,218
270	583,23	550,9	4.111	1,593	2,257
280	574,48	573,7	4,194	1,627	2,298
290	565,69	596,8	4,276	1,662	2,340
300	556,89	620,5	4,356	1,700	2,383
310	548,08	644,5	4,435	1,739	2,426
320	539,26	669,0	4,512	1,779	2,469
330	530,44	693,9	4,589	1,820	2,512
340	521,62	719,2	4,665	1,861	2,554
350	512,81	745,0	4,739	1,903	2,597
360	504,01	771,2	4,813	1,944	2,638
370	495,21	797,7	4,886	1,986	2,680
380	486,43	824,7	4,958	2,027	2,720
390	477,65	852,1	5,029	2,068	2,760
400	468,89	879,9	5,099	2,108	2,799
410	460,16	908,1	5,169	2,148	2,836
420	451,45	936,7	5,238	2,188	2,873
430	442,80	965,6	5,306	2,227	2,909
440	434,19	994,8	5,373	2,265	2,944
450	425,66	1024,5	5,440	2,303	2 978
460	417,22	1054,4	5,505	2,341	3,010
470	408,87	1084,7	5,570	2,378	3,042
480	400,65	1115,2	5,635	2,415	3,072
490	392,55	1146,1	5,698	2,451	3,102
500	384,60	1177,3	5,761	2,487	3,130
520	369,16	1240,4	5,885	2,559	3,185
540	354,42	1304,6	6,006	2,628	3,237
560	340,40	1369,9	6,125	2,695	3,286
580	327,14	1436,0	6,241	2,759	3,331
600	314,64	1503,1	6,355	2,821	3,374
620	302,89	1571,0	6,466	2,880	3,414
640	291,88	1639,6	6,575	2,938	3,453
660	281,57	1709,1	6,682	2,995	3,492
680	271,94	1779,4	6,787	3,056	3,535
700	262,94	1850,5	6,890	3,120	3,582

$p=45$

T	ρ	h	s	c_v	c_p
100	733,88	210,3	2,104	—	—
110	725,17	227,5	2,273	—	—
120	716,24	246,5	2,438	—	—
130	707,24	266,2	2,596	—	—
140	698,24	286,2	2,744	1,445	2,003
150	689,31	306,3	2,882	1,426	2,003
160	680,49	326,3	3,011	1,413	2,003

Table II.7 (*Continued*)

T	ρ	h	s	c_v	c_p
		$p=45$			
170	671,80	346,3	3,133	1,410	2,007
180	663,22	366,4	3,248	1,415	2,015
190	654,75	386,6	3,357	1,424	2,026
200	646,35	407,0	3,461	1,437	2,040
210	637,99	427,5	3,561	1,453	2,059
220	629,65	448,2	3,658	1,471	2,081
230	621,31	469,1	3,751	1,491	2,108
240	612,95	490,3	3,841	1,514	2,138
250	604,56	511,9	3,929	1,539	2,171
260	596,13	533,8	4,015	1,567	2,208
270	587,68	556,0	4,099	1,597	2,246
280	579,20	578,7	4,181	1,630	2,287
290	570,70	601,8	4,262	1,666	2,328
300	562,20	625,3	4,342	1,703	2,370
310	553,70	649,2	4,420	1,742	2,412
320	545,22	673,5	4,498	1,782	2,455
330	536,76	698,3	4,574	1,823	2,497
340	528,32	723,4	4,649	1,864	2,538
350	519,91	749,0	4,723	1,906	2,579
360	511,53	775,0	4,796	1,948	2,620
370	503,19	801,4	4,869	1,990	2,659
380	494,88	828,2	4,940	2,032	2,699
390	486,61	855,4	5,011	2,073	2,737
400	478,37	882,9	5,080	2,114	2,775
410	470,18	910,9	5,149	2,155	2,811
420	462,04	939,2	5,218	2,195	2,847
430	453,96	967,8	5,285	2,234	2,883
440	445,94	996,8	5,352	2,273	2,917
450	438,00	1026,1	5,418	2,312	2,950
460	430,13	1055,8	5,483	2,350	2,983
470	422,36	1085,8	5,547	2,387	3,015
480	414,69	1116,1	5,611	2,425	3,046
490	407,13	1146,7	5,674	2,461	3,076
500	399,70	1177,6	5,737	2,498	3,105
520	385,23	1240,3	5,860	2,569	3,162
540	371,33	1304,1	5,980	2,639	3,216
560	358,05	1368,9	6,098	2,706	3,267
580	345,39	1434,7	6,213	2,770	3,315
600	333,37	1501,5	6,326	2,831	3,361
620	321,99	1569,1	6,437	2,890	3,404
640	311,23	1637,6	6,546	2,947	3,446
660	301,08	1707,0	6,653	3,005	3,488
680	291,51	1777,2	6,757	3,064	3,533
700	282,50	1848,3	6,861	3,129	3,582
		$p=50$			
100	735,29	216,4	2,096	—	—
110	726,83	233,5	2,265	—	—
120	718,09	252,4	2,429	—	—
130	709,21	272,1	2,587	—	—
140	700,31	292,1	2,735	1,456	2,002
150	691,46	312,1	2,873	1,433	2,002

Table II.7 (*Continued*)

T	ρ	h	s	c_v	$\dot{c}_p$
		$p=50$			
160	682,72	332,1	3,002	1,420	2,002
170	674,13	352,2	3,124	1,416	2,005
180	665,67	372,2	3,238	1,421	2,013
190	657,33	392,4	3,347	1,430	2,023
200	649,08	412,7	3,452	1,443	2,037
210	640,90	433,2	3,551	1,459	2,054
220	632,75	453,8	3,647	1,476	2,075
230	624,61	474,7	3,740	1,496	2,101
240	616,46	495,8	3,830	1,519	2,130
250	608,30	517,3	3,918	1,543	2,162
260	600,10	539,1	4,003	1,571	2,198
270	591,89	561,3	4,087	1,601	2,236
280	583,65	583,8	4,169	1,634	2,276
290	575,42	606,8	4,250	1,669	2,317
300	567,18	630,2	4,329	1,706	2,359
310	558,96	654,0	4,407	1,744	2,401
320	550,77	678,2	4,484	1,784	2,443
330	542,61	702,8	4,560	1,825	2,484
340	534,49	727,9	4,634	1,867	2,525
350	526,41	753,3	4,708	1,909	2,565
360	518,39	779,2	4,781	1,952	2,605
370	510,41	805,4	4,853	1,994	2,644
380	502,49	832,0	4,924	2,036	2,682
390	494,63	859,1	4,994	2,078	2,720
400	486,82	886,4	5,063	2,119	2,756
410	479,08	914,2	5,132	2,160	2,792
420	471,39	942,3	5,199	2,200	2,828
430	463,77	970,7	5,266	2,240	2,862
440	456,23	999,5	5,333	2,280	2,896
450	448,76	1028,7	5,398	2,319	2,929
460	441,37	1058,1	5,463	2,357	2,962
470	434,07	1087,9	5,527	2,396	2,994
480	426,87	1118,0	5,590	2,433	3,025
490	419,77	1148,4	5,653	2,470	3,056
500	412,77	1179,1	5,715	2,507	3,085
520	399,13	1241,4	5,837	2,579	3,143
540	385,98	1304,8	5,957	2,649	3,199
560	373,35	1369,3	6,074	2,716	3,252
580	361,27	1434,9	6,189	2,780	3,302
600	349,73	1501,4	6,302	2,841	3,349
620	338,73	1568,8	6,412	2,900	3,394
640	328,28	1637,1	6,521	2,957	3,438
660	318,34	1706,3	6,627	3,014	3,482
680	308,92	1776,4	6,732	3,073	3,529
700	299,99	1847,5	6,835	3,137	3,580
		$p=55$			
100	736,69	222,4	2,089	—	—
110	728,50	239,5	2,257	—	—
120	719,96	258,3	2,420	—	—
130	711,20	278,0	2,578	—	—
140	702,37	298,0	2,726	1,466	2,000
150	693,59	318,0	2,864	1,441	2,000

Table II.7 (*Continued*)

T	p	h	s	c_v	c_p
			$p=55$		
160	684,92	338,0	2,993	1,426	2,001
170	676,41	358,0	3,114	1,422	2,004
180	668,06	378,1	3,229	1,426	2,011
190	659,84	398,2	3,338	1,436	2,021
200	651,74	418,5	3,442	1,449	2,034
210	643,71	438,9	3,542	1,464	2,050
220	635,74	459,5	3,637	1,482	2,070
230	627,78	480,3	3,730	1,501	2,094
240	619,83	501,4	3,820	1,523	2,123
250	611,87	522,8	3,907	1,548	2,155
260	603,89	544,5	3,992	1,575	2,190
270	595,89	566,6	4,075	1,605	2,227
280	587,88	589,0	4,157	1,637	2,267
290	579,88	611,9	4,237	1,672	2,308
300	571,88	635,2	4,316	1,708	2,349
310	563,90	658,9	4,394	1,747	2,391
320	555,96	683,0	4,471	1,787	2,432
330	548,06	707,6	4,546	1,828	2,474
340	540,21	732,5	4,621	1,869	2,514
350	532,42	757,8	4,694	1,911	2,554
360	524,69	783,6	4,766	1,954	2,593
370	517,03	809,7	4,838	1,997	2,632
380	509,43	836,2	4,909	2,039	2,669
390	501,90	863,1	4,979	2,081	2,706
400	494,45	890,3	5,047	2,123	2,742
410	487,07	917,9	5,116	2,164	2,778
420	479,76	945,9	5,183	2,205	2,813
430	472,52	974,2	5,250	2,246	2,847
440	465,37	1002,8	5,315	2,286	2,880
450	458,29	1031,8	5,381	2,325	2,913
460	451,30	1061,1	5,445	2,364	2,946
470	444,40	1090,7	5,509	2,403	2,977
480	437,59	1120,6	5,572	2,441	3,009
490	430,87	1150,9	5,634	2,478	3,040
500	424,26	1181,4	5,696	2,515	3,070
520	411,33	1243,4	5,817	2,588	3,129
540	398,85	1306,5	5,936	2,658	3,185
560	386,82	1370,8	6,053	2,725	3,239
580	375,26	1436,1	6,168	2,790	3,290
600	364,18	1502,4	6,280	2,851	3,339
620	353,58	1569,6	6,390	2,909	3,385
640	343,44	1637,7	6,499	2,966	3,430
660	333,77	1706,8	6,605	3,023	3,475
680	324,54	1776,8	6,709	3,082	3,524
700	315,75	1847,8	6,812	3,145	3,577
			$p=60$		
100	738,08	228,5	2,081	—	—
110	730,19	245,5	2,249	—	—
120	721,86	264,3	2,412	—	—
130	713,22	283,9	2,569	—	—
140	704,45	303,8	2,717	1,476	1,997

Table II.7 (*Continued*)

T	ρ	h	s	c_v	c_p

$p = 60$

150	695,71	323,8	2,855	1,448	1,999
160	687,10	343,8	2,984	1,431	2,000
170	678,66	363,8	3,105	1,427	2,003
180	670,40	383,9	3,220	1,431	2,010
190	662,29	404,0	3,329	1,441	2,019
200	654,32	424,3	3,433	1,454	2,031
210	646,43	444,6	3,532	1,469	2,046
220	638,62	465,2	3,628	1,487	2,066
230	630,83	486,0	3,720	1,506	2,089
240	623,06	507,0	3,809	1,528	2,116
250	615,29	528,3	3,896	1,552	2,147
260	607,51	550,0	3,981	1,579	2,182
270	599,71	572,0	4,064	1,608	2,219
280	591,91	594,3	4,146	1,640	2,258
290	584,11	617,1	4,226	1,674	2,299
300	576,33	640,3	4,304	1,710	2,340
310	568,57	663,9	4,382	1,748	2,382
320	560,84	688,0	4,458	1,788	2,424
330	553,17	712,4	4,533	1,829	2,465
340	545,56	737,3	4,607	1,871	2,505
350	538,01	762,5	4,681	1,913	2,545
360	530,53	788,2	4,753	1,956	2,584
370	523,13	814,2	4,824	1,998	2,622
380	515,81	840,6	4,895	2,041	2,659
390	508,57	867,4	4,964	2,084	2,696
400	501,40	894,5	5,033	2,126	2,731
410	494,32	922,0	5,101	2,167	2,767
420	487,33	949,8	5,168	2,209	2,801
430	480,41	978,0	5,234	2,250	2,835
440	473,59	1006,5	5,300	2,290	2,868
450	466,84	1035,4	5,364	2,330	2,901
460	460,19	1064,5	5,429	2,369	2,933
470	453,62	1094,0	5,492	2,408	2,965
480	447,14	1123,8	5,555	2,447	2,996
490	440,75	1153,9	5,617	2,485	3,027
500	434,46	1184,4	5,678	2,522	3,057
520	422,17	1246,1	5,799	2,595	3,117
540	410,27	1309,0	5,918	2,666	3,174
560	398,79	1373,0	6,035	2,734	3,228
580	387,73	1438,1	6,149	2,798	3,280
600	377,08	1504,2	6,261	2,859	3,330
620	366,86	1571,3	6,371	2,918	3,377
640	357,05	1639,3	6,479	2,974	3,423
660	347,65	1708,2	6,585	3,031	3,469
680	338,64	1778,1	6,689	3,090	3,518
700	330,03	1849,0	6,792	3,153	3,572

$p = 65$

100	739,48	234,5	2,074	—	—
110	731,92	251,5	2,241	—	—
120	723,81	270,2	2,404	—	—
130	715,28	289,8	2,561	—	—

Table II.7 (*Continued*)

T	ρ	h	s	c_v	c_p
		$p = 65$			
140	706,55	309,7	2,708	1,486	1,995
150	697,83	329,6	2,846	1,454	1,998
160	689,26	349,6	2,975	1,436	1,999
170	680,88	369,6	3,096	1,431	2,003
180	672,69	389,7	3,211	1,435	2,009
190	664,68	409,8	3,320	1,445	2,018
200	656,82	430,1	3,423	1,458	2,029
210	649,07	450,4	3,523	1,474	2,044
220	641,40	470,9	3,618	1,491	2,062
230	633,78	491,7	3,710	1,510	2,084
240	626,17	512,6	3,800	1,532	2,110
250	618,58	533,9	3,886	1,556	2,141
260	610,98	555,5	3,971	1,582	2,175
270	603,37	577,4	4,054	1,611	2,212
280	595,75	599,7	4,135	1,642	2,251
290	588,14	622,4	4,215	1,676	2,291
300	580,55	645,5	4,293	1,712	2,333
310	572,98	669,1	4,370	1,750	2,374
320	565,46	693,0	4,446	1,789	2,416
330	557,99	717,4	4,521	1,830	2,457
340	550,58	742,2	4,595	1,872	2,497
350	543,24	767,3	4,668	1,914	2,537
360	535,98	792,9	4,740	1,957	2,576
370	528,81	818,8	4,811	2,000	2,614
380	521,72	845,2	4,881	2,043	2,651
390	514,71	871,9	4,951	2,085	2,687
400	507,80	898,9	5,019	2,128	2,723
410	500,97	926,3	5,087	2,170	2,758
420	494,24	954,1	5,154	2,211	2,792
430	487,60	982,2	5,220	2,253	2,826
440	481,04	1010,6	5,285	2,293	2,859
450	474,58	1039,3	5,350	2,334	2,891
460	468,21	1068,4	5,414	2,373	2,923
470	461,92	1097,8	5,477	2,413	2,955
480	455,73	1127,5	5,539	2,452	2,986
490	449,63	1157,5	5,601	2,490	3,017
500	443,62	1187,8	5,662	2,528	3,048
520	431,88	1249,4	5,783	2,602	3,107
540	420,51	1312,1	5,902	2,673	3,165
560	409,52	1376,0	6,018	2,741	3,220
580	398,90	1440,9	6,132	2,805	3,272
600	388,67	1506,8	6,243	2,867	3,322
620	378,82	1573,7	6,353	2,925	3,370
640	369,33	1641,6	6,461	2,982	3,416
660	360,21	1710,4	6,567	3,039	3,463
680	351,45	1780,1	6,671	3,097	3,513
700	343,03	1850,9	6,773	3,160	3,568
		$p = 70$			
100	740,88	240,6	2,068	—	—
110	733,69	257,6	2,234	—	—
120	725,84	276,2	2,396	—	—

Table II.7 (*Continued*)

T	ρ	h	s	c_v	c_p

$p=70$

130	717,40	295,7	2,552	—	—
140	708,68	315,5	2,699	1,495	1,992
150	699,96	335,5	2,837	1,459	1,997
160	691,40	355,4	2,966	1,440	1,999
170	683,06	375,5	3,087	1,435	2,003
180	674,94	395,5	3,202	1,439	2,009
190	667,01	415,6	3,311	1,449	2,017
200	659,26	435,9	3,414	1,462	2,028
210	651,63	456,2	3,514	1,478	2,041
220	644,09	476,7	3,609	1,495	2,058
230	636,62	497,4	3,701	1,514	2,080
240	629,17	518,3	3,790	1,536	2,105
250	621,74	539,5	3,877	1,559	2,135
260	614,31	561,0	3,961	1,585	2,168
270	606,87	582,9	4,043	1,613	2,205
280	599,43	605,1	4,124	1,644	2,243
290	592,00	627,8	4,204	1,678	2,284
300	584,58	650,8	4,282	1,713	2,325
310	577,19	674,3	4,359	1,751	2,367
320	569,84	698,1	4,435	1,790	2,409
330	562,55	722,4	4,509	1,831	2,450
340	555,32	747,1	4,583	1,872	2,491
350	548,17	772,3	4,656	1,915	2,531
360	541,09	797,8	4,728	1,957	2,570
370	534,11	823,6	4,799	2,000	2,608
380	527,22	849,9	4,869	2,043	2,645
390	520,42	876,5	4,938	2,086	2,681
400	513,72	903,5	5,006	2,129	2,716
410	507,11	930,9	5,074	2,171	2,751
420	500,60	958,5	5,140	2,213	2,758
430	494,19	986,6	5,206	2,255	2,819
440	487,87	1014,9	5,271	2,296	2,852
450	481,64	1043,6	5,336	2,336	2,884
460	475,51	1072,6	5,400	2,376	2,916
470	469,47	1101,9	5,463	2,416	2,947
480	463,52	1131,5	5,525	2,455	2,979
490	457,66	1161,5	5,587	2,494	3,009
500	451,90	1191,7	5,648	2,532	3,040
520	440,64	1253,1	5,768	2,606	3,099
540	429,73	1315,7	5,886	2,678	3,157
560	419,19	1379,4	6,002	2,747	3,212
580	408,99	1444,2	6,116	2,812	3,265
600	399,14	1510,0	6,227	2,873	3,315
620	389,64	1576,8	6,337	2,932	3,363
640	380,47	1644,5	6,444	2,989	3,410
660	371,64	1713,2	6,550	3,045	3,458
680	363,12	1782,9	6,654	3,104	3,508
700	354,91	1853,6	6,757	3,167	3,563

$p=75$

110	735,54	263,6	2,228	—	—
120	727,96	282,1	2,389	—	—
130	719,60	301,5	2,544	—	—

Table II.7 (*Continued*)

T	ρ	h	s	c_v	c_p
		$p=75$			
140	710,85	321,3	2,691	1,502	1,989
150	702,10	341,3	2,828	1,463	1,996
160	693,54	361,3	2,957	1,443	2,000
170	685,22	381,3	3,078	1,437	2,004
180	677,14	401,3	3,193	1,442	2,010
190	669,29	421,5	3,302	1,452	2,017
200	661,63	441,7	3,406	1,466	2,027
210	654,11	462,0	3,505	1,482	2,039
220	646,70	482,5	3,600	1,499	2,056
230	639,36	503,1	3,692	1,518	2,076
240	632,06	524,0	3,781	1,539	2,100
250	624,78	545,2	3,867	1,562	2,129
260	617,51	566,6	3,951	1,587	2,162
270	610,23	588,4	4,033	1,615	2,198
280	602,95	610,6	4,114	1,646	2,237
290	595,68	633,2	4,193	1,679	2,277
300	588,42	656,1	4,271	1,714	2,319
310	581,19	679,5	4,348	1,751	2,361
320	574,01	703,3	4,424	1,790	2,403
330	566,87	727,6	4,498	1,831	2,444
340	559,81	752,2	4,572	1,872	2,485
350	552,82	777,3	4,644	1,914	2,525
360	545,91	802,7	4,716	1,957	2,564
370	539,10	828,6	4,787	2,000	2,602
380	532,38	854,8	4,857	2,043	2,640
390	525,76	881,4	4,926	2,086	2,676
400	519,24	908,3	4,994	2,129	2,711
410	512,82	935,6	5,061	2,172	2,746
420	506,50	963,2	5,128	2,214	2,780
430	500,28	991,2	5,194	2,256	2,813
440	494,16	1019,5	5,259	2,297	2,846
450	488,14	1048,1	5,323	2,338	2,878
460	482,21	1077,0	5,387	2,378	2,910
470	476,38	1106,3	5,450	2,418	2,942
480	470,64	1135,9	5,512	2,458	2,973
490	464,99	1165,8	5,573	2,497	3,004
500	459,44	1195,9	5,634	2,535	3,034
520	448,60	1257,2	5,755	2,610	3,094
540	438,10	1319,7	5,872	2,682	3,151
560	427,95	1383,3	5,988	2,751	3,207
580	418,14	1447,9	6,102	2,817	3,260
600	408,65	1513,6	6,213	2,879	3,310
620	399,48	1580,3	6,322	2,938	3,358
640	390,61	1648,0	6,430	2,994	3,405
660	382,05	1716,5	6,535	3,051	3,453
680	373,78	1786,1	6,639	3,110	3,504
700	365,80	1856,7	6,741	3,173	3,559

Table II.7 (*Continued*)

T	ρ	h	s	c_v	c_p
		$p=80$			
110	737,48	269,7	2,221	—	—
120	730,24	288,1	2,381	—	—
130	721,91	307,4	2,536	—	—
140	713,09	327,1	2,682	—	—
150	704,26	347,1	2,819	1,466	1,996
160	695,66	367,1	2,948	1,444	2,001
170	687,35	387,1	3,070	1,439	2,005
180	679,31	407,2	3,185	1,444	2,011
190	671,52	427,3	3,293	1,455	2,017
200	663,94	447,5	3,397	1,469	2,026
210	656,52	467,8	3,496	1,485	2,038
220	649,23	488,3	3,591	1,502	2,053
230	642,01	508,9	3,683	1,521	2,073
240	634,85	529,8	3,772	1,542	2,096
250	627,72	550,9	3,858	1,565	2,124
260	620,59	572,3	3,942	1,590	2,156
270	613,47	594,0	4,024	1,617	2,192
280	606,34	616,1	4,104	1,647	2,230
290	599,22	638,6	4,183	1,680	2,271
300	592,11	661,5	4,261	1,715	2,312
310	585,03	684,8	4,337	1,752	2,355
320	577,98	708,6	4,413	1,790	2,397
330	571,00	732,8	4,487	1,830	2,439
340	564,07	757,4	4,561	1,872	2,480
350	557,23	782,4	4,633	1,914	2,520
360	550,47	807,8	4,705	1,956	2,560
370	543,80	833,6	4,775	1,999	2,598
380	537,24	859,7	4,845	2,043	2,635
390	530,77	886,3	4,914	2,086	2,672
400	524,41	913,2	4,982	2,129	2,707
410	518,15	940,4	5,049	2,172	2,742
420	511,99	968,0	5,116	2,214	2,776
430	505,94	995,9	5,182	2,256	2,809
440	499,99	1024,2	5,247	2,298	2,842
450	494,14	1052,8	5,311	2,339	2,874
460	488,39	1081,7	5,374	2,380	2,906
470	482,74	1110,9	5,437	2,420	2,938
480	477,18	1140,4	5,499	2,460	2,969
490	471,72	1170,3	5,561	2,499	2,999
500	466,35	1200,4	5,622	2,537	3,030
520	455,87	1261,6	5,742	2,613	3,089
540	445,74	1324,0	5,859	2,686	3,147
560	435,95	1387,5	5,975	2,755	3,202
580	426,47	1452,1	6,088	2,821	3,255
600	417,31	1517,7	6,199	2,883	3,306
620	408,45	1584,3	6,309	2,942	3,354
640	399,87	1651,8	6,416	2,999	3,401
660	391,58	1720,3	6,521	3,056	3,449
680	383,55	1789,8	6,625	3,115	3,500
700	375,79	1860,3	6,727	3,178	3,555

Table II.7 (*Continued*)

T	ρ	h	s	c_v	c_p
			$p=85$		
130	724,39	313,2	2,527	—	—
140	715,40	332,9	2,673	—	—
150	706,45	352,8	2,811	1,467	1,997
160	697,79	372,8	2,940	1,444	2,003
170	689,46	392,9	3,061	1,439	2,007
180	681,44	413,0	3,176	1,445	2,012
190	673,71	433,1	3,285	1,456	2,018
200	666,20	453,4	3,389	1,471	2,026
210	658,87	473,7	3,488	1,488	2,037
220	651,68	494,1	3,583	1,505	2,051
230	644,58	514,7	3,674	1,524	2,070
240	637,55	535,5	3,763	1,545	2,092
250	630,55	556,6	3,849	1,567	2,120
260	623,56	577,9	3,933	1,592	2,151
270	616,58	599,6	4,015	1,619	2,186
280	609,59	621,7	4,095	1,648	2,224
290	602,61	644,1	4,173	1,680	2,264
300	595,64	666,9	4,251	1,715	2,306
310	588,70	690,2	4,327	1,751	2,349
320	581,79	713,9	4,402	1,790	2,391
330	574,93	738,0	4,477	1,830	2,434
340	568,14	762,6	4,550	1,871	2,475
350	561,42	787,6	4,622	1,913	2,516
360	554,80	812,9	4,694	1,955	2,556
370	548,26	838,7	4,764	1,998	2,594
380	541,82	864,8	4,834	2,041	2,632
390	535,49	891,3	4,903	2,085	2,669
400	529,27	918,2	4,971	2,128	2,704
410	523,15	945,4	5,038	2,171	2,739
420	517,13	972,9	5,105	2,213	2,773
430	511,23	1000,8	5,170	2,256	2,806
440	505,42	1029,1	5,235	2,297	2,839
450	499,73	1057,6	5,299	2,339	2,872
460	494,13	1086,5	5,363	2,380	2,903
470	488,63	1115,7	5,425	2,420	2,935
480	483,23	1145,2	5,488	2,460	2,966
490	477,93	1175,0	5,549	2,500	2,996
500	472,71	1205,1	5,610	2,539	3,027
520	462,56	1266,3	5,730	2,615	3,086
540	452,75	1328,6	5,847	2,688	3,144
560	443,27	1392,0	5,963	2,758	3,199
580	434,10	1456,5	6,076	2,824	3,252
600	425,24	1522,1	6,187	2,887	3,302
620	416,66	1588,6	6,296	2,946	3,351
640	408,35	1656,1	6,403	3,003	3,398
660	400,31	1724,5	6,508	3,060	3,446
680	392,53	1793,9	6,612	3,120	3,496
700	384,98	1864,4	6,714	3,183	3,552

Table II.7 (*Continued*)

T	ρ	h	s	c_v	c_p
		$p=90$			
130	727,11	319,0	2,519	—	—
140	717,82	338,6	2,664	—	—
150	708,68	358,6	2,802	1,465	1,999
160	699,92	378,6	2,931	1,443	2,005
170	691,55	398,7	3,053	1,439	2,010
180	683,54	418,8	3,168	1,445	2,014
190	675,84	439,0	3,277	1,457	2,020
200	668,40	459,2	3,380	1,473	2,027
210	661,15	479,5	3,480	1,490	2,037
220	654,06	499,9	3,575	1,508	2,050
230	647,07	520,5	3,666	1,527	2,067
240	640,16	541,3	3,755	1,547	2,089
250	633,28	562,3	3,840	1,569	2,115
260	626,43	583,6	3,924	1,594	2,146
270	619,58	605,3	4,005	1,620	2,180
280	612,73	627,2	4,085	1,649	2,218
290	605,88	649,6	4,164	1,681	2,259
300	599,04	672,4	4,241	1,715	2,300
310	592,22	695,6	4,317	1,751	2,343
320	585,44	719,3	4,392	1,789	2,386
330	578,70	743,4	4,467	1,828	2,429
340	572,03	767,9	4,450	1,869	2,471
350	565,43	792,8	4,612	1,911	2,512
360	558,92	818,1	4,683	1,954	2,552
370	552,50	843,8	4,754	1,997	2,591
380	546,18	869,9	4,823	2,040	2,629
390	539,96	896,4	4,892	2,083	2,666
400	533,86	923,2	4,960	2,126	2,702
410	527,86	950,4	5,027	2,169	2,737
420	521,97	978,0	5,094	2,212	2,771
430	516,19	1005,9	5,159	2,255	2,805
440	510,51	1034,1	5,224	2,297	2,837
450	504,95	1062,6	5,288	2,338	2,870
460	499,48	1091,5	5,352	2,379	2,901
470	494,12	1120,6	5,414	2,420	2,933
480	488,85	1150,1	5,476	2,461	2,964
490	483,69	1179,9	5,538	2,500	2,994
500	478,61	1210,0	5,599	2,539	3,025
520	468,74	1271,1	5,718	2,616	3,084
540	459,21	1333,4	5,836	2,689	3,141
560	450,00	1396,7	5,951	2,760	3,197
580	441,11	1461,2	6,064	2,826	3,250
600	432,51	1526,7	6,175	2,889	3,300
620	424,20	1593,2	6,284	2,949	3,348
640	416,14	1660,6	6,391	3,007	3,395
660	408,34	1729,0	6,497	3,064	3,443
680	400,78	1798,4	6,600	3,123	3,494
700	393,45	1868,8	6,702	3,187	3,549

Table II.7 (*Continued*)

T	ρ	h	s	c_v	c_p
			p=95		
140	720,39	344,3	2,655	—	—
150	710,95	364,3	2,793	1,461	2,002
160	702,05	384,3	2,922	1,440	2,009
170	693,62	404,4	3,044	1,437	2,013
180	685,61	424,6	3,159	1,444	2,017
190	677,94	444,8	3,268	1,458	2,022
200	670,55	465,0	3,372	1,474	2,028
210	663,37	485,4	3,471	1,491	2,037
220	656,37	505,8	3,566	1,510	2,049
230	649,49	526,4	3,658	1,529	2,065
240	642,68	547,1	3,746	1,549	2,086
250	635,93	568,1	3,832	1,571	2,111
260	629,19	589,3	3,915	1,595	2,141
270	622,47	610,9	3,997	1,621	2,175
280	615,75	632,9	4,076	1,650	2,213
290	609,02	655,2	4,155	1,681	2,253
300	602,31	677,9	4,232	1,715	2,295
310	595,61	701,1	4,308	1,750	2,338
320	588,94	724,7	4,383	1,788	2,381
330	582,32	748,7	4,457	1,827	2,425
340	575,75	773,2	4,530	1,868	2,467
350	569,26	798,1	4,602	1,909	2,509
360	562,85	823,3	4,673	1,952	2,549
370	556,54	849,0	4,743	1,995	2,589
380	550,32	875,1	4,813	2,038	2,627
390	544,21	901,6	4,882	2,081	2,664
400	538,21	928,4	4,950	2,124	2,700
410	532,32	955,6	5,017	2,168	2,735
420	526,53	983,1	5,083	2,210	2,770
430	520,86	1011,0	5,149	2,253	2,803
440	515,30	1039,2	5,213	2,295	2,836
450	509,85	1067,7	5,278	2,337	2,868
460	504,50	1096,5	5,341	2,379	2,900
470	499,25	1125,7	5,404	2,419	2,932
480	494,10	1155,2	5,466	2,460	2,963
490	489,06	1185,0	5,527	2,500	2,993
500	484,10	1215,0	5,588	2,539	3,024
520	474,47	1276,1	5,708	2,616	3,083
540	465,19	1338,3	5,825	2,690	3,140
560	456,23	1401,7	5,940	2,761	3,195
580	447,58	1466,1	6,053	2,828	3,248
600	439,23	1531,6	6,164	2,891	3,298
620	431,14	1598,0	6,273	2,951	3,346
640	423,32	1665,4	6,380	3,009	3,393
660	415,74	1733,8	6,485	3,067	3,441
680	408,39	1803,1	6,589	3,126	3,491
700	401,26	1873,5	6,691	3,190	3,547

Table II.7 (*Continued*)

T	ρ	h	s	c_v	c_p
			$p=100$		
140	723,17	349,9	2,6₊5	—	—
150	713,29	369,9	2,783	1,453	2,008
160	704,19	390,0	2,913	1,435	2,014
170	695,68	410,2	3,036	1,433	2,018
180	687,64	430,4	3,151	1,443	2,021
190	679,99	450,6	3,260	1,457	2,025
200	672,64	470,9	3,364	1,475	2,030
210	665,54	491,2	3,463	1,493	2,038
220	658,62	511,6	3,558	1,511	2,049
230	651,83	532,2	3,650	1,531	2,064
240	645,13	552,9	3,738	1,551	2,083
250	638,48	573,9	3,824	1,573	2,108
260	631,87	595,1	3,907	1,596	2,137
270	625,26	616,6	3,988	1,622	2,170
280	618,66	638,5	4,068	1,650	2,207
290	612,06	660,8	4,146	1,681	2,247
300	605,46	683,5	4,223	1,714	2,289
310	598,87	706,6	4,299	1,749	2,333
320	592,31	730,1	4,373	1,786	2,377
330	585,79	754,1	4,447	1,825	2,420
340	579,33	778,5	4,520	1,866	2,463
350	572,93	803,4	4,592	1,907	2,506
360	566,62	828,6	4,663	1,949	2,547
370	560,40	854,3	4,733	1,992	2,586
380	554,27	880,4	4,803	2,035	2,625
390	548,26	906,8	4,872	2,079	2,663
400	542,35	933,6	4,940	2,122	2,699
410	536,55	960,8	5,007	2,165	2,734
420	530,86	988,3	5,073	2,208	2,769
430	525,29	1016,2	5,138	2,251	2,803
440	519,82	1044,3	5,203	2,293	2,836
450	514,46	1072,9	5,267	2,335	2,868
460	509,21	1101,7	5,331	2,377	2,900
470	504,07	1130,9	5,393	2,418	2,931
480	499,03	1160,3	5,456	2,459	2,962
490	494,08	1190,1	5,517	2,499	2,993
500	489,23	1220,2	5,578	2,539	3,023
520	479,82	1281,2	5,697	2,616	3,082
540	470,75	1343,5	5,815	2,690	3,140
560	462,01	1406,8	5,930	2,761	3,195
580	453,58	1471,2	6,043	2,829	3,247
600	445,44	1536,7	6,154	2,892	3,297
620	437,57	1603,1	6,263	2,953	3,345
640	429,95	1670,5	6,370	3,011	3,392
660	422,57	1738,8	6,475	3,069	3,439
680	415,42	1808,1	6,578	3,128	3,490
700	408,48	1878,4	6,680	3,192	3,545

Table II.8 Thermodynamic Properties of Propane in the Single-Phase Region

T	w	μ	k	α/α_0	ν/ν_0
			$p=0{,}01$		
110	1792	—0,64	227701	0,159	25874
120	1771	—0,60	216867	0,176	27012
130	1733	—0,58	205896	0,192	27826
140	1688	—0,57	192188	0,210	27989
150	1638	—0,57	179246	0,229	28090
160	1582	—0,56	164610	0,249	27795
170	1520	—0,55	149700	0,272	27298
180	1456	—0,53	134981	0,296	26648
190	204,7	38,21	1,18	1,013	1,01
210	209,6	43,52	1,17	1,015	1,01
200	214,4	43,57	1,17	1,014	1,01
220	219,0	41,11	1,16	1,013	1,01
230	223,5	37,59	1,16	1,012	1,01
240	227,9	33,79	1,15	1,011	1,01
250	232,2	30,08	1,15	1,009	1,01
260	236,3	26,67	1,14	1,008	1,01
270	240,4	23,61	1,14	1,007	1,01
280	244,4	20,92	1,13	1,006	1,00
290	248,3	18,58	1,13	1,006	1,00
300	252,1	16,56	1,13	1,005	1,00
310	255,9	14,82	1,12	1,004	1,00
320	259,6	13,32	1,12	1,004	1,00
330	263,3	12,03	1,12	1,004	1,00
340	266,9	10,92	1,11	1,003	1,00
350	270,4	9,96	1,11	1,003	1,00
360	273,9	9,13	1,11	1,003	1,00
370	277,4	8,41	1,10	1,002	1,00
380	280,8	7,78	1,10	1,002	1,00
390	284,2	7,22	1,10	1,002	1,00
400	287,5	6,73	1,10	1,002	1,00
410	290,8	6,30	1,09	1,002	1,00
420	294,1	5,91	1,09	1,002	1,00
430	297,3	5,57	1,09	1,002	1,00
440	300,5	5,26	1,09	1,001	1,00
450	303,6	4,98	1,09	1,001	1,00
460	306,8	4,72	1,09	1,001	1,00
470	309,9	4,49	1,08	1,001	1,00
480	312,9	4,28	1,08	1,001	1,00
490	316,0	4,08	1,08	1,001	1,00
500	319,0	3,90	1,08	1,001	1,00
520	324,9	3,58	1,08	1,001	1,00
540	330,7	3,30	1,07	1,001	1,00
560	336,5	3,05	1,07	1,001	1,00
580	342,2	2,84	1,07	1,001	1,00
600	347,7	2,65	1,07	1,001	1,00
620	353,2	2,47	1,07	1,001	1,00
640	358,6	2,32	1,07	1,001	1,00
660	364,0	2,17	1,06	1,001	1,00
680	369,2	2,04	1,06	1,001	1,00
700	374,4	1,91	1,06	1,001	1,00

Table II.8 (*Continued*)

T	w	μ	k	α/α_0	γ/γ_0
		$p=0,05$			
110	1793	—0,64	45565	0,159	5176
120	1772	—0,60	43856	0,176	5461
130	1733	—0,58	41352	0,192	5587
140	1689	—0,57	38636	0,210	5626
150	1638	—0,57	35866	0,229	5619
160	1582	—0,56	32933	0,249	5560
170	1521	—0,55	29955	0,271	5461
180	1456	—0,53	27025	0,296	5334
190	1390	—0,52	24205	0,323	5183
200	1324	—0,50	21594	0,352	5026
210	1258	—0,47	19153	0,385	4854
220	216,4	42,49	1,15	1,071	1,05
230	221,2	38,50	1,15	1,063	1,04
240	225,8	34,40	1,15	1,056	1,04
250	230,3	30,52	1,14	1,049	1,03
260	234,6	26,99	1,14	1,042	1,03
270	238,8	23,86	1,13	1,037	1,03
280	243,0	21,12	1,13	1,032	1,02
290	247,0	18,74	1,13	1,028	1,02
300	250,9	16,69	1,12	1,025	1,02
310	254,8	14,93	1,12	1,022	1,02
320	258,6	13,41	1,12	1,020	1,01
330	262,3	12,11	1,11	1,018	1,01
340	266,0	10,99	1,11	1,016	1,01
350	269,6	10,02	1,11	1,015	1,01
360	273,1	9,18	1,10	1,013	1,01
370	276,7	8,45	1,10	1,012	1,01
380	280,1	7,81	1,10	1,011	1,01
390	283,5	7,25	1,10	1,010	1,01
400	286,9	6,76	1,09	1,010	1,01
410	290,2	6,32	1,09	1,009	1,01
420	293,5	5,93	1,09	1,008	1,01
430	296,8	5,59	1,09	1,008	1,01
440	300,0	5,27	1,09	1,007	1,01
450	303,2	4,99	1,09	1,007	1,00
460	306,4	4,73	1,08	1,007	1,00
470	309,5	4,50	1,08	1,006	1,00
480	312,6	4,28	1,08	1.006	1,00
490	315,6	4,09	1,08	1,006	1,00
500	318,7	3,90	1,08	1,005	1,00
520	324,6	3,58	1,08	1,005	1,00
540	330,5	3,30	1,07	1,004	1,00
560	336,3	3,05	1,07	1,004	1,00
580	342,0	2,84	1,07	1,004	1,00
600	347,6	2,64	1,07	1,003	1,00
620	353,1	2,47	1,07	1,003	1,00
640	358,5	2,31	1,07	1,003	1,00
660	363,9	2,17	1,06	1,003	1,00
680	369,1	2,03	1,06	1,003	1,00
700	374,3	1,90	1,06	1,002	1,00

Table II.8 (*Continued*)

T	w	μ	k	α/α_0	γ/γ_0
			$p=0,1$		
110	1794	—0,64	22810	0,158	2590
120	1772	—0,60	22133	0,176	2756
130	1733	—0,58	20689	0,192	2795
140	1689	—0,57	19340	0,210	2815
150	1639	—0,57	17942	0,229	2810
160	1582	—0,56	16475	0,249	2780
170	1521	—0,55	14987	0,271	2731
180	1457	—0,53	13523	0,296	2668
190	1391	—0,52	12122	0,322	2594
200	1325	—0,50	10789	0,352	2510
210	1259	—0,47	9586	0,385	2428
220	1194	—0,44	8462	0,421	2338
230	1130	—0,41	7431	0,462	2244
240	223,0	35,23	1,14	1,118	1,08
250	227,8	31,10	1,13	1,102	1,07
260	232,4	27,42	1,13	1,088	1,06
270	236,8	24,19	1,13	1,077	1,05
280	241,1	21,38	1,12	1,067	1,05
290	245,3	18,95	1,12	1,059	1,04
300	249,4	16,86	1,12	1,052	1,04
310	253,4	15,07	1,11	1,046	1,03
320	257,3	13,53	1,11	1,041	1,03
330	261,1	12,21	1,11	1,036	1,02
340	264,9	11,07	1,11	1,033	1,02
350	268,6	10,09	1,10	1,030	1,02
360	272,2	9,24	1,10	1,027	1,02
370	275,7	8,50	1,10	1,025	1,02
380	279,3	7,86	1,10	1,023	1,02
390	282,7	7,29	1,09	1,021	1,01
400	286,2	6,79	1,09	1,019	1,01
410	289,5	6,35	1,09	1,018	1,01
420	292,9	5,96	1,09	1,017	1,01
430	296,2	5,61	1,09	1,016	1,01
440	299,4	5,29	1,09	1,015	1,01
450	302,7	5,00	1,08	1,014	1,01
460	305,8	4,74	1,08	1,013	1,01
470	309,0	4,51	1,08	1,012	1,01
480	312,1	4,29	1,08	1,012	1,01
490	315,2	4,09	1,08	1,011	1,01
500	318,3	3,91	1,08	1,011	1,01
520	324,3	3,58	1,08	1,010	1,01
540	330,2	3,30	1,07	1,009	1,01
560	336,0	3,05	1,07	1,008	1,01
580	341,7	2,84	1,07	1,007	1,01
600	347,4	2,64	1,07	1,007	1,01
620	352,9	2,47	1,07	1,006	1,01
640	358,4	2,31	1,07	1,006	1,00
660	363,8	2,16	1,06	1,005	1,00
680	369,1	2,03	1,06	1,005	1,00
700	374,2	1,90	1,06	1,005	1,00

Table II.8 (*Continued*)

T	w	μ	k	α/α_0	γ/γ_0
		$p = 0,101325$			
110	1794	—0,64	22513	0,158	2557
120	1772	—0,60	21841	0,176	2719
130	1734	—0,58	20419	0,192	2758
140	1689	—0,57	19088	0,210	2779
150	1639	—0,57	17709	0,229	2774
160	1582	—0,56	16260	0,249	2744
170	1521	—0,55	14791	0,271	2696
180	1457	—0,53	13347	0,296	2633
190	1391	—0,52	11964	0,322	2561
200	1325	—0,50	10648	0,352	2477
210	1259	—0,47	9460	0,385	2396
220	1194	—0,44	8359	0,421	2307
230	1130	—0,41	7334	0,462	2215
240	222,9	35,25	1,14	1,120	1,08
250	227,7	31,12	1,13	1,104	1,07
260	252,3	27,43	1,13	1,090	1,06
270	236,8	24,19	1,13	1,078	1,05
280	241,1	21,38	1,12	1,068	1,05
290	245,3	18,95	1,12	1,059	1,04
300	249,4	16,87	1,12	1,052	1,04
310	253,4	15,07	1,11	1,046	1,03
320	257,3	13,54	1,11	1,041	1,03
330	261,1	12,21	1,11	1,037	1,02
340	264,8	11,08	1,11	1,033	1,02
350	268,5	10,09	1,10	1,030	1,02
360	272,2	9,24	1,10	1,027	1,02
370	275,7	8,50	1,10	1,025	1,02
380	279,2	7,86	1,10	1,023	1,02
390	282,7	7,29	1,09	1,021	1,01
400	286,1	6,79	1,09	1,020	1,01
410	289,5	6,35	1,09	1,018	1,01
420	292,9	5,96	1,09	1,017	1,01
430	296,2	5,61	1,09	1,016	1,01
440	299,4	5,29	1,09	1,015	1,01
450	302,6	5,00	1,08	1,014	1,01
460	305,8	4,74	1,08	1,013	1,01
470	309,0	4,51	1,08	1,013	1,01
480	312,1	4,29	1,08	1,012	1,01
490	315,2	4,09	1,08	1,011	1,01
500	318,3	3,91	1,08	1,011	1,01
520	324,3	3,58	1,08	1,010	1,01
540	330,2	3,30	1,07	1,009	1,01
560	336,0	3,05	1,07	1,008	1,01
580	341,7	2,84	1,07	1,008	1,01
600	374,4	2,64	1,07	1,007	1,01
620	352,9	2,47	1,07	1,006	1,01
640	358,4	2,31	1,07	1,006	1,00
660	363,8	2,16	1,06	1,005	1,00
680	369,1	2,03	1,06	1,005	1,00
700	374,2	1,90	1,06	1,005	1,00
		$p = 0,15$			
110	1795	—0,64	15224	0,158	1728
120	1773	—0,60	14713	0,176	1831
130	1734	—0,58	13801	0,192	1864

Table II.8 (*Continued*)

T	w	μ	k	α/α_0	γ/γ_0
		$p=0{,}15$			
140	1690	—0,57	12994	0,210	1891
150	1639	—0,57	11969	0,229	1874
160	1583	—0,56	10990	0,249	1854
170	1522	—0,55	9997	0,271	1822
180	1457	—0,53	9021	0,296	1779
190	1391	—0,52	8087	0,322	1730
200	1325	—0,50	7202	0,352	1675
210	1259	—0,47	6396	0,384	1619
220	1194	—0,45	5647	0,421	1559
230	1130	—0,41	4960	0,462	1497
240	1067	—0,38	4329	0,508	1432
250	225,2	31,73	1,12	1,161	1,11
260	230,1	27,87	1,12	1,139	1,09
270	234,7	24,53	1,12	1,120	1,08
280	239,3	21,64	1,12	1,104	1,07
290	243,6	19,16	1,11	1,090	1,06
300	247,8	17,04	1,11	1,079	1,05
310	252,0	15,22	1,11	1,070	1,05
320	256,0	13,65	1,11	1,062	1,04
330	259,9	12,31	1,10	1,056	1,04
340	263,7	11,16	1,10	1,050	1,03
350	267,5	10,17	1,10	1,045	1,03
360	271,2	9,30	1,10	1,041	1,03
370	274,8	8,56	1,10	1,037	1,02
380	278,4	7,90	1,09	1,034	1,02
390	281,9	7,33	1,09	1,032	1,02
400	285,4	6,83	1,09	1,029	1,02
410	288,8	6,38	1,09	1,027	1,02
420	292,2	5,98	1,09	1,025	1,02
430	295,6	5,62	1,09	1,024	1,02
440	298,9	5,31	1,08	1,022	1,02
450	302,1	5,02	1,08	1,021	1,01
460	305,3	4,76	1,08	1,020	1,01
470	308,5	4,52	1,08	1,019	1,01
480	311,7	4,30	1,08	1,018	1,01
490	314,8	4,10	1,08	1,017	1,01
500	317,9	3,91	1,08	1,018	1,01
520	323,9	3,58	1,07	1,015	1,01
540	329,9	3,30	1,07	1,013	1,01
560	335,8	3,05	1,07	1,012	1,01
580	341,5	2,83	1,07	1,011	1,01
600	347,2	2,64	1,07	1,010	1,01
620	352,8	2,46	1,07	1,009	1,01
640	358,3	2,31	1,07	1,009	1,01
660	363,7	2,16	1,06	1,008	1,01
680	369,0	2,03	1,06	1,008	1,01
700	374,2	1,90	1,06	1,007	1,01
		$p=0{,}2$			
110	1796	—0,64	11430	0,158	1297
120	1774	—0,60	11024	0,176	1372
130	1734	—0,58	10357	0,192	1398
140	1690	—0,57	9679	0,210	1408
150	1640	—0,57	8982	0,229	1406
160	1583	—0,56	8247	0,249	1391

Table II.8 (*Continued*)

T	w	μ	k	α/α_0	γ/γ_0
			$p=0,2$		
170	1522	—0,55	7503	0,271	1367
180	1458	—0,53	6770	0,295	1335
190	1392	—0,52	6070	0,322	1298
200	1325	—0,50	5443	0,352	1265
210	1260	—0,47	4801	0,384	1215
220	1195	—0,45	4240	0,421	1170
230	1131	—0,41	3724	0,461	1123
240	1068	—0,38	3251	0,507	1074
250	222,5	32,39	1,12	1,227	1,15
260	227,7	28,36	1.11	1,194	1,13
270	232,6	24,89	1,11	1,166	1,11
280	237,3	21,93	1,11	1,143	1,10
290	241,9	19,39	1,11	1,124	1,08
300	246,2	17,22	1,11	1,109	1,07
310	250,5	15,37	1,10	1,096	1,06
320	254,6	13,78	1,10	1,085	1,06
330	258,7	12,42	1,10	1,076	1,05
340	262,6	11,25	1,10	1,068	1,05
350	266,4	10,24	1,10	1,061	1,04
360	270,2	9,37	1,09	1,055	1,04
370	273,9	8,61	1,09	1,051	1,03
380	277,6	7,95	1,09	1,046	1,03
390	281,1	7,37	1,09	1,043	1,03
400	284,7	6,86	1,09	1,040	1,03
410	288,1	6,41	1,09	1,037	1,02
420	291,6	6,00	1,09	1,034	1,02
430	295,0	5,64	1,08	1,032	1,02
440	298,3	5,32	1,08	1,030	1,02
450	301,6	5,03	1,08	1,028	1,02
460	304,8	4,77	1,08	1,027	1,02
470	308,1	4,53	1,08	1,025	1,02
480	311,2	4,31	1,08	1,024	1,02
490	314,4	4,10	1,08	1,023	1,02
500	317,5	3,92	1,08	1,021	1,02
520	323,6	3,59	1,07	1,019	1,01
540	329,6	3,30	1,07	1,018	1,01
560	335,5	3,05	1,07	1,016	1,01
580	341,3	2,83	1,07	1,015	1,01
600	347,0	2,64	1,07	1,014	1,01
620	352,6	2,46	1,07	1,013	1,01
640	358,1	2,30	1,07	1,012	1,01
660	363,6	2,16	1,06	1,011	1,01
680	368,9	2,02	1,06	1,010	1,01
700	374,1	1,89	1,06	1,009	1,01
			$p=0,3$		
110	1798	—0,64	7631	0,158	865,5
120	1775	—0,60	7349	0,175	913,9
130	1735	—0,58	6913	0,192	932,9
140	1691	—0,57	6465	0,210	940,1
150	1640	—0,57	5995	0,229	938,0
160	1584	—0,56	5505	0,249	927,9
170	1523	—0,55	5008	0,271	911,5
180	1459	—0,54	4516	0,295	889,8
190	1393	—0,52	4054	0,322	866,3

Table II.8 (*Continued*)

T	w	μ	k	α/α_0	γ/γ_0
			$p=0,3$		
200	1326	—0,50	3612	0,351	838,8
210	1261	—0,47	3207	0,384	810,6
220	1196	—0,45	2832	0,420	780,7
230	1132	—0,41	2488	0,461	749,5
240	1069	—0,37	2173	0,507	717,0
250	1006	—0,34	1883	0,559	683,0
260	222,5	29,43	1,10	1,322	1,21
270	228,1	25,69	1,10	1,271	1,18
280	233,3	22,53	1,10	1,231	1,15
290	238,2	19,87	1,10	1,199	1,13
300	242,9	17,60	1,10	1,172	1,11
310	247,5	15,68	1,09	1,151	1,10
320	251,9	14,04	1,09	1,133	1,09
330	256,1	12,63	1,09	1,118	1,08
340	260,3	11,43	1,09	1,105	1,07
350	264,3	10,39	1,09	1,095	1,06
360	268,2	9,50	1,09	1,086	1,06
370	272,1	8,72	1,09	1,078	1,05
380	275,8	8,04	1,09	1,071	1,05
390	279,5	7,45	1,08	1,065	1,04
400	283,2	6,92	1,08	1,060	1,04
410	286,7	6,46	1,08	1,056	1,04
420	290,3	6,05	1,08	1,052	1,03
430	293,7	5,68	1,08	1,049	1,03
440	297,1	5,36	1,08	1,046	1,03
450	300,4	5,06	1,08	1,043	1,03
460	303,8	4,79	1,08	1,040	1,03
470	307,1	4,54	1,08	1,038	1,03
480	310,3	4,32	1,08	1,036	1,03
490	313,5	4,12	1,07	1,034	1,02
500	316,7	3,93	1,07	1,032	1,02
520	322,9	3,59	1,07	1,029	1,02
540	329,0	3,30	1,07	1,027	1,02
560	335,0	3,05	1,07	1,024	1,02
580	340,9	2,83	1,07	1,022	1,02
600	346,6	2,63	1,07	1,020	1,02
620	352,3	2,46	1,07	1,019	1,02
640	357,9	2,30	1,06	1,017	1,01
660	363,3	2,15	1,06	1,016	1,01
680	368,7	2,01	1,06	1,015	1,01
700	374,0	1,89	1,06	1,014	1,01
			$p=0,4$		
110	1799	—0,64	5736	0,158	650,1
120	1776	—0,60	5516	0,175	685,6
130	1736	—0,58	5192	0,192	700,3
140	1692	—0,57	4892	0,210	711,0
150	1641	—0,57	4501	0,228	703,9
160	1585	—0,56	4133	0,249	696,4
170	1524	—0,55	3761	0,271	684,1
180	1459	—0,54	3394	0,295	668,3
190	1394	—0,52	3044	0,322	649,9
200	1327	—0,50	2714	0,351	629,6
210	1262	—0,47	2409	0,383	608,5
220	1197	—0,45	2128	0,420	586,1

Table II.8 (*Continued*)

T	ω	μ	k	α/α₀	γ/γ,,

$$p = 0{,}4$$

T	ω	μ	k	α/α_0	$\gamma/\gamma_{,,}$
230	1133	—0,41	1870	0,460	562,7
240	1070	—0,38	1634	0,506	538,4
250	1007	—0,34	1416	0,558	512,9
260	943,9	—0,29	1216	0,619	486,3
270	223,2	26,59	1,08	1,398	1,25
280	228,9	23,21	1,08	1,334	1,21
290	234,4	20,39	1,08	1,284	1,18
300	239,5	18,01	1,08	1,244	1,16
310	244,4	16,01	1,08	1,212	1,14
320	249,0	14,31	1,08	1,185	1,12
330	253,5	12,86	1,08	1,164	1,11
340	257,9	11,61	1,08	1,146	1,10
350	262,1	10,55	1,08	1,130	1,09
360	266,2	9,63	1,08	1,118	1,08
370	270,2	8,83	1,08	1,107	1,07
380	274,1	8,13	1,08	1,097	1,06
390	277,9	7,53	1,08	1,089	1,06
400	281,7	6,99	1,08	1,082	1,05
410	285,3	6,52	1,08	1,076	1,05
420	288,9	6,10	1,08	1,071	1,05
430	292,5	5,72	1,08	1,066	1,04
440	296,0	5,39	1,08	1,062	1,04
450	299,4	5,09	1,08	1,058	1,04
460	302,8	4,81	1,07	1,054	1,04
470	306,2	4,56	1,07	1,051	1,04
480	309,5	4,34	1,07	1,048	1,03
490	312,7	4,13	1,07	1,046	1,03
500	315,9	3,94	1,07	1,043	1,03
520	322,2	3,59	1,07	1,039	1,03
540	328,4	3,30	1,07	1,036	1,03
560	334,5	3,05	1,07	1,032	1,02
580	340,4	2,82	1,07	1,030	1,02
600	346,2	2,63	1,07	1,027	1,02
620	352,0	2,45	1,07	1,025	1,02
640	357,6	2,29	1,06	1,023	1,02
660	363,1	2,14	1,06	1,022	1,02
680	368,5	2,01	1,06	1,020	1,02
700	373,9	1,88	1,06	1,018	1,02

$$p = 0{,}5$$

T	ω	μ	k	α/α_0	$\gamma/\gamma_{,,}$
110	1801	—0,64	4599	0,158	520,9
120	1777	—0,60	4417	0,175	548,8
130	1737	—0,58	4158	0,192	560,6
140	1693	—0,57	3888	0,210	564,8
150	1642	—0,57	3605	0,228	563,4
160	1586	—0,56	3310	0,249	557,4
170	1525	—0,55	3012	0,271	547,6
180	1459	—0,54	2714	0,295	534,8
190	1393	—0,52	2435	0,322	520,1
200	1328	—0,50	2174	0,351	504,1
210	1263	—0,47	1931	0,383	487,2
220	1198	—0,45	1706	0,419	469,3

Table II.8 (*Continued*)

T	w	μ	k	a/a_0	γ/γ_0
			$p=0,5$		
230	1134	—0,42	1500	0,459	450,6
240	1071	—0,38	1310	0,505	431,2
250	1008	—0,34	1136	0,557	410,8
260	945,3	—0,29	975,9	0,617	389,7
270	881,5	—0,24	828,0	0,690	367,4
280	224,3	23,96	1,07	1,456	1,28
290	230,3	20,96	1,07	1,382	1,24
300	235,9	18,46	1,07	1,325	1,21
310	241,1	16,36	1,07	1,280	1,18
320	246,1	14,59	1,07	1,243	1,16
330	250,9	13,09	1,08	1,213	1,14
340	255,4	11,81	1,08	1,189	1,12
350	259,8	10,71	1,08	1,169	1,11
360	264,1	9,76	1,08	1,152	1,10
370	268,3	8,94	1,08	1,137	1,09
380	272,3	8,23	1,08	1,125	1,08
390	276,3	7,60	1,07	1,114	1,07
400	280,1	7,06	1,07	1,105	1,07
410	283,9	6,57	1,07	1,097	1,06
420	287,6	6,15	1,07	1,090	1,06
430	291,3	5,76	1,07	1,083	1,06
440	294,8	5,42	1,07	1,078	1,05
450	298,3	5,11	1,07	1,073	1,05
460	301,8	4,83	1,07	1,069	1,05
470	305,2	4,58	1,07	1,065	1,04
480	308,6	4,35	1,07	1,061	1,04
490	311,9	4,14	1,07	1,058	1,04
500	315,2	3,94	1,07	1,055	1,04
520	321,6	3,60	1,07	1,049	1,04
540	327,8	3,30	1,07	1,045	1,03
560	334,0	3,05	1,07	1,041	1,03
580	340,0	2,82	1,07	1,037	1,03
600	345,9	2,62	1,07	1,034	1,03
620	351,7	2,44	1,06	1,031	1,03
640	357,3	2,28	1,06	1,029	1,02
660	362,9	2,14	1,06	1,027	1,02
680	368,4	2,00	1,06	1,025	1 02
700	373,7	1,87	1,06	1,023	1,02
			$p=0,6$		
110	1803	—0,64	3840	0,158	434,7
120	1778	—0,60	3686	0,175	457,6
130	1738	—0,58	3469	0,192	467,5
140	1693	—0,57	3243	0,209	470,9
150	1643	—0,57	3007	0,228	469,8
160	1587	—0,56	2762	0,249	464,8
170	1526	—0,55	2513	0,271	456,6
180	1461	—0,54	2269	0,295	446,1
190	1395	—0,52	2034	0,321	433,7
200	1329	—0,50	1815	0,350	420,5
210	1264	—0,47	1612	0,383	406,3
220	1199	—0,45	1424	0,419	391,4
230	1135	—0,42	1252	0,459	375,9

Table II.8 (*Continued*)

T	w	μ	k	α/α_0	γ/γ_0
		$p=0{,}6$			
240	1072	−0,38	1095	0,504	359,8
250	1010	−0,34	949,5	0,556	342,9
260	946,8	−0,29	816,0	0,616	325,2
270	883,1	−0,24	692,7	0,688	306,7
280	818,1	−0,17	579,9	0,777	287,1
290	226,0	21,59	1,06	1,497	1,30
300	232,0	18,94	1,06	1,417	1,26
310	237,7	16,74	1,06	1,356	1,22
320	243,0	14,89	1,06	1,307	1,19
330	248,1	13,33	1,07	1,268	1,17
340	252,9	12,00	1,07	1,236	1,15
350	257,6	10,87	1,07	1,209	1,13
360	262,0	9,90	1,07	1,188	1,12
370	266,3	9,05	1,07	1,169	1,11
380	270,5	8,32	1,07	1,153	1,10
390	274,6	7,68	1,07	1,140	1,09
400	278,6	7,12	1,07	1,128	1,08
410	282,5	6,63	1,07	1,118	1,08
420	286,3	6,19	1,07	1,109	1,07
430	290,0	5,80	1,07	1,102	1,07
440	293,7	5,45	1,07	1,095	1,06
450	297,3	5,14	1,07	1,089	1,06
460	300,8	4,86	1,07	1,083	1,06
470	304,3	4,60	1,07	1,078	1,05
480	307,7	4,36	1,07	1,074	1,05
490	311,1	4,15	1,07	1,070	1,05
500	314,4	3,95	1,07	1,066	1,05
520	320,9	3,60	1,07	1,059	1,04
540	327,3	3,30	1,07	1,054	1,04
560	333,5	3,04	1,07	1,049	1,04
580	339,5	2,82	1,06	1,045	1,03
600	345,5	2,62	1,06	1,041	1,03
620	351,3	2,44	1,06	1,038	1,03
640	357,1	2,28	1,06	1,035	1,03
660	362,7	2,13	1,06	1,032	1,03
680	368,2	1,99	1,06	1,030	1,03
700	373,6	1,86	1,06	1,028	1,02
		$p=0{,}7$			
110	1805	−0,64	3299	0,158	373,2
120	1780	−0,60	3163	0,175	392,6
130	1739	−0,58	2977	0,192	401,0
140	1694	−0,57	2783	0,209	403,9
150	1644	−0,57	2580	0,228	402,9
160	1588	−0,56	2370	0,248	398,6
170	1526	−0,55	2157	0,270	391,6
180	1462	−0,54	1947	0,294	382,6
190	1396	−0,52	1747	0,321	372,1
200	1330	−0,50	1558	0,350	360,7
210	1265	−0,47	1384	0,382	348,5
220	1200	−0,45	1223	0,418	335,8
230	1137	−0,42	1076	0,458	322,5
240	1074	−0,38	940,5	0,503	308,7

Table II.8 (*Continued*)

T	w	μ	k	α/α_0	ν/γ_0
			$p=0,7$		
250	1011	−0,34	816,2	0,555	294,3
260	948,2	−0,29	701,8	0,615	279,2
270	884,7	−0,24	596,2	0,686	263,3
280	819,9	−0,17	498,6	0,774	246,7
290	221,4	22,30	1,04	1,634	1,37
300	228,0	19,47	1,05	1,524	1,32
310	234,2	17,14	1,05	1,441	1,27
320	239,9	15,21	1,05	1,377	1,23
330	245,3	13,59	1,06	1,327	1,20
340	250,4	12,21	1,06	1,286	1,18
350	255,2	11,04	1,06	1,253	1,16
360	259,9	10,03	1,06	1,226	1,14
370	264,4	9,17	1,06	1,203	1,13
380	268,7	8,42	1,06	1,184	1,12
390	272,9	7,76	1,06	1,167	1,11
400	277,0	7,19	1,07	1,153	1,10
410	281,0	6,69	1,07	1,140	1,09
420	285,0	6,24	1,07	1,130	1,09
430	288,8	5,84	1,07	1,120	1,08
440	292,5	5,48	1,07	1,112	1,08
450	296,2	5,17	1,07	1,105	1,07
460	299,8	4,88	1,07	1,098	1,07
470	303,3	4,61	1,07	1,092	1,06
480	306,8	4,38	1,07	1,087	1,06
490	310,2	4,16	1,07	1,082	1,06
500	313,6	3,96	1,07	1,077	1,05
520	320,2	3,61	1,07	1,070	1,05
540	326,7	3,30	1,06	1,063	1,05
560	333,0	3,04	1,06	1,057	1,04
580	339,1	2,81	1,06	1,052	1,04
600	345,1	2,61	1,06	1,048	1,04
620	351,0	2,43	1,06	1,044	1,04
640	356,8	2,27	1,06	1,041	1,03
660	362,5	2,12	1,06	1,037	1,03
680	368,0	1,98	1,06	1,035	1,03
700	373,5	1,86	1,06	1,032	1,03
			$p=0,8$		
110	1806	−0,64	2892	0,158	327,0
120	1781	−0,60	2771	0,175	343,7
130	1740	−0,58	2608	0,192	351,1
140	1695	−0,57	2438	0,209	353,6
150	1645	−0,57	2260	0,228	352,8
160	1588	−0,56	2076	0,248	349,0
170	1527	−0,55	1889	0,270	342,9
180	1463	−0,54	1706	0,294	335,0
190	1397	−0,52	1531	0,320	325,8
200	1331	−0,50	1366	0,350	315,8
210	1266	−0,47	1213	0,382	305,2
220	1201	−0,45	1072	0,418	294,1
230	1138	−0,42	943,4	0,458	282,5
240	1075	−0,38	825,0	0,502	270,4
250	1012	−0,34	716,2	0,554	257,8

Table II.8 (*Continued*)

T	w	μ	k	α/α_0	γ/γ
			$p=0,8$		
260	949,6	−0,30	616,1	0,613	244,6
270	886,3	−0,24	523,7	0,684	230,8
280	821,7	−0,17	438,4	0,772	216,3
290	754,9	−0,09	359,5	0,883	200,8
300	223,8	20,05	1,03	1,648	1,38
310	230,4	17,58	1,04	1,539	1,32
320	236,6	15,55	1,04	1,456	1,28
330	242,3	13,85	1,05	1,392	1,24
340	247,7	12,43	1,05	1,341	1,21
350	252,8	11,21	1,05	1,300	1,19
360	257,7	10,18	1,06	1,267	1,17
370	262,4	9,29	1,06	1,239	1,15
380	266,9	8,52	1,06	1,215	1,14
390	271,3	7,84	1,06	1,195	1,13
400	275,5	7,26	1,06	1,178	1,12
410	279,6	6,74	1,06	1,164	1,11
420	283,6	6,29	1,06	1,151	1,10
430	287,5	5,88	1,06	1,140	1,09
440	291,4	5,52	1,06	1,130	1,09
450	295,1	5,19	1,06	1,121	1,08
460	298,8	4,90	1,06	1,113	1,08
470	302,4	4,63	1,06	1,106	1,07
480	305,9	4,39	1,06	1,100	1,07
490	309,4	4,17	1,06	1,094	1,07
500	312,9	3,97	1,06	1,089	1,06
520	319,6	3,61	1,06	1,080	1,06
540	326,1	3,30	1,06	1,072	1,05
560	332,5	3,04	1,06	1,065	1,05
580	338,7	2,81	1,06	1,060	1,05
600	344,8	2,61	1,06	1,055	1,04
620	350,7	2,42	1,06	1,050	1,04
640	356,6	2,26	1,06	1,046	1,04
660	362,3	2,11	1,06	1,043	1,04
680	367,9	1,98	1,06	1,039	1,03
700	373,4	1,85	1,06	1,037	1,03
			$p=0,9$		
110	1808	−0,64	2576	0,158	291,0
120	1782	−0,60	2466	0,175	305,8
130	1741	−0,58	2321	0,192	312,3
140	1696	−0,57	2169	0,209	314,5
150	1646	−0,57	2011	0,228	313,8
160	1589	−0,56	1848	0,248	310,4
170	1528	−0,55	1682	0,270	305,0
180	1464	−0,54	1519	0,294	298,0
190	1398	−0,52	1362	0,320	289,8
200	1332	−0,50	1216	0,349	280,9
210	1267	−0,48	1080	0,381	271,5
220	1202	−0,45	955,1	0,417	261,6
230	1139	−0,42	840,4	0,457	251,4
240	1076	−0,38	735,2	0,502	240,6
250	1014	−0,34	638,5	0,553	229,5
260	951,1	−0,30	549,5	0,612	217,8

Table II.8 (*Continued*)

T	w	μ	k	α/τ_0	γ/γ_0

$p=0,9$

270	887,9	−0,24	467,4	0,683	205,5
280	823,4	−0,17	391,5	0,769	192,6
290	756,9	−0,09	321,4	0,880	179,0
300	219,2	20,70	1,01	1,796	1,45
310	226,5	18,06	1,02	1,651	1,38
320	233,2	15,91	1,03	1,545	1,32
330	239,3	14,14	1,04	1,464	1,28
340	245,0	12,65	1,04	1,401	1,25
350	250,4	11,39	1,05	1,351	1,22
360	255,5	10,32	1,05	1,310	1,19
370	260,4	9,41	1,05	1,276	1,17
380	265,0	8,61	1,05	1,248	1,16
390	269,5	7,93	1,05	1,225	1,14
400	273,9	7,33	1,06	1,205	1,13
410	278,1	6,80	1,06	1,187	1,12
420	282,3	6,33	1,06	1 172	1,11
430	286,3	5,92	1,06	1,159	1,11
440	290,2	5,55	1,06	1,148	1,10
450	294,0	5,22	1,06	1,138	1,09
460	297,8	4,92	1,06	1,129	1,09
470	301,5	4,65	1,06	1,121	1,08
480	305,1	4,40	1,06	1,113	1,08
490	308,6	4,18	1,06	1,107	1,07
500	312,1	3,97	1,06	1,101	1,07
520	318,9	3,61	1,06	1,090	1,06
540	325,5	3,30	1,06	1,081	1,06
560	332,0	3,04	1,06	1,074	1,06
580	338,3	2,80	1,06	1,067	1,05
600	344,4	2,60	1,06	1,062	1,05
620	350,4	2,42	1,06	1 056	1,05
640	356,3	2,26	1,06	1,052	1,04
660	362,1	2,11	1,06	1,048	1,04
680	367,7	1,97	1,06	1,044	1,04
700	373,2	1,84	1,06	1,041	1,04

$p=1,0$

110	1810	−0,64	2323	0,158	262,3
120	1783	−0,60	2224	0,175	275,5
130	1742	−0,58	2091	0,192	281,3
140	1697	−0,57	1955	0,209	283,2
150	1646	−0,57	1812	0,228	282,5
160	1590	−0,56	1665	0,248	279,5
170	1529	−0,55	1515	0,270	274,6
180	1465	−0,54	1368	0,294	268,3
190	1399	−0,52	1228	0,320	261,0
200	1333	−0,50	1096	0,349	253,0
210	1268	−0,48	973,8	0,381	244,6
220	1203	−0,45	861,3	0,417	235,7
230	1140	−0,42	758,0	0,456	226,4
240	1077	−0,38	663,3	0,501	216,8
250	1015	−0,34	576,2	0,552	206,8
260	952,5	−0,30	496,2	0,611	196,3
270	889,5	−0,24	422,3	0,681	185,3

Table II.8 (*Continued*)

T	w	μ	k	α/α_0	γ/γ_0

$$p = 1{,}0$$

T	w	μ	k	α/α_0	γ/γ_0
280	825,2	—0,18	354,0	0,767	173,7
290	758,9	—0,09	291,0	0,876	161,5
300	689,7	0,02	232,8	1,023	148,4
310	222,4	18,58	1,01	1,781	1,44
320	229,6	16,30	1,02	1,644	1,38
330	236,1	14,43	1,03	1,544	1,32
340	242,2	12,88	1,03	1,466	1,28
350	247,9	11,58	1,04	1,405	1,25
360	253,2	10,47	1,04	1,356	1,22
370	258,3	9,53	1,04	1,316	1,20
380	263,2	8,71	1,05	1,283	1,18
390	267,8	8,01	1,05	1,256	1,16
400	272,3	7,39	1,05	1,232	1,15
410	276,7	6,85	1,05	1,212	1,14
420	280,9	6,38	1,05	1,195	1,13
430	285,0	5,96	1,06	1,180	1,12
440	289,0	5,58	1,06	1,166	1,11
450	293,0	5,24	1,06	1,155	1,10
460	296,8	4,94	1,06	1,144	1,10
470	300,5	4,66	1,06	1,135	1,09
480	304,2	4,42	1,06	1,127	1,09
490	307,8	4,19	1,06	1,119	1,08
500	311,3	3,98	1,06	1,113	1,08
520	318,3	3,61	1,06	1,101	1,07
540	325,0	3,30	1,06	1,091	1,07
560	331,5	3,03	1,06	1,082	1,06
580	337,8	2,80	1,06	1,075	1,06
600	344,1	2,59	1,06	1,068	1,05
620	350,1	2,41	1,06	1,063	1,05
640	356,1	2,25	1,06	1,058	1,05
660	361,9	2,10	1,06	1,053	1,04
680	367,6	1,96	1,06	1,049	1,04
700	373,1	1,83	1,06	1,045	1,04

$$p = 1{,}2$$

T	w	μ	k	α/α_0	γ/γ_0
110	1813	—0,64	1944	0,158	219,2
120	1786	—0,60	1858	0,175	230,0
130	1744	—0,58	1747	0,192	234,7
140	1699	—0,57	1632	0,209	236,3
150	1648	—0,57	1514	0,228	235,7
160	1592	—0,56	1390	0,248	233,2
170	1531	—0,55	1266	0,269	229,1
180	1466	—0,54	1143	0,293	223,9
190	1401	—0,52	1026	0,319	217,8
200	1335	—0,50	916,2	0,348	211,2
210	1270	—0,48	814,3	0,380	204,1
220	1206	—0,45	720,5	0,416	196,7
230	1142	—0,42	634,4	0,455	189,1
240	1080	—0,38	555,5	0,499	181,1
250	1018	—0,34	482,9	0,550	172,8
260	955,3	—0,30	416,2	0,608	164,1
270	892,6	—0,25	354,6	0,677	154,9
280	828,7	—0,18	297,8	0,762	145,4

Table II.8 (*Continued*)

T	w	μ	k	α/α_0	γ/γ
			$p=1,2$		
290	762,9	−0,10	245,3	0,869	135,2
300	694,3	0,01	196,8	1,012	124,5
310	213,4	19,80	0,98	2,118	1,59
320	221,9	17,18	0,99	1,890	1,49
330	229,5	15,09	1,00	1,732	1,42
340	236,4	13,38	1,01	1,616	1,36
350	242,7	11,97	1,02	1,528	1,32
360	248,6	10,79	1,03	1,459	1,28
370	254,1	9,78	1,03	1,404	1,25
380	259,3	8,92	1,04	1,359	1,22
390	264,3	8,18	1,04	1,322	1,20
400	269,1	7,53	1,04	1,291	1,19
410	273,7	6,97	1,04	1,264	1,17
420	278,2	6,47	1,05	1,242	1,16
430	282,5	6,03	1,05	1,222	1,15
440	286,7	5,64	1,05	1,205	1,14
450	290,8	5,29	1,05	1,190	1,13
460	294,8	4,98	1,05	1,177	1,12
470	298,7	4,70	1,05	1,165	1,11
480	302,5	4,44	1,05	1,155	1,11
490	306,2	4,21	1,06	1,145	1,10
500	309,9	3,99	1,06	1,137	1,10
520	317,0	3,62	1,06	1,122	1,09
540	323,8	3,30	1,06	1,110	1,08
560	330,5	3,03	1,06	1,099	1,07
580	337,0	2,79	1,06	1,090	1,07
600	343,4	2,58	1,06	1,082	1,06
620	349,5	2,40	1,06	1,075	1,06
640	355,6	2,24	1,06	1,069	1,06
660	361,5	2,09	1,06	1,063	1,05
680	367,2	1,95	1,06	1,059	1,05
700	372,9	1,82	1,06	1,054	1,05
			$p=1,4$		
110	1817	−0,64	1672	0,158	188,4
120	1788	−0,60	1597	0,175	197,5
130	1746	−0,58	1501	0,191	201,5
140	1700	−0,57	1402	0,209	202,8
150	1650	−0,57	1300	0,227	202,3
160	1593	−0,56	1194	0,247	200,1
170	1532	−0,55	1087	0,269	196,6
180	1468	−0,54	982,5	0,293	192,1
190	1403	−0,52	882,0	0,319	186,9
200	1337	−0,50	787,7	0,348	181,3
210	1272	−0,48	700,3	0,379	175,2
220	1208	−0,45	619,9	0,415	168,9
230	1144	−0,42	546,1	0,454	162,4
240	1082	−0,39	478,4	0,498	155,6
250	1020	−0,35	416,2	0,548	148,5
260	958,1	−0,30	359,0	0,606	141,0
270	895,7	−0,25	306,3	0,674	133,3
280	832,1	−0,18	257,6	0,757	125,1
290	766,8	−0,10	212,6	0,863	116,5

Table II.8 (*Continued*)

T	w	μ	k	α/α_0	γ/γ_0

			$p=1,4$		
300	698,8	0,00	171,1	1,002	107,3
310	627,0	0,15	132,9	1,198	97,47
320	213,4	18.23	0,96	2,226	1,64
330	222,3	15.84	0,98	1,974	1,53
340	230,1	13.94	0,99	1,800	1,45
350	257,2	12,40	1,00	1,674	1,39
360	243.7	11,12	1,01	1,578	1,34
370	249,7	10,04	1,02	1,503	1,30
380	255,4	9,13	1,02	1,444	1,27
390	260,8	8,35	1,03	1,395	1,25
400	265,9	7,67	1,03	1,354	1,22
410	270,8	7,08	1,04	1,321	1,20
420	275,5	6,56	1,04	1,292	1,19
430	280,0	6.11	1,04	1,267	1,17
440	284,4	5,70	1,04	1,246	1,16
450	288,7	5,34	1,05	1,227	1,15
460	292.8	5,02	1,05	1,211	1,14
470	296,8	4,73	1,05	1,196	1,13
480	300,8	4,46	1,05	1,183	1,13
490	304,6	4,22	1,05	1,172	1,12
500	308,4	4,01	1,05	1,162	1,11
520	315,7	3,62	1,05	1,144	1,10
540	322,7	3,30	1,06	1,129	1,09
560	329,6	3,02	1,06	1,116	1,09
580	336,2	2,78	1,06	1,105	1,08
600	342,7	2,57	1,06	1,096	1,07
620	349,0	2,39	1,06	1,087	1,07
640	355,1	2,22	1,06	1,080	1,07
660	361,1	2,07	1,06	1,074	1,06
680	366,9	1,93	' 06	1,068	1,06
700	372.6	1,80	1 06	1,063	1,05

			$p=1,6$		
110	1820	—0.64	1469	0,158	165,3
120	1790	—0,60	1401	0,175	173,1
130	1748	—0,58	1316	0,191	176,5
140	1702	—0,57	1230	0,209	177,7
150	1651	—0,57	1140	0,227	177.2
160	1595	—0,56	1047	0,247	175,3
170	1534	—0,55	953,8	0,269	172,2
180	1470	—0,54	861,8	0,292	168,3
190	1404	—0,52	773,9	0,318	163,8
200	1339	—0,50	691,3	0,347	158,8
210	1274	—0,48	614,9	0,379	153,6
220	1210	—0,45	544,5	0,414	148,1
230	1147	—0,42	479,9	0,453	142.4
240	1084	—0,39	420,6	0,496	136,4
250	1023	—0,35	366,2	0,546	130,2
260	960.9	—0,30	316,2	0,603	123,8
270	898.7	—0.25	270,0	0,671	117,0
280	835.5	—0,19	227,4	0,753	109,9
290	770.6	—0,11	188,1	0,856	102,4

Table II.8 (*Continued*)

T	w	μ	k	α/α_0	γ/γ_0
			$p=1,6$		
300	703,2	—0,01	151,8	0,992	94,45
310	632,2	0,13	118,4	1,182	85,93
320	555,6	0,34	87,73	1,471	76,60
330	214,4	16,72	0,95	2,296	1,66
340	223,4	14,57	0,97	2,031	1,56
350	231,4	12,87	0,98	1,849	1,48
360	238,6	11,48	0,99	1,717	1,41
370	245,2	10,32	1,00	1,616	1,36
380	251,4	9,35	1,01	1,538	1,32
390	257,1	8,52	1,02	1,475	1,29
400	262,6	7,81	1,02	1,424	1,26
410	267,8	7,19	1,03	1,381	1,24
420	272,7	6,65	1,03	1,345	1,22
430	277,5	6,18	1,03	1,315	1,20
440	282,1	5,76	1,04	1,289	1,19
450	286,5	5,39	1,04	1,266	1,18
460	290,8	5,06	1,04	1,246	1,16
470	295,0	4,76	1,04	1,229	1,15
480	299,1	4,49	1,05	1,213	1,15
490	303,0	4,24	1,05	1,199	1,14
500	306,9	4,02	1,05	1,187	1,13
520	314,4	3,63	1,05	1,166	1,12
540	321,7	3,30	1,05	1,148	1,11
560	328,6	3,02	1,06	1,133	1,10
580	335,4	2,77	1,06	1,120	1,09
600	342,0	2,56	1,06	1,109	1,09
620	348,4	2,37	1,06	1,100	1,08
640	354,6	2,21	1,06	1,091	1,07
660	360,7	2,06	1,06	1,084	1,07
680	366,6	1,92	1,06	1,077	1,07
700	372,4	1,79	1,06	1,071	1,06
			$p=1,8$		
110	1824	—0,64	1311	0,158	147,3
120	1793	—0,60	1249	0,175	154,1
130	1750	—0,58	1172	0,191	157,1
140	1704	—0,57	1095	0,208	158,1
150	1653	—0,57	1015	0,227	157,7
160	1597	—0,56	993,1	0,247	156,0
170	1536	—0,55	849,8	0,268	153,3
180	1472	—0,54	768,0	0,292	149,8
190	1406	—0,52	689,8	0,318	145,8
200	1341	—0,50	616,4	0,346	141,4
210	1276	—0,48	548,4	0,378	136,7
220	1212	—0,45	485,8	0,413	131,8
230	1149	—0,42	428,3	0,451	126,8
240	1087	—0,39	375,7	0,495	121,5
250	1025	—0,35	327,3	0,544	116,0
260	963,7	—0,31	282,8	0,601	110,3
270	901,7	—0,25	241,8	0,667	104,3
280	838,9	—0,19	204,0	0,748	98,05
290	774,4	—0,11	169,0	0,850	91,44

Table II.8 (*Continued*)

T	w	μ	k	α/α_0	γ/γ_0

$$p=1,8$$

T	w	μ	k	α/α_0	γ/γ_0
300	707,6	—0,01	136,8	0,983	84,44
310	637,4	0,12	107,2	1,166	76,95
320	561,9	0,32	79,96	1,443	68,77
330	205,5	17,80	0,92	2,753	1,83
340	216,2	15,30	0,95	2,331	1,68
350	225,3	13,38	0,96	2,065	1,57
360	233,3	11,86	0,98	1,881	1,49
370	240,6	10,61	0,99	1,746	1,43
380	247,2	9,58	1,00	1,644	1,38
390	253,4	8,70	1,01	1,564	1,34
400	259,2	7,95	1,01	1,499	1.31
410	264,7	7,31	1,02	1,446	1,28
420	270,0	6,75	1,02	1,402	1,25
430	275,0	6,25	1,03	1,365	1,23
440	279,7	5,82	1,03	1,334	1,22
450	284,4	5,44	1,03	1,306	1,20
460	288,8	5,09	1,04	1,283	1,19
470	293,2	4,78	1,04	1,262	1,18
480	297,4	4,51	1,04	1,244	1,17
490	301,5	4,26	1,04	1,227	1,16
500	305,5	4,03	1,05	1,213	1,15
520	313,2	3,63	1,05	1,188	1,13
540	320,6	3,29	1,05	1,167	1,12
560	327,7	3,01	1,05	1,150	1,11
580	334,6	2,76	1,06	1.135	1,10
600	341,3	2,55	1,06	1,123	1,10
620	347,8	2,36	1,06	1,112	1,09
640	354,2	2,19	1,06	1,102	1,08
660	360,3	2,04	1,06	1,094	1,08
680	366 3	1,90	1,06	1,087	1,07
700	372,2	1,78	1,06	1,080	1,07

$$p=2,0$$

T	w	μ	k	α/α_0	γ/γ_0
110	1827	—0,64	1184	0,158	132,9
120	1795	—0,60	1127	0,175	139,0
130	1752	—0,58	1058	0,191	141,6
140	1705	—0,57	987,8	0,208	142,5
150	1654	—0,57	915,8	0,227	142,1
160	1598	—0,56	841,6	0,247	140,5
170	1537	—0,55	766,6	0,268	138,1
180	1473	—0,54	693,0	0,292	135,0
190	1408	—0,52	622,6	0,317	131,4
200	1342	—0,50	556,4	0,346	127,4
210	1278	—0,48	495,2	0,377	123,2
220	1214	—0,45	438,8	0,412	118,8
230	1151	—0,42	387,1	0,450	114,3
240	1089	—0,39	339,7	0,493	109,6
250	1028	—0,35	296,2	0,542	104,7
260	966,4	—0,31	256,2	0,598	99,56
270	904,7	—0,26	219,2	0,664	94,20
280	842,2	—0,19	185,2	0,744	88,58
290	778,2	—0,12	153,8	0,844	82,68
300	711,9	—0,02	124,8	0,973	76,43

Table II.8 (*Continued*)

T	w	μ	k	α/α_0	γ/γ_0
			$p=2,0$		
310	642.4	0,11	98,16	1.152	69,75
320	568.1	0,30	73,72	1,416	62,49
330	485,8	0,61	51,29	1,863	54,36
340	208,2	16,15	0,92	2,739	1,83
350	218,7	13,96	0.94	2,336	1,68
360	227,7	12,28	0,96	2,078	1.58
370	235,7	10,93	0,97	1,897	1,50
380	243,0	9,81	0,99	1.764	1,44
390	249.6	8,89	1,00	1,662	1,39
400	255,8	8,10	1,00	1,582	1,35
410	261,7	7,42	1,01	1,517	1,32
420	267,2	6,84	1,02	1,464	1,29
430	272,4	6,33	1,02	1,419	1,27
440	277,4	5,88	1,03	1,381	1,25
450	282,2	5,48	1,03	1,349	1,23
460	286,9	5,13	1,03	1,321	1,21
470	291,4	4,81	1,04	1,297	1,20
480	295,7	4,53	1,04	1,275	1,19
490	299.9	4,27	1,04	1,256	1,18
500	304,1	4,04	1,04	1,239	1,17
520	312,0	3,63	1,05	1,211	1,15
540	319,6	3,29	1,05	1,187	1,14
560	326,8	3,00	1,05	1,167	1,12
580	333,9	2,75	1,05	1,151	1,11
600	340,7	2,54	1,06	1,137	1,11
620	347,3	2,35	1,06	1,124	1,10
640	353,7	2,18	1,06	1,114	1,09
660	359,9	2,03	1,06	1,104	1,09
680	366,0	1,89	1.06	1,096	1.08
700	371,9	1,76	1,06	1.088	1,08
			$p=2,2$		
110	1830	−0,64	1081	0.158	121,1
120	1797	−0,60	1027	0,175	126,5
130	1754	−0,58	963,7	0,191	128,9
140	1707	−0,57	899,8	0,208	129,7
150	1656	−0,57	834,3	0,227	129,3
160	1600	−0,56	766,7	0,246	127,9
170	1539	−0,55	698,5	0,268	125,7
180	1475	−0,54	631,6	0,291	122,8
190	1410	−0,52	567,4	0,317	119,6
200	1344	−0,50	507,4	0,345	116,0
210	1279	−0,48	451,6	0,376	112,2
220	1216	−0,45	400,4	0,411	108.2
230	1153	−0,42	353,4	0,449	104,1
240	1091	−0,39	310,3	0,492	99,84
250	1030	−0,35	270,7	0.540	95,39
260	969.1	−0,31	234,3	0.596	90,76
270	907,7	−0,26	200,8	0,661	85,91
280	845,5	−0,20	169,8	0.740	80,83
290	781,9	−0,12	141,3	0,838	75,50
300	716.2	−0,03	115.0	0.964	69,87

Table II.8 (*Continued*)

T	w	μ	k	α/α_0	γ/γ_0
		$p=2,2$			
310	647,4	0,10	90,77	1,137	63,86
320	574,1	0,28	68,60	1,391	57,34
330	493,6	0,57	48,30	1,810	50,09
340	199,1	17,19	0,89	3,334	2,01
350	211,6	14,63	0,92	2,691	1,82
360	221,8	12,73	0,94	2,319	1,68
370	230,7	11,26	0,96	2,075	1,58
380	238,6	10,06	0,97	1,902	1,51
390	245,8	9,08	0,98	1,773	1,45
400	252,4	8,24	0,99	1,673	1,40
410	258,6	7,54	1,00	1,593	1,36
420	264,4	6,93	1,01	1,529	1,33
430	269,9	6,40	1,01	1,476	1,30
440	275,1	5,93	1,02	1,431	1,28
450	280,1	5,53	1,02	1,393	1,26
460	284,9	5,16	1,03	1,361	1,24
470	289,6	4,84	1,03	1,332	1,22
480	294,1	4,55	1,03	1,307	1,21
490	298,4	4,28	1,04	1,286	1,20
500	302,6	4,05	1,04	1,266	1,18
520	310,8	3,63	1,04	1,234	1,17
540	318,5	3,29	1,05	1,207	1,15
560	326,0	2,99	1,05	1,185	1,14
580	333,1	2,74	1,05	1,166	1,13
600	340,0	2,53	1,05	1,150	1,12
620	346,7	2,34	1,06	1,136	1,11
640	353,3	2,17	1,06	1,124	1,10
660	359,6	2,02	1,06	1,114	1,09
680	365,7	1,88	1,06	1,105	1,09
700	371,7	1,75	1,06	1,096	1,08
		$p=2,4$			
110	1833	—0,64	994,2	0,158	111,3
120	1800	—0,60	944,2	0,174	116,2
130	1755	—0,58	885,3	0,191	118,3
140	1709	—0,57	826,6	0,208	119,0
150	1658	—0,57	766,3	0,226	118,6
160	1601	—0,56	704,4	0,246	117,4
170	1540	—0,55	641,8	0,267	115,3
180	1477	—0,54	580,4	0,291	112,8
190	1411	—0,52	521,6	0,316	109,7
200	1346	—0,50	466,5	0,344	106,5
210	1281	—0,48	415,4	0,375	103,0
220	1218	—0,45	368,4	0,410	99,37
230	1155	—0,42	325,3	0,448	95,62
240	1094	—0,39	285,8	0,490	91,71
250	1033	—0,35	249,5	0,538	87,65
260	971,8	—0,31	216,1	0,594	83,42
270	910,7	—0,26	185,3	0,658	79,00
280	848,8	—0,20	157,0	0,736	74,37
290	785,5	—0,13	130,8	0,832	69,52
300	720,3	—0,03	106,7	0,956	64,39
310	652,3	0,09	84,60	1,124	58,94

Table II.8 (*Continued*)

T	w	μ	k	α/α_0	γ/γ_0
		$p=2,4$			
320	580,0	0,27	64,32	1,367	53,04
330	501,1	0,54	45,77	1,761	46,52
340	188,6	18,54	0,85	4,307	2,26
350	203,7	15,40	0,89	3,178	1,97
360	215,5	13,24	0,92	2,622	1,80
370	225,4	11,61	0,94	2,286	1,67
380	234,0	10,32	0,96	2,060	1,58
390	241,8	9,27	0,97	1,896	1,51
400	248,9	8,39	0,98	1,773	1,45
410	255,4	7,65	0,99	1,676	1,41
420	261,6	7,02	1,00	1,599	1,37
430	267,3	6,47	1,01	1,536	1,33
440	272,8	5,99	1,01	1,484	1,31
450	278,0	5,57	1,02	1,439	1,28
460	283,0	5,19	1,02	1,402	1,26
470	287,8	4,86	1,03	1,369	1,25
480	292,4	4,56	1,03	1,341	1,23
490	296,9	4,30	1,03	1,316	1,22
500	301,3	4,05	1,04	1,294	1,20
520	309,6	3,63	1,04	1,257	1,18
540	317,5	3,28	1,05	1,227	1,16
560	325,1	2,99	1,05	1,202	1,15
580	332,4	2,73	1,05	1,181	1,14
600	339,4	2,51	1,05	1,164	1,13
620	346,2	2,32	1,06	1,149	1,12
640	352,8	2,15	1,06	1,135	1,11
660	359,2	2,00	1,06	1,124	1,10
680	365,4	1,86	1,06	1,114	1,10
700	371,5	1,73	1,06	1,105	1,09
		$p=2,6$			
110	1837	—0,64	921,1	0,158	103,0
120	1802	—0,60	873,8	0,174	107,4
130	1757	—0,58	819,0	0,191	109,3
140	1710	—0,57	764,5	0,208	110,0
150	1659	—0,57	708,8	0,226	109,6
160	1603	—0,56	651,6	0,246	108,5
170	1542	—0,55	593,8	0,267	106,6
180	1478	—0,54	537,0	0,290	104,2
190	1413	—0,52	482,7	0,316	101,4
200	1348	—0,50	431,9	0,344	98,41
210	1283	—0,48	384,7	0,375	95,21
220	1219	—0,45	341,1	0,409	91,79
230	1157	—0,43	301,5	0,447	88,42
240	1096	—0,39	264,9	0,489	84,80
250	1035	—0,36	231,5	0,537	81,10
260	974,4	—0,31	200,7	0,591	77,21
270	913,6	—0,26	172,3	0,655	73,15
280	852,0	—0,20	146,1	0,731	68,91
290	789,2	—0,13	122,0	0,826	64,45
300	724,4	—0,04	99,78	0,947	59,76
310	657,0	0,08	79,37	1,111	54,77
320	585,7	0,25	60,68	1,345	49,40

Table II.8 (*Continued*)

T	w	μ	k	α/α_0	γ/γ_0
			$p=2,6$		
330	508,3	0,51	43,62	1,717	43,48
340	420,0	0,97	28,04	2,434	36,68
350	194,9	16,35	0,86	3,897	2,17
360	208,7	13,82	0,90	3,017	1,93
370	219,9	12,00	0,93	2,543	1,77
380	229,4	10,60	0,95	2,243	1,66
390	237,7	9,47	0,96	2,036	1,57
400	245,3	8,55	0,97	1,883	1,51
410	252,3	7,77	0,98	1,767	1,45
420	258,7	7,11	0,99	1,675	1,41
430	264,8	6,54	1,00	1,600	1,37
440	270,5	6,04	1,01	1,539	1,34
450	275,9	5,61	1,01	1,488	1,31
460	281,1	5,23	1,02	1,445	1,29
470	286,0	4,89	1,02	1,407	1,27
480	290,8	4,58	1,03	1,375	1,25
490	295,4	4,31	1,03	1,347	1,24
500	299,9	4,06	1,03	1,322	1,22
520	308,4	3,63	1,04	1,281	1,20
540	316,5	3,28	1,04	1,247	1,18
560	324,3	2,98	1,05	1,220	1,16
580	331,7	2,72	1,05	1,197	1,15
600	338,8	2,50	1,05	1,177	1,14
620	345,7	2,31	1,05	1,161	1,13
640	352,4	2,14	1,06	1,146	1,12
660	358,9	1,99	1,06	1,134	1,11
680	365,2	1,85	1,06	1,122	1,10
700	371,3	1,72	1,06	1,113	1,10
			$p=2,8$		
110	1840	−0,64	858,4	0,158	95,86
120	1804	−0,60	813,5	0,174	99,90
130	1759	−0,58	762,2	0,190	101,6
140	1712	−0,57	711,4	0,208	102,2
150	1661	−0,57	659,6	0,226	101,9
160	1604	−0,56	606,3	0,245	100,8
170	1544	−0,55	552,6	0,267	99,08
180	1480	−0,54	499,9	0,290	96,86
190	1415	−0,52	449,5	0,315	94,29
200	1350	−0,50	402,2	0,343	91,50
210	1285	−0,48	358,4	0,374	88,54
220	1222	−0,46	318,0	0,408	85,45
230	1160	−0,43	280,9	0,446	82,21
240	1098	−0,39	247,2	0,488	78,94
250	1038	−0,36	216,1	0,535	75,48
260	977,1	−0,31	187,5	0,589	71,89
270	916,5	−0,27	161,1	0,652	68,14
280	855,2	−0,21	136,8	0,727	64,22
290	792,7	−0,14	114,4	0,820	60,11
300	728,5	−0,05	93,81	0,939	55,78
310	661,7	0,07	74,88	1,098	51,19
320	591,3	0,24	57,55	1,323	46,26
330	515,3	0,48	41,76	1,676	40,86

Table II.8 (*Continued*)

T	w	μ	k	η/η_0	γ/γ_0
			$p=2,8$		
340	429,7	0,91	27,37	2,329	34,71
350	184,6	17,55	0,83	5,100	2,43
360	201,3	14,48	0,88	3,555	2,09
370	214,0	12,42	0,91	2,862	1,89
380	224,5	10,89	0,93	2,460	1,75
390	233,6	9,68	0,95	2,194	1,65
400	241,7	8,70	0,96	2,006	1,57
410	249,1	7,89	0,98	1,865	1,50
420	255,9	7,20	0,99	1,756	1,45
430	262,2	6,61	0,99	1,669	1,41
440	268,2	6,10	1,00	1,598	1,37
450	273,8	5,65	1,01	1,539	1.34
460	279,2	5,26	1,01	1,489	1,32
470	284,3	4,91	1,02	1,447	1,29
480	289,2	4,60	1,02	1,411	1,27
490	294,0	4,32	1,03	1,379	1,26
500	298,5	4,07	1,03	1,351	1,24
520	307,3	3,63	1,04	1,305	1,22
540	315,6	3,27	1,04	1,268	1,19
560	323,4	2,97	1,05	1,237	1,18
580	331,0	2,71	1,05	1,212	1,16
600	338,2	2,49	1,05	1,191	1,15
620	345,2	2,30	1,05	1,173	1,14
640	352,0	2,13	1,06	1,157	1,13
660	358,6	1,97	1,06	1,143	1,12
680	364,9	1,84	1,06	1,131	1,11
700	371,1	1,71	1,06	1,120	1,10
			$p=3,0$		
110	1843	—0,64	804,0	0,158	89,68
120	1806	—0,60	761,3	0,174	93,39
130	1761	—0,58	712,9	0,190	94,99
140	1713	—0,57	665,3	0,207	95,51
150	1662	—0,57	616,8	0,226	95,21
160	1606	—0,56	567,1	0,245	94,19
170	1545	—0,55	516,9	0,266	92,58
180	1482	—0,54	467,7	0,289	90,50
190	1417	—0,52	420,6	0,315	88,12
200	1351	—0,50	376,5	0,342	85,51
210	1287	—0,48	335,5	0,373	82,76
220	1224	—0,46	297,9	0,407	79,89
230	1162	—0,43	263,4	0,444	76,91
240	1100	—0,39	231,8	0,486	73,82
250	1040	—0,36	202,7	0,533	70,62
260	979,7	—0,32	176,0	0,587	67,28
270	919,3	—0,27	151,4	0,649	63,79
280	858,4	—0,21	128,8	0,723	60,15
290	796,3	—0,14	107,9	0,815	56,34
300	732,5	—0,05	88,63	0,931	52,33
310	666,4	0,06	70,97	1,086	48,08
320	596,8	0,22	54,82	1,303	43,54
330	522,2	0,46	40,12	1,637	38,58

Table II.8 (*Continued*)

T	w	μ	k	α/α_0	γ/γ_0
			$p=3,0$		
340	438,8	0,85	26,78	2,237	32,98
350	335,8	1,75	14,50	3,829	26,20
360	193,1	15,26	0,85	4,340	2,30
370	207,7	12,88	0,89	3,270	2,02
380	219,4	11,20	0,92	2,718	1,85
390	229,3	9,90	0,94	2,377	1,72
400	238,0	8,86	0,95	2,143	1,63
410	245,8	8,01	0,97	1,973	1,56
420	253,0	7,29	0,98	1,843	1,50
430	259,6	6,68	0,99	1,741	1,45
440	265,9	6,15	1,00	1,659	1,41
450	271,7	5,69	1,00	1,592	1,37
460	277,3	5,29	1,01	1,535	1,34
470	282,6	4,93	1,02	1,488	1,32
480	287,6	4,61	1,02	1,447	1,30
490	292,5	4,33	1,02	1,411	1,28
500	297,2	4,07	1,03	1,381	1,26
520	306,2	3,63	1,03	1,329	1,23
540	314,6	3,27	1,04	1,288	1,21
560	322,6	2,96	1,04	1,255	1,19
580	330,3	2,70	1,05	1,228	1,17
600	337,7	2,48	1,05	1,204	1,16
620	344,8	2,28	1,05	1,185	1,15
640	351,6	2,11	1,06	1,168	1,14
660	358,2	1,96	1,06	1,153	1,13
680	364,7	1,82	1,06	1,140	1,12
700	370,9	1,69	1,06	1,128	1,11
			$p=3,2$		
110	1846	—0,64	756,4	0 158	84,27
120	1808	—0,60	715,5	0,174	87,69
130	1762	—0,58	669,7	0,190	89,16
140	1715	—0,57	624,9	0,207	89,63
150	1664	—0,57	579,4	0,225	89,35
160	1608	—0,56	532,8	0,245	88,40
170	1547	—0,55	485,7	0,266	86,89
180	1483	—0,54	439,5	0,289	84,94
190	1418	—0,52	395,4	0,314	82,71
200	1353	—0,50	353,9	0,342	80,28
210	1289	—0,48	315,6	0,372	77,70
220	1226	—0,46	280,3	0,406	75,01
230	1164	—0,43	247,9	0,443	72,24
240	1103	—0,40	218,2	0 485	69,35
250	1042	—0,36	191,0	0,531	66,35
260	982,3	—0,32	166,0	0,584	63,24
270	922,2	—0,27	142,9	0,646	59,99
280	861,5	—0,21	121,7	0,720	56,59
290	799,8	—0,14	102,1	0,809	53,04
300	736,5	—0,06	84,08	0,923	49,30
310	670,9	0,06	67,55	1,074	45,36
320	602,2	0,21	52,43	1,284	41,15
330	528,8	0,43	38,68	1,602	36,57
340	447,6	0,80	26,23	2,155	31,44

Table II.8 (*Continued*)

T	w	μ	k	α/α_0	ν/ν_0
			$p=3,2$		
350	350,6	1,57	14,89	3,478	25,30
360	183,7	16,21	0,82	5,622	2,56
370	201,0	13,40	0,87	3,814	2,18
380	214,1	11,52	0,90	3,034	1,96
390	224,9	10,12	0,93	2,586	1,81
400	234,2	9,02	0,94	2,296	1,70
410	242,5	8,12	0,96	2,091	1,61
420	250,1	7,38	0,97	1,937	1,55
430	257,1	6,74	0,98	1,819	1,49
440	263,6	6,20	0,99	1,724	1,44
450	269,6	5,73	1,00	1,647	1,41
460	275,4	5,31	1,01	1,584	1,37
470	280,9	4,95	1,01	1,530	1,35
480	286,1	4,63	1,02	1,484	1,32
490	291,1	4,34	1,02	1,445	1,30
500	295,9	4,08	1,03	1,411	1,28
520	305,1	3,63	1,03	1,354	1,25
540	313,7	3,26	1,04	1,309	1,22
560	321,8	2,95	1,04	1,273	1,20
580	329,6	2,69	1,05	1,243	1,18
600	337,1	2,47	1,05	1,218	1,17
620	344,3	2,27	1,05	1,197	1,16
640	351,2	2,10	1,06	1,178	1,14
660	357,9	1,95	1,06	1,162	1,13
680	364,4	1,81	1,06	1,148	1,13
700	370,7	1,68	1,06	1,136	1,12
			$p=3,4$		
110	1849	—0,64	714,5	0,158	79,50
120	1811	—0,60	675,1	0,174	82,66
130	1764	—0,58	631,7	0,190	84,02
140	1717	—0,57	589,3	0,207	84,45
150	1665	—0,57	546,4	0,225	84,18
160	1609	—0,56	502,5	0,245	83,28
170	1548	—0,55	458,1	0,266	81,86
180	1485	—0,54	414,6	0,289	80,03
190	1420	—0,52	373,1	0,314	77,94
200	1355	—0,50	334,1	0,341	75,65
210	1291	—0,48	298,0	0,372	73,23
220	1228	—0,46	264,7	0,405	70,72
230	1166	—0,43	234,3	0,442	68,11
240	1105	—0,40	206,3	0,483	65,40
250	1045	—0,36	180,7	0,530	62,59
260	984,9	—0,32	157,1	0,582	59,67
270	925,0	—0,27	135,5	0,643	56,63
280	864,6	—0,22	115,5	0,716	53,45
290	803,2	—0,15	97,03	0,804	50,12
300	740,4	—0,06	80,07	0,916	46,63
310	675,4	0,05	64,53	1,062	42,96
320	607,5	0,20	50,31	1,265	39,03
330	535,2	0,41	37,40	1,568	34,79
340	455,9	0,76	25,73	2,082	30,06
350	363,2	1,44	15,16	3,223	24,51

Table II.8 (*Continued*)

T	w	μ	k	a/a_0	v/v,
			$p=3,4$		
360	172,3	17,44	0,79	8,217	2,93
370	193,7	13,99	0,85	4,579	2,37
380	208,5	11,88	0,89	3,427	2,09
390	220,3	10,36	0,91	2,835	1,90
400	230,4	9,19	0,93	2,470	1,77
410	239,2	8,24	0,95	2,221	1,67
420	247,2	7,46	0,96	2,039	1,60
430	254,5	6,81	0,98	1,901	1,53
440	261,3	6,25	0,99	1,793	1,48
450	267,6	5,76	0,99	1,706	1,44
460	273,5	5,34	1,00	1,634	1,40
470	279,2	4,97	1,01	1,574	1,37
480	284,5	4,64	1,01	1,523	1,35
490	289,7	4,35	1,02	1,479	1,32
500	294,6	4,08	1,02	1,441	1,30
520	304,0	3,63	1,03	1,379	1,27
540	312,8	3,26	1,04	1,330	1,24
560	321,1	2,95	1,04	1,291	1,22
580	329,0	2,68	1,05	1,258	1,20
600	336,6	2,46	1,05	1,231	1,18
620	343,8	2,26	1,05	1,208	1,17
640	350,8	2,09	1,05	1,189	1,15
660	357,6	1,93	1,06	1,172	1,14
680	364,2	1,80	1,06	1,157	1,13
700	370,5	1,67	1,06	1,144	1,12
			$p=3,6$		
110	1852	−0,64	677,1	0,158	75,25
120	1813	−0,60	639,2	0,174	78,19
130	1766	−0,58	597,8	0,190	79,44
140	1718	−0,58	557,7	0,207	79,84
150	1667	−0,57	517,1	0,225	79,58
160	1611	−0,56	475,5	0,244	78,74
170	1550	−0,55	433,6	0,265	77,40
180	1486	−0,54	392,6	0,288	75,68
190	1422	−0,52	353,3	0,313	73,70
200	1357	−0,51	316,4	0,341	71,54
210	1293	−0,48	282,3	0,371	69,26
220	1230	−0,46	250,9	0,404	66,89
230	1168	−0,43	222,1	0,441	64,43
240	1107	−0,40	195,7	0,482	61,89
250	1047	−0,36	171,5	0,528	59,25
260	987,5	−0,32	149,3	0,580	56,50
270	927,8	−0,27	128,8	0,640	53,64
280	867,7	−0,22	109,9	0,712	50,65
290	806,6	−0,15	92,51	0,799	47,53
300	744,2	−0,07	76,50	0,908	44,26
310	679,8	0,04	61,84	1,051	40,82
320	612,7	0,18	48,42	1,248	37,15
330	541,5	0,39	36,25	1,537	33,19
340	464,0	0,72	25,26	2,015	28,82
350	374,9	1,33	15,37	3,015	23,76

Table II.8 (*Continued*)

T	w	μ	k	α/α_0	ν/ν
			$p=3,6$		
360	253,5	3,30	6,19	7,649	16,94
370					
380	202,6	12,25	0,87	3,931	2,23
390	215,6	10,60	0,90	3,129	2,01
400	226,5	9,35	0,92	2,667	1,85
410	235,9	8,36	0.94	2,365	1,74
420	244,3	7,55	0,96	2,150	1,65
430	251,9	6,87	0,97	1,990	1,58
440	259,0	6,30	0,98	1,866	1,52
450	265,5	5,80	0,99	1,767	1,47
460	271,7	5,37	1,00	1,686	1,43
470	277,5	4,99	1,01	1,619	1,40
480	283,0	4,65	1,01	1,562	1,37
490	288,3	4,35	1,02	1,514	1,34
500	295,4	4,09	1,02	1,473	1,32
520	302,9	3,63	1,03	1,405	1,28
540	311,9	3,25	1,04	1,351	1,25
560	320,3	2,94	1,04	1,309	1,23
580	328,4	2,67	1 05	1,274	1,21
600	336,0	2,44	1 05	1,245	1,19
620	343,4	2,25	1 05	1,220	1,17
640	350,5	2,07	1 05	1,199	1,16
660	357,3	1,92	1,06	1,181	1,15
680	363,9	1,78	1,06	1,165	1,14
700	370,3	1,66	1,06	1,151	1,13
			$p=3,8$		
110	1855	—0,64	643,6	0,157	71,45
120	1815	—0,60	607,0	0.174	74,18
130	1768	—0,58	567,5	0,190	75,35
140	1720	—0,58	529,3	0,207	75,72
150	1668	—0,57	490,8	0,225	75,47
160	1612	—0,56	451,4	0,244	74,67
170	1551	—0,55	411,7	0,265	73,40
180	1488	—0,54	372,8	0,288	71,78
190	1423	—0,52	335,6	0,313	69,90
200	1358	—0,51	300,6	0,340	67,86
210	1294	—0,48	268,3	0,370	65,71
220	1232	—0,46	238,5	0,403	63,47
230	1170	—0,43	211,2	0,440	61,15
240	1109	—0,40	186,2	0,481	58,75
250	1049	—0,36	163,3	0,526	56,26
260	990,0	—0,32	142,2	0,578	53,66
270	930,6	—0,28	122,8	0,638	50,96
280	870,7	—0,22	104,9	0,708	48,15
290	810,0	—0,16	88,46	0,794	45,21
300	748,0	—0,07	73,30	0,901	42,13
310	684,2	0,03	59,41	1,041	38,89
320	617,7	0,17	46,72	1.231	35,46
330	547,6	0,37	35,21	1,507	31,76
340	471,8	0,68	24,83	1,955	27,70
350	385,8	1,24	15,52	2,843	23,05

Table II.8 (*Continued*)

T	w	μ	k	α/α_0	ν/ν_t
			$p=3,8$		
360	275,1	2,77	7,06	6,039	17,10
370	176.3	15,42	0,81	7,794	2,94
380	196,3	12,65	0,86	4,602	2,41
390	210,8	10,85	0,89	3,484	2,12
400	222,5	9,52	0,92	2,894	1,94
410	232,5	8,48	0,94	2,524	1,81
420	241,4	7,64	0,95	2,270	1,71
430	249,4	6,94	0,97	2,084	1,63
440	256,7	6,34	0,98	1,942	1,56
450	263,5	5,83	0,99	1,830	1,51
460	269,9	5,39	0,99	1,740	1,47
470	275,9	5,00	1,00	1,665	1,43
480	281,5	4,66	1,01	1,603	1,40
490	286,9	4,36	1,01	1,550	1,37
500	292,1	4,09	1,02	1,504	1,34
520	301,9	3,62	1,03	1,430	1,30
540	311,0	3,24	1,03	1,373	1,27
560	319,6	2,93	1,04	1,327	1,24
580	327,7	2,66	1,04	1,289	1,22
600	335,5	2,43	1,05	1,258	1,20
620	343,0	2,23	1,05	1,232	1,18
640	350,1	2,06	1,05	1,210	1,17
660	357,1	1,91	1,06	1,190	1,16
680	363,7	1,77	1.06	1,174	1,15
700	370,2	1,64	1,06	1,159	1.14
			$p=4,0$		
110	1858	—0,64	613,5	0,157	68,03
120	1817	—0,60	578,1	0,174	70,58
130	1769	—0,58	540,2	0,190	71,67
140	1721	—0,58	503,8	0,207	72,01
150	1670	—0 57	467,2	0,225	71,77
160	1614	—0,56	429,7	0,244	71,01
170	1553	—0,55	392,0	0,265	69,80
180	1490	—0,54	355,0	0,287	68,26
190	1425	—0,53	319,6	0,312	66,49
200	1360	—0,51	286,4	0,339	64,55
210	1296	—0,48	255,6	0,369	62,51
220	1233	—0,46	227,4	0,402	60,39
230	1172	—0,43	201,4	0,439	58,20
240	1111	—0,40	177,7	0,479	55,92
250	1052	—0,36	155,9	0,525	53,56
260	992,5	—0,32	135,9	0,576	51,11
270	933,3	—0,28	117,4	0,635	48,56
280	873,7	—0,22	100,5	0,705	45,90
290	813,4	—0,16	84,81	0,789	43,12
300	751,8	—0,08	70,42	0,894	40,22
310	688,4	0,02	57,22	1,030	37,16
320	622,7	0,16	45,18	1,215	33,93
330	553,5	0,35	34,27	1,479	30,47
340	479,2	0,64	24,44	1,899	26,68
350	396,1	1,15	15,64	2,696	22,39

Table II.8 (*Continued*)

T	w	μ	k	α/α_0	γ/γ_0

$$p=4,0$$

T	w	μ	k	α/α_0	γ/γ_0
360	293,3	2,41	7,75	5,111	17,09
370	165,3	16,24	0,79	12,52	3,42
380	189,6	13,07	0,84	5,542	2,61
390	205,7	11,10	0,88	3,919	2,26
400	218,4	9,69	0,91	3,155	2,03
410	229,1	8,60	0,93	2,702	1,88
420	238,4	7,72	0,95	2,401	1,76
430	246,8	7,00	0,96	2,185	1,68
440	254,4	6,39	0,97	2,024	1,60
450	261,5	5,86	0,98	1,897	1,55
460	268,1	5,41	0,99	1,796	1,50
470	274,2	5,02	1,00	1,714	1,46
480	280,1	4,67	1,01	1,645	1,42
490	285,6	4,36	1,01	1,587	1,39
500	290,9	4,09	1,02	1,537	1,36
520	300,9	3,62	1,03	1,457	1,32
540	310,2	3,24	1,03	1,394	1,28
560	318,9	2,92	1,04	1,345	1,25
580	327,1	2,65	1,04	1,305	1,23
600	335,0	2,42	1,05	1,272	1,21
620	342,6	2,22	1,05	1,244	1,19
640	349,8	2,05	1,05	1,220	1,18
660	356,8	1,89	1,06	1,200	1,17
680	363,5	1,76	1,06	1,182	1,15
700	370,0	1,63	1,06	1,166	1,14

$$p=4,2$$

T	w	μ	k	α/α_0	γ/γ_0
110	1861	—0,64	586,2	0,157	64,93
120	1819	—0,60	551,9	0,174	67,32
130	1771	—0,58	515,6	0,190	68,33
140	1723	—0,58	480,8	0,206	68,65
150	1671	—0,57	445,8	0,224	68,42
160	1615	—0,56	410,1	0,244	67 69
170	1555	—0,55	374,1	0,264	66,55
180	1491	—0,54	338,9	0,287	65,08
190	1427	—0,53	305,2	0,312	63,39
200	1362	—0 51	273,5	0,339	61,56
210	1298	—0,49	244,2	0,369	59,62
220	1235	—0,46	217,3	0,402	57,60
230	1174	—0,43	192,6	0,438	55,52
240	1114	—0,40	170,0	0,478	53,36
250	1054	—0,37	149,2	0,523	51,12
260	995,0	—0,33	130,1	0,574	48,80
270	936,0	—0,28	112,6	0,632	46,38
280	876,7	—0,23	96,40	0,701	43,86
290	816,7	—0,16	81,50	0,784	41,23
300	755,5	—0,08	67,80	0,888	38,48
310	692,7	0,02	55,24	1,020	35,60
320	627,6	0,15	43,79	1,199	32,55
330	559,4	0,33	33,40	1,453	29,29
340	486 5	0,61	24,06	1,849	25,75
350	405,8	1,08	15,73	2,570	21,77

Table II.8 (*Continued*)

T	w	μ	k	a/a_0	γ/γ_0
			$p=4,2$		
360	309,2	2,14	8,31	4,493	16,99
370	150,7	16,50	0,81	41,79	4,51
380	182,5	13,49	0,83	6,950	2,87
390	200,5	11,36	0,87	4,466	2,41
400	214,3	9,85	0,90	3,459	2,14
410	225,7	8,71	0,92	2,901	1,96
420	235,5	7,80	0,94	2,544	1,83
430	244,3	7,05	0,95	2,294	1,73
440	252,2	6,43	0,97	2,109	1,65
450	259,5	5,89	0,98	1,967	1,58
460	266,3	5,43	0,99	1,855	1,53
470	272,6	5,03	1,00	1,763	1,49
480	278,6	4,68	1,00	1,688	1,45
490	284,3	4,37	1,01	1,624	1,41
500	289,7	4,09	1,02	1,570	1,38
520	299,9	3,62	1,02	1,483	1,34
540	309,3	3,23	1,03	1,416	1,30
560	318,2	2,91	1,04	1,363	1,27
580	326,6	2,64	1,04	1,320	1,24
600	334,5	2,41	1,05	1,285	1,22
620	342,2	2,21	1,05	1,255	1,20
640	349,5	2,04	1,05	1,230	1,19
660	356,5	1,88	1,06	1,209	1,17
680	363,3	1,74	1,06	1,190	1,16
700	369,9	1,62	1,06	1,173	1,15
			$p=4,24746$		
110	1862,1	—0,64	580,2	0,157	64,24
120	1819,5	—0,60	546,0	0,174	66,59
130	1771,3	—0,58	510,0	0,190	67,59
140	1723,0	—0,58	475,6	0,206	67,90
150	1671,6	—0,57	441,0	0,224	67,67
160	1615,4	—0,56	405,7	0,244	66,95
170	1554,9	—0,55	370,1	0,264	65,82
180	1491,5	—0,54	335,3	0,287	64,37
190	1426,9	—0,53	301,9	0,312	62,70
200	1362,3	—0,51	270,7	0,339	60,89
210	1298,4	—0,49	241,7	0,368	58,97
220	1235,8	—0,46	215,0	0,401	56,98
230	1174,4	—0,43	190,6	0,438	54,92
240	1114,1	—0,40	168,2	0,478	52,79
250	1054,6	—0,37	147,7	0,522	50,58
260	995,6	—0,33	128,8	0,573	48,28
270	936,7	—0,28	111,5	0,632	45,89
280	877,4	—0,23	95,50	0,700	43,40
290	817,5	—0,16	80,76	0,783	40,80
300	756,4	—0,09	67,22	0,886	38,09
310	693,7	0,01	54,80	1,018	35,24
320	628,7	0,15	43,48	1,195	32,24
330	560,7	0,33	33,21	1,447	29.03
340	488,2	0,60	23,98	1,837	25,54
350	408,1	1,06	15,75	2,543	21,63

Table II.8 (*Continued*)

T	w	μ	k	α/α_0	v/v_0
			$p=4,24746$		
360	312,7	2,08	8,43	4,375	16,96
370	145,7	15,36	0,88	162,3	5,42
380	180,7	13,58	0,83	7,393	2,95
390	199,3	11,42	0,87	4,616	2,44
400	213,3	9,89	0,90	3,539	2,16
410	224,8	8,74	0,92	2,952	1,98
420	234,8	7,82	0,94	2,580	1,84
430	243,7	7,07	0,95	2,321	1,74
440	251,7	6,44	0,97	2,131	1,66
450	259,0	5,90	0,98	1,985	1,59
460	265,8	5,44	0,99	1,869	1,54
70	272,2	5,04	1,00	1,775	1,49
180	278,3	4,68	1,00	1,698	1,45
490	284,0	4,37	1,01	1,633	1,42
500	289,4	4,09	1,01	1,578	1,39
520	299,7	3,62	1,02	1,489	1,34
540	309,1	3,23	1,03	1,421	1,30
560	318,0	2,91	1,04	1,367	1,27
580	326,4	2,63	1,04	1,324	1,25
600	334,4	2,40	1,05	1,288	1,22
620	342,1	2,21	1,05	1,258	1,21
640	349,4	2,03	1,05	1,233	1,19
660	356,5	1,88	1,06	1,211	1,17
680	363,3	1,74	1,06	1,192	1,16
700	369,8	1,61	1,06	1,175	1,15
			$p=4,4$		
110	1864	—0,64	561,5	0,157	62,12
120	1821	—0,60	528,0	0,173	64,35
130	1773	—0,58	493,1	0,189	65,30
140	1724	—0,58	459,8	0,206	65,60
150	1673	—0,57	426,3	0,224	65,37
160	1617	—0,56	392,2	0,243	64,68
170	1556	—0,55	357,9	0,264	63,59
180	1493	—0,54	324,2	0,287	62,19
190	1428	—0,53	292,0	0,311	60,58
200	1364	—0,51	261,8	0,338	58,83
210	1300	—0,49	233,8	0,368	56,99
220	1237	—0,46	208,1	0,401	55,07
230	1176	—0,43	184,5	0,437	53,09
240	1116	—0,40	162,9	0,477	51,03
250	1056	—0,37	143,1	0,521	48,91
260	997,5	—0,33	124,9	0,572	46,69
270	938,7	—0,28	108,2	0,630	44,40
280	879,7	—0,23	92,71	0,698	42,00
290	820,0	—0,17	78,50	0,779	39,51
300	759,2	—0,09	65,42	0,881	36,90
310	696,8	0,01	53,44	1,011	34,17
320	632,4	0,14	42,51	1,184	31,29
330	565,0	0,31	32,61	1,428	28,21
340	493,5	0,58	23,72	1,802	24,89
350	415,1	1,01	15,79	2,460	21,19

Table II.8 (*Continued*)

T	w	μ	k	a/a_0	v/γ_0
			$p=4,4$		
360	323,4	1,93	8,78	4,047	16,83
370	189,3	6,23	2,43	21,25	10.30
380	174,9	13,84	0,82	9.273	3.21
390	195,2	11,60	0,86	5,167	2.58
400	210,1	10,01	0,89	3,817	2,25
410	222,2	8,82	0,91	3,125	2,04
420	232,6	7,88	0,93	2,700	1,89
430	241,7	7,11	0,95	2,410	1.78
440	250,0	6,47	0,96	2.200	1,69
450	257,5	5,92	0,97	2,041	1.62
460	264.5	5,45	0,98	1,916	1.56
470	271,0	5,05	0,99	1,815	1,52
480	277,2	4,69	1.00	1,732	1,47
490	283,0	4,37	1,01	1,663	1,44
500	288,5	4,09	1,01	1,604	1,41
520	298,9	3,61	1,02	1.510	1,36
540	308,5	3,22	1,03	1,438	1,31
560	317,5	2,90	1,04	1,381	1,28
580	326,0	2,63	1,04	1,336	1,25
600	334,1	2,39	1,05	1.298	1,23
620	341,8	2,20	1,05	1.267	1,21
640	349,2	2,02	1,05	1.240	1,20
660	356,3	1,87	1,06	1,218	1.18
680	363,1	1,73	1,06	1,198	1,17
700	369,7	1,60	1,06	1,181	1,16
			$p=4,6$		
100	—	—	—	—	—
110	1867	—0,64	538,8	0,157	59,54
120	1823	—0,60	506,3	0,173	61,64
130	1774	—0,58	472,6	0,189	62,53
140	1726	—0,58	440,6	0,206	62.81
150	1674	—0,57	408,6	0,224	62,59
160	1618	—0,56	375,9	0,243	61.93
170	1558	—0,55	343,0	0,264	60,88
180	1494	—0,54	310,8	0,286	59,55
190	1430	—0,53	280,0	0,311	58.02
200	1365	—0,51	251,1	0,338	56.35
210	1302	—0,49	224,4	0,367	54.59
220	1239	—0,46	199,7	0,400	52,76
230	1178	—0,43	177,2	0,436	50.86
240	1118	—0,40	156,5	0,475	48,91
250	1059	—0,37	137,6	0,520	46,88
260	1000	—0,33	120,1	0,570	44,77
270	941,4	—0,29	104,1	0,627	42.58
280	882,6	—0,23	89,34	0,694	40,31
290	823,2	—0,17	75,75	0,775	37.93
300	762,8	—0,09	63,24	0,874	35,45
310	700,9	0,00	51,79	1,001	32.86
320	637,1	0,13	41,34	1,170	30.13
330	570,6	0,30	31,89	1,405	27,23
340	500,3	0,55	23,39	1,758	24.09
350	423,9	0,95	15,84	2,362	20,64

Table II.8 (*Continued*)

T	w	μ	k	α/α_0	γ/γ_0
			$p=4{,}6$		
360	336,4	1,76	9,19	3,706	16,64
370	220,2	4,52	3,34	11,12	11,27
380	167,1	13,95	0,83	13,66	3,70
390	189,8	11,80	0,85	6,092	2,78
400	205,9	10,16	0,88	4,242	2,38
410	218,7	8,92	0,91	3,378	2,14
420	229,6	7,95	0,93	2,870	1,96
430	239,2	7,16	0,95	2,535	1,84
440	247,8	6,50	0,96	2,296	1,74
450	255,6	5,95	0,97	2,118	1,66
460	262,8	5,47	0,98	1,979	1,60
470	269,5	5,06	0,99	1,868	1,55
480	275,8	4,69	1,00	1,777	1,50
490	281,7	4,37	1,01	1,702	1,46
500	287,4	4,09	1,01	1,638	1,43
520	298,0	3,61	1,02	1,537	1,37
540	307,7	3,21	1,03	1,460	1,33
560	316,9	2,89	1,04	1,400	1,30
580	325,5	2,61	1,04	1,351	1,27
600	333,6	2,38	1,05	1,311	1,24
620	341,4	2,18	1,05	1,278	1,22
640	348,9	2,01	1,05	1,250	1,20
660	356,0	1,86	1,06	1,227	1,19
680	362,9	1,72	1,06	1,206	1,17
700	369,6	1,59	1,06	1,188	1,16
			$p=4{,}8$		
110	1870	—0,64	518,0	0,157	57,18
120	1825	—0,60	486,3	0,173	59,16
130	1776	—0,58	453,8	0,189	59,99
140	1727	—0,58	423,0	0,206	60,25
150	1676	—0,57	392,3	0,224	60,04
160	1619	—0,56	361,0	0,243	59,40
170	1559	—0,55	329,4	0,263	58,41
180	1496	—0,54	298,6	0,286	57,13
190	1431	—0,53	269,0	0,310	55,66
200	1367	—0,51	241,3	0,337	54,07
210	1303	—0,49	215,7	0,366	52,38
220	1241	—0,46	192,1	0,399	50,63
230	1180	—0,44	170,4	0,435	48,83
240	1120	—0,40	150,6	0,474	46,96
250	1061	—0,37	132,5	0,518	45,02
260	1002	—0,33	115,8	0,568	43,01
270	944,1	—0,29	100,4	0,624	40,92
280	885,5	—0,24	86,25	0,691	38,75
290	826,4	—0,17	73,22	0,770	36,49
300	766,4	—0,10	61,24	0,868	34,13
310	705,0	—0,01	50,27	0,992	31,66
320	641,7	0,12	40,27	1,156	29,07
330	576,1	0,28	31,21	1,382	26,32
340	506,9	0,52	23,08	1,718	23,31
350	432,4	0,90	15,86	2,275	20,12

Table II.8 (*Continued*)

T	w	μ	k	α/α_0	γ/γ_0
		$p=4,8$			
360	348,5	1,61	9,53	3,436	16,43
370	243,3	3,66	4,05	8,027	11,76
380	160,5	13,26	0,88	22,93	4,45
390	184,4	11,93	0,85	7,341	3,03
400	201,7	10,28	0,88	4,749	2,52
410	215,3	9,01	0,90	3,663	2,23
420	226,8	8,01	0,92	3,057	2,04
430	236,7	7,20	0,94	2,669	1,90
440	245,6	6,53	0,96	2,398	1,79
450	253,7	5,97	0,97	2,198	1,71
460	261,1	5,48	0,98	2,044	1,64
470	268,0	5,06	0,99	1,923	1,58
480	274,4	4,70	1,00	1,824	1,53
490	280,5	4,37	1,00	1,742	1,49
500	286,3	4,09	1,01	1,673	1,45
520	297,1	3,60	1,02	1,564	1,39
540	307,0	3,20	1,03	1,482	1,35
560	316,2	2,88	1,04	1,418	1,31
580	324,9	2,60	1,04	1,367	1,28
600	333,2	2,37	1,05	1,325	1,25
620	341,1	2,17	1,05	1,290	1,23
640	348,6	2,00	1,05	1,260	1,21
660	355,8	1,84	1,06	1,235	1,20
680	362,8	1,71	1,06	1,214	1,18
700	369,5	1,58	1,06	1,195	1,17
		$p=5,0$			
110	1873	--0,64	498,9	0,157	55,01
120	1827	—0,60	467,9	0,173	56,87
130	1777	—0,58	436,5	0,189	57,66
140	1729	—0,58	406,8	0,206	57,89
150	1677	—0,57	377,3	0,224	57,69
160	1621	—0,56	347,2	0,243	57,08
170	1561	—0,55	316,9	0,263	56,13
180	1497	—0,54	287,3	0,285	54,91
190	1433	—0,53	258,9	0,310	53,50
200	1369	—0,51	232,3	0,336	51,97
210	1305	—0,49	207,7	0,366	50,36
220	1243	—0,46	185,0	0,398	48,68
230	1182	—0,44	164,2	0,434	46,95
240	1122	—0,41	145,2	0,473	45,16
250	1063	—0,37	127,8	0,516	43,31
260	1005	—0,33	111,7	0,566	41,39
270	946,7	—0,29	96,99	0,622	39,39
280	888,4	—0,24	83,40	0,687	37,32
290	829,6	—0,18	70,90	0,766	35,16
300	769,9	—0,10	59,40	0,862	32,91
310	708,9	—0,01	48,87	0,983	30,56
320	646,3	0,11	39,27	1,143	28,09
330	581,4	0,27	30,59	1,361	25,48
340	513,3	0,49	22,79	1,680	22,68
350	440,5	0,85	15,88	2,197	19,63

Table II.8 (*Continued*)

T	w	μ	k	α/α_0	ν/ν_0
		$p=5,0$			
360	359,7	1,49	9.83	3,215	16,20
370	262,5	3,11	4,64	6,471	12,04
380	160,3	10,98	1,07	32,95	5,64
390	179.3	11,92	0,86	9,048	3,32
400	197,6	10,37	0.87	5,358	2,68
410	212,0	9,08	0,90	3,986	2,34
420	223,9	8,07	0.92	3,262	2,12
430	234,3	7,24	0,94	2,813	1,96
440	243,5	6,56	0.95	2,506	1,84
450	251,8	5,99	0,97	2,282	1,75
460	259,4	5,50	0,98	2,112	1,67
470	266,5	5,07	0,99	1,979	1,61
480	273,1	4,70	1,00	1,871	1,56
490	279,3	4,37	1,00	1.783	1,51
500	285,2	4,08	1,01	1,709	1,47
520	296,2	3,59	1,02	1,592	1,41
540	306,2	3,19	1.03	1,504	1,36
560	315,6	2,87	1,04	1,436	1,32
580	324,4	2,59	1,04	1,382	1,29
600	332,8	2,36	1,05	1,338	1.26
620	340,7	2,16	1,05	1,301	1,24
640	348,3	1,98	1,05	1,270	1,22
660	355,6	1,83	1,06	1,244	1,20
680	362,6	1,69	1.06	1,221	1,19
700	369,3	1,57	1.06	1,202	1,18
		$p=5,5$			
110	1880	−0,64	457,1	0,157	50,26
120	1832	−0,60	427,8	0,173	51,88
130	1781	−0,58	398,7	0,189	52,56
140	1732	−0,58	371,5	0,205	52,76
150	1681	−0,57	344,6	0,223	52,57
160	1624	−0,56	317,2	0,242	52,02
170	1564	−0,56	289,6	0,262	51,15
180	1501	−0,54	262,6	0,284	50,04
190	1437	−0,53	236.8	0,309	48,77
200	1373	−0,51	212,6	0,335	47,39
210	1310	−0,49	190.2	0,364	45,93
220	1247	−0,46	169,6	0,396	44,42
230	1187	−0,44	150,7	0,431	42,86
240	1127	−0,41	133,4	0,470	41,25
250	1069	−0,37	117,5	0,512	39,58
260	1011	−0,34	102,9	0,561	37,85
270	953,2	−0,29	89,52	0,615	36,06
280	895,5	−0,24	77,18	0,679	34,19
290	837,4	−0,19	65,81	0,755	32,25
300	778,6	−0,12	55,37	0.847	30,24
310	718,7	−0,03	45,80	0,962	28,14
320	657,4	0,08	37,09	1,111	25.94
330	594,2	0,23	29,20	1,312	23,63
340	528,6	0,44	22,13	1,595	21,17
350	459,5	0,74	15,87	2,032	18,53

Table II.8 (*Continued*)

T	w	μ	k	$\alpha/\alpha_.$	γ/γ_0

$p=5,5$

T	w	μ	k	$\alpha/\alpha_.$	γ/γ_0
360	384,9	1,25	10,42	2,802	15,62
370	300,9	2,30	5,78	4,601	12,31
380	203,3	5,55	2,16	12,96	8,15
390	172,3	10,58	0,97	15,12	4,39
400	188,7	10,27	0,89	7,427	3,18
410	204,1	9,15	0,90	4,984	2,65
420	217,2	8,14	0,92	3,860	2,34
430	228,5	7,31	0,93	3,217	2,13
440	238,4	6,61	0,95	2,801	1,98
450	247,3	6,02	0,96	2,509	1,86
460	255,4	5,51	0,97	2,293	1,77
470	262,9	5,08	0,98	2,127	1,69
480	269,9	4,70	0,99	1,995	1,63
490	276,4	4,37	1,00	1,888	1,58
500	282,6	4,07	1,01	1,800	1,53
520	294,0	3,57	1,02	1,662	1,46
540	304,5	3,17	1,03	1,561	1,40
560	314,2	2,84	1,04	1,482	1,36
580	323,3	2,56	1,04	1,421	1,32
600	331,8	2,33	1,05	1,371	1,29
620	340,0	2,13	1,05	1,329	1,26
640	347,7	1,95	1,05	1,295	1,24
660	355,2	1,80	1,06	1,266	1,22
680	362,3	1,66	1,06	1,241	1,21
700	369,1	1,54	1,06	1,219	1,19

$p=6,0$

T	w	μ	k	$\alpha/\alpha_.$	γ/γ_0
110	1887	−0,64	422,2	0,157	46,30
120	1837	−0,60	394,3	0,173	47,71
130	1785	−0,58	367,2	0,188	48,30
140	1736	−0,58	342,1	0,205	48,47
150	1684	−0,57	317,3	0,222	48,30
160	1628	−0,56	292,1	0,241	47,79
170	1568	−0,56	266,8	0,261	47,00
180	1505	−0,54	242,1	0,283	45,99
190	1441	−0,53	218,4	0,307	44,83
200	1377	−0,51	196,2	0,334	43,57
210	1314	−0,49	175,6	0,362	42,24
220	1252	−0,47	156,7	0,394	40,86
230	1192	−0,44	139,4	0,428	39,44
240	1132	−0,41	123,5	0,466	37,98
250	1074	−0,38	108,9	0,509	36,47
260	1017	−0,34	95,58	0,556	34,90
270	959,6	−0,30	83,29	0,609	33,27
280	902,5	−0,25	71,98	0,671	31,58
290	845,0	−0,19	61,56	0,745	29,82
300	787,0	−0,13	51,99	0,833	28,00
310	728,1	−0,04	43,23	0,943	26,11
320	668,1	0,06	35,24	1,083	24,13
330	606,5	0,20	28,02	1,267	22,07
340	543,0	0,39	21,55	1,523	19,88
350	477,0	0,65	15,83	1,899	17,56

Table II.8 (*Continued*)

T	w	μ	k	a/a_0	v/v_0

$p=6,0$

360	407,1	1,07	10,85	2,511	15,05
370	331,5	1,83	6,62	3,709	12,29
380	248,3	3,57	3,25	7,033	9,12
390	184,2	7,56	1,34	14,88	5,72
400	184,6	9,38	0,97	9,921	3,84
410	198,2	8,93	0,92	6,244	3,04
420	211,5	8,09	0,92	4,582	2,60
430	223,3	7,29	0.93	3,687	2,33
440	233,8	6,61	0,95	3,133	2,13
450	243,2	6,02	0,96	2,758	1,99
460	251,8	5,51	0,97	2,488	1,87
470	259,7	5,07	0,98	2,284	1,78
480	267,0	4,69	0,99	2,126	1,71
490	273,8	4,35	1,00	1,998	1,65
500	280,3	4,05	1,01	1,894	1,59
520	292,1	3,55	1,02	1,734	1,51
540	302,9	3,14	1,03	1,617	1,44
560	312,9	2,81	1,04	1,529	1,39
580	322,2	2,53	1,04	1,459	1,35
600	331,0	2,30	1,05	1,403	1,32
620	339,3	2,10	1.05	1,357	1,29
640	347,2	1,92	1,06	1,319	1,26
660	354,8	1,77	1,06	1,287	1,24
680	362,0	1,63	1,06	1,259	1,22
700	368,9	1,51	1,06	1,235	1,21

$p=6,5$

110	1894	—0,64	392,6	0,157	42,94
120	1841	—0,60	365,9	0,172	44,18
130	1789	—0,58	340,4	0,188	44,70
140	1739	—0,58	317,1	0,204	44,85
150	1687	—0,57	294,2	0,222	44,68
160	1631	--0,57	270,9	0,241	44,21
170	1571	—0,56	247,6	0,261	43,49
180	1509	—0,54	224,7	0,282	42,56
190	1445	—0,53	202,8	0,306	41,49
200	1381	—0,51	182,3	0,332	40,33
210	1318	—0,49	163,3	0,361	39,12
220	1257	—0,47	145,8	0,392	37,85
230	1196	—0,44	129,8	0,426	36,55
240	1137	—0,41	115,1	0,463	35,21
250	1080	—0,38	101,7	0,505	33,83
260	1023	—0,34	89,36	0,551	32,39
270	965,9	—0,30	78,01	0,604	30,91
280	909,3	—0,26	67,57	0,664	29,36
290	852,5	—0,20	57,95	0,735	27,77
300	795,2	—0,14	49,12	0,820	26,11
310	737,3	—0,06	41,03	0,924	24,39
320	678,3	0,04	33,66	1,056	22,60
330	618,3	0,17	27,00	1,227	20,73
340	556,7	0,34	21,04	1,459	18,77
350	493,2	0,58	15,76	1,788	16,71

Table II.8 (*Continued*)

T	w	μ	k	α/α_0	γ/γ
			$p=6,5$		
360	427,1	0,93	11,17	2,294	14,50
370	357,5	1,52	7,27	3,171	12,13
380	283,6	2,66	4,11	5,049	9,54
390	214,4	5,10	1,96	9,638	6,76
400	189,7	7,72	1,16	10,89	4,60
410	196,1	8,27	0,99	7,502	3,49
420	207,7	7,82	0,95	5,377	2,90
430	219,4	7,17	0,95	4,204	2,54
440	230,1	6,54	0,95	3,495	2,30
450	239,8	5,97	0,96	3,026	2,12
460	248,7	5,48	0,97	2,695	1,98
470	256,9	5,04	0,98	2,450	1,88
480	264,4	4,66	0,99	2,261	1,79
490	271,5	4,32	1,00	2,112	1,72
500	278,2	4,02	1,01	1,991	1,66
520	290,4	3,52	1,02	1,807	1,56
540	301,6	3,11	1,03	1,674	1,48
560	311,8	2,78	1,04	1,575	1,43
580	321,4	2,50	1,04	1,497	1,38
600	330,3	2,26	1,05	1,435	1,34
620	338,8	2,06	1,05	1,384	1,31
640	346,8	1,89	1,06	1,342	1,28
660	354,5	1,74	1,06	1,307	1,26
680	361,8	1,60	1,06	1,277	1,24
700	368,8	1,48	1,07	1,251	1,22
			$p=7,0$		
110	1900	—0,64	367,2	0,156	40,05
120	1846	—0,60	341,6	0,172	41,15
130	1793	—0,58	317,5	0,188	41,61
140	1743	—0,58	295,7	0,204	41,73
150	1691	—0,57	274,3	0,221	41,58
160	1635	—0,57	252,7	0,240	41,15
170	1575	—0,56	231,0	0,260	40,48
180	1512	—0,55	209,8	0,281	39,62
190	1449	—0,53	189,4	0,305	38,63
200	1385	—0,51	170,4	0,331	37,56
210	1322	—0,49	152,7	0,359	36,44
220	1261	—0,47	136,4	0,390	35,27
230	1201	—0,44	121,6	0,423	34,07
240	1142	—0,42	108,0	0,460	32,84
250	1085	—0,38	95,47	0,501	31,57
260	1028	—0,35	84,01	0,547	30,25
270	972,0	—0,31	73,47	0,598	28,88
280	916,0	—0,26	63,77	0,657	27,46
290	859,8	—0,21	54,85	0,725	26,00
300	803,3	—0,15	46,65	0,807	24,48
310	746,1	—0,07	39,14	0,907	22,90
320	688,3	0,02	32,30	1,031	21,27
330	629,5	0,14	26,11	1,191	19,58
340	569,6	0,30	20,57	1,403	17,81
350	508,4	0,51	15,67	1,695	15,95

Table II.8 (*Continued*)

T	w	μ	k	α/σ_0	γ/γ
			$p=7,0$		
360	445,4	0,82	11,40	2,123	13,99
370	380,3	1,30	7,78	2,806	11,92
380	312,9	2,12	4,81	4,054	9,71
390	247,4	3,68	2,63	6,619	7,40
400	205,5	5,95	1,47	9,466	5,31
410	199,5	7,22	1,11	8,170	3,99
420	206,9	7,31	1.01	6,093	3,24
430	217,1	6,91	0,98	4,724	2,78
440	227,4	6,39	0,97	3,868	2,48
450	237,1	5,88	0,97	3,303	2,26
460	246,2	5,41	0.98	2,909	2,10
470	254,5	4,98	0,99	2,620	1,98
480	262,3	4,61	1,00	2,400	1,87
490	269,6	4,28	1,00	2,227	1,79
500	276,4	3,98	1,01	2.089	1,72
520	289,0	3,48	1,02	1,880	1,61
540	300,4	3,08	1,03	1,731	1,53
560	310,9	2,74	1.04	1,620	1,46
580	320,6	2,47	1,05	1,534	1,41
600	329,8	2,23	1,05	1,466	1,37
620	338,4	2,03	1,06	1,411	1,33
640	346,5	1,86	1,06	1,365	1,30
660	354,3	1,71	1,06	1,327	1,28
680	361,7	1,58	1,07	1,294	1,25
700	368,8	1,45	1,07	1.266	1,24
			$p=7,5$		
110	1907	—0,64	345,1	0,156	37,55
120	1850	—0,60	320,4	0,172	38,52
130	1796	—0,58	297,6	0,187	38,92
140	1746	—0,58	277,1	0,204	39,03
150	1694	—0,57	257,1	0,221	38,89
160	1638	—0,57	236,9	0,239	38,49
170	1578	—0,56	216,7	0,259	37,86
180	1516	—0,55	196,8	0,281	37,07
190	1452	—0,53	177,8	0,304	36,15
200	1389	—0,51	160,0	0,329	35,16
210	1326	—0,49	143,5	0,357	34,11
220	1265	—0,47	128,3	0,388	33,03
230	1206	—0,45	114,4	0,421	31,92
240	1147	—0,42	101,7	0,458	30,78
250	1090	—0,39	90,07	0,498	29,60
260	1034	—0,35	79,37	0,542	28,38
270	978,1	—0,31	69,53	0,592	27,12
280	922,5	—0,27	60,48	0,650	25,81
290	866,9	—0,22	52,15	0,716	24,46
300	811,1	—0,16	44,49	0,795	23,06
310	754,8	—0,08	37,48	0,890	21,61
320	697,9	0,01	31,10	1,008	20,11
330	640,3	0,12	25,32	1,158	18,56
340	582,0	0,26	20,15	1,353	16,95
350	522,7	0,46	15,57	1,615	15,27

Table II.8 (*Continued*)

T	w	μ	k	α/α_0	γ/γ
			$p=7,5$		
360	462,3	0,73	11,58	1,984	13,51
370	400,7	1,12	8,19	2,538	11,67
380	338,1	1,75	5,39	3,448	9,75
390	277,0	2,84	3,24	5,072	7,77
400	228,6	4,53	1,89	7,407	5,89
410	209,1	6,03	1,30	7,910	4,47
420	209,8	6,59	1,10	6,518	3,59
430	217,2	6,50	1,03	5,166	3,04
440	226,3	6,14	1,00	4,219	2,67
450	235,6	5,72	0,99	3,575	2,42
460	244,5	5,29	0,99	3,121	2,23
470	252,8	4,90	1,00	2,790	2,08
480	260,7	4,54	1,00	2,538	1,96
490	268,1	4,22	1,01	2,343	1,87
500	275,0	3,93	1,01	2,186	1,79
520	287,9	3,44	1,02	1,953	1,66
540	299,5	3,04	1,03	1,787	1,57
560	310,2	2,71	1,04	1,665	1,50
580	320,1	2,43	1,05	1,571	1,44
600	329,3	2,20	1,05	1,497	1,39
620	338,1	2,00	1,06	1,437	1,36
640	346,4	1,83	1,06	1,387	1,32
660	354,2	1,68	1,06	1,346	1,30
680	361,7	1,55	1,07	1,311	1,27
700	368,9	1,43	1,07	1,281	1,25
			$p=8,0$		
110	1913	—0,64	325,8	0,156	35,35
120	1855	—0,60	301,9	0,172	36,22
130	1800	—0,58	280,2	0,187	36,57
140	1749	—0,58	260,9	0,203	36,67
150	1697	—0,57	242,0	0,220	36,53
160	1641	—0,57	223,1	0,239	36,16
170	1582	—0,56	204,1	0,258	35,58
180	1520	—0,55	185,5	0,280	34,83
190	1456	—0,53	167,7	0,303	33,98
200	1393	—0,52	150,9	0,328	33,05
210	1331	—0,50	135,5	0,356	32,08
220	1270	—0,47	121,2	0,386	31,07
230	1210	—0,45	108,2	0,419	30,04
240	1152	—0,42	96,26	0,455	28,98
250	1095	—0,39	85,33	0,494	27,88
260	1039	—0,35	75,30	0,538	26,75
270	984,0	—0,32	66,08	0,587	25,58
280	928,9	—0,27	57,59	0,643	24,36
290	873,9	—0,22	49,77	0,708	23,11
300	818,7	—0,16	42,60	0,784	21,81
310	763,2	—0,09	36,02	0,875	20,48
320	707,2	—0,01	30,04	0,987	19,09
330	650,8	0,10	24,62	1,128	17,67
340	593,8	0,23	19,77	1,308	16,19
350	536,2	0,41	15,47	1,545	14,66

Table II.8 (*Continued*)

T	w	μ	k	a/a_0	γ/γ_0
			$p=8,0$		
360	478,0	0,65	11,72	1,869	13,07
370	419,3	0,98	8,52	2,332	11,41
380	360,5	1,49	5,86	3,036	9,71
390	303,2	2,30	3,78	4,169	7,97
400	253,7	3,54	2,33	5,840	6,30
410	224,3	4,92	1,55	7,010	4,90
420	216,8	5,76	1,23	6,527	3,93
430	220,0	5,96	1,10	5,443	3,30
440	227,1	5,80	1,05	4,508	2,87
450	235,3	5,49	1,02	3,820	2,58
460	243,7	5,14	1,01	3,321	2,35
470	251,9	4,78	1,01	2,953	2,19
480	259,7	4,45	1,01	2,673	2,05
490	267,1	4,15	1,02	2,455	1,94
500	274,1	3,87	1,02	2,282	1,86
520	287,0	3,39	1,03	2,024	1,72
540	298,8	2,99	1,04	1,842	1,61
560	309,6	2,67	1,05	1,709	1,53
580	319,7	2,39	1,05	1,607	1,47
600	329,1	2,16	1,06	1,526	1,42
620	337,9	1,97	1,06	1,462	1,38
640	346,3	1,80	1,06	1,408	1,34
660	354,3	1,65	1,07	1,364	1,31
680	361,8	1,52	1,07	1,327	1,29
700	369,1	1,40	1,07	1,295	1,26
			$p=8,5$		
110	1919	−0,64	308,6	0,156	33,41
120	1859	−0,60	285,5	0,171	34,18
130	1803	−0,58	264,8	0,187	34,50
140	1752	−0,58	246,5	0,203	34,58
150	1700	−0,57	228,7	0,220	34,45
160	1645	−0,57	210,9	0,238	34,10
170	1585	−0,56	193,0	0,258	33,56
180	1523	−0,55	175,5	0,279	32,86
190	1460	−0,53	158,7	0,302	32,06
200	1397	−0,52	142,9	0,327	31,19
210	1335	−0,50	128,4	0,354	30,28
220	1274	−0,47	115,0	0,384	29,34
230	1215	−0,45	102,7	0,416	28,38
240	1157	−0,42	91,44	0,452	27,38
250	1100	−0,39	81,15	0,491	26,36
260	1045	−0,36	71,70	0,534	25,31
270	989,8	−0,32	63,02	0,582	24,21
280	935,2	−0,28	55,03	0,636	23,08
290	880,7	−0,23	47,67	0,699	21,92
300	826,2	−0,17	40,92	0,773	20,71
310	771,4	−0,11	34,73	0,860	19,47
320	716,3	−0,02	29,09	0,967	18,19
330	660,9	0,08	23,99	1,099	16,87
340	605,1	0,20	19,42	1,267	15,51
350	549,0	0,36	15,36	1,483	14,10

Table II.8 (*Continued*)

T	w	μ	k	α/α_0	ν/ν_0
			$p=8,5$		
360	492,8	0,58	11,82	1,771	12,65
370	436,5	0,87	8,79	2,167	11,16
380	380,6	1,29	6,27	2.735	9,62
390	326,5	1,92	4,25	3,582	8,08
400	277,9	2,85	2,77	4,792	6,59
410	243,1	4,01	1,85	5,976	5,26
420	227,5	4,93	1,40	6,169	4,26
430	225,6	5,36	1,20	5,505	3,56
440	229,8	5,39	1,11	4,694	3,08
450	236,6	5,21	1,06	4,016	2,74
460	244,1	4,94	1.04	3,496	2,49
470	251,8	4,64	1,03	3,102	2,30
480	259,4	4,34	1,03	2,799	2,15
490	266,6	4,05	1,03	2,562	2,03
500	273,6	3,79	1,03	2,373	1,93
520	286,6	3,33	1.04	2,092	1,77
540	298,4	2,94	1,05	1,896	1,66
560	309,4	2,62	1,05	1,751	1,57
580	319,5	2,35	1,06	1,641	1,50
600	329,0	2,13	1.06	1,555	1,45
620	337,9	1,93	1,07	1,486	1,40
640	346,4	1,77	1,07	1,429	1,36
660	354,4	1,62	1,07	1,382	1,33
680	362,0	1,49	1,07	1,342	1,30
700	369,3	1,37	1,07	1,309	1,28
			$p=9,0$		
110	1925	—0,64	293,4	0,155	31,68
120	1863	—0,60	270,9	0,171	32,37
130	1806	—0,58	251,1	0,186	32,65
140	1755	—0,58	233,7	0,202	32,73
150	1703	—0,57	216,9	0,220	32,60
160	1648	—0,57	200,0	0,238	32,27
170	1589	—0,56	183,1	0,257	31,76
180	1527	—0,55	166,5	0,278	31,11
190	1464	—0,53	150,7	0,301	30,35
200	1401	—0,52	135,8	0,323	29,54
210	1339	—0,50	122,0	0,352	28,68
220	1278	—0,48	109,4	0,382	27,80
230	1219	—0,45	97,76	0.414	26,90
240	1162	—0,42	87,14	0.449	25,97
250	1105	—0,39	77,42	0,488	25,01
260	1050	—0,36	68,50	0,530	24,02
270	995,6	—0,32	60.29	0,577	23,00
280	941,4	—0,28	52.74	0,630	21,94
290	887,4	—0,24	45,80	0,691	20,85
300	833,5	—0,18	39,41	0,762	19,73
310	779,4	—0,12	33.57	0,847	18,57
320	725,1	—0,04	28,24	0,948	17,38
330	670,6	0,06	23,42	1,073	16,15
340	616,0	0,17	19.09	1,230	14,89
350	561,3	0,32	15,25	1,428	13,60

Table II.8 (*Continued*)

T	w	μ	k	α/α₀	ν/ν₀

$$p=9,0$$

T	w	μ	k	α/α₀	ν/ν₀
360	506,7	0,52	11,90	1,687	12,27
370	452,5	0,78	9,02	2,031	10,90
380	399,0	1,13	6,61	2,504	9,51
390	347,5	1,64	4,66	3,168	8,12
400	300,4	2,37	3,18	4,080	6,77
410	263,1	3,29	2,18	5,088	5,55
420	241,4	4,18	1,61	5,609	4,55
430	233,9	4,74	1,32	5,362	3,81
440	234,7	4,93	1,19	4,758	3,28
450	239,4	4,88	1,12	4,145	2,90
460	245,7	4,70	1,08	3,634	2,62
470	252,7	4,46	1,06	3,229	2,41
480	259,8	4,20	1,05	2,912	2,24
490	266,8	3,94	1,05	2,660	2,11
500	273,6	3,70	1,05	2,458	2,00
520	286,5	3,26	1,05	2,157	1,83
540	298,4	2,89	1,05	1,946	1,70
560	309,3	2,58	1,06	1,792	1,61
580	319,5	2,31	1,06	1,674	1,53
600	329,1	2,09	1,07	1,582	1,48
620	338,1	1,90	1,07	1,509	1,43
640	346,6	1,73	1,07	1,449	1,38
660	354,7	1,59	1,08	1,399	1,35
680	362,4	1,46	1,08	1,357	1,32
700	369,7	1,34	1,08	1,321	1,29

$$p=9,5$$

T	w	μ	k	α/α₀	ν/ν₀
110	1931	−0,64	279,7	0,155	30,13
120	1867	−0,60	257,8	0,171	30,75
130	1810	−0,58	238,8	0,186	31,00
140	1758	−0,58	222,2	0,202	31,06
150	1706	−0,57	206,3	0,219	30,94
160	1651	−0,57	190,3	0,237	30,63
170	1592	−0,56	174,3	0,256	30,15
180	1530	−0,55	158,6	0,277	29,54
190	1467	−0,53	143,5	0,299	28,83
200	1404	−0,52	129,4	0,324	28,06
210	1343	−0,50	116,4	0,351	27,25
220	1282	−0,48	104,4	0,380	26,42
230	1223	−0,45	93,36	0,412	25,57
240	1166	−0,43	83,30	0,446	24,70
250	1110	−0,40	74,08	0,484	23,80
260	1055	−0,36	65,62	0,526	22,87
270	1001	−0,33	57,85	0,572	21,91
280	947,5	−0,29	50,69	0,624	20,92
290	894,0	−0,24	44,11	0,683	19,90
300	840,6	−0,19	38,06	0,752	18,84
310	787,2	−0,13	32,52	0,833	17,76
320	733,6	−0,05	27,47	0,931	16,65
330	680,0	0,04	22,90	1,049	15,51
340	626,4	0,15	18,79	1,195	14,33
350	573,0	0,29	15,15	1,379	13,13

Table II.8 (*Continued*)

T	w	μ	k	α/α_0	γ/γ_0

			$p=9,5$		
360	519,9	0,46	11,96	1,613	11,90
370	467,4	0,69	9,21	1,917	10,65
380	416,1	1,00	6,90	2,320	9,38
390	366,7	1,43	5,01	2,860	8,12
400	321,2	2,01	3,55	3,574	6,89
410	283,1	2,75	2,50	4,397	5,76
420	257,2	3,55	1,84	5,009	4,79
430	244,6	4,15	1,48	5,071	4,03
440	241,6	4,45	1,29	4,701	3,48
450	243,8	4,52	1,18	4,198	3,07
460	248,6	4,42	1,13	3,726	2,76
470	254,6	4,25	1,10	3,329	2,52
480	261,1	4,04	1,08	3,006	2,34
490	267,7	3,81	1,07	2,746	2,19
500	274,3	3,59	1,07	2,535	2,07
520	286,9	3,18	1,06	2,217	1,88
540	298,6	2,83	1,06	1,994	1,75
560	309,6	2,53	1,07	1,830	1,65
580	319,8	2,27	1,07	1,706	1,57
600	329,4	2,05	1,07	1,609	1,50
620	338,4	1,86	1,08	1,531	1,45
640	347,0	1,70	1,08	1,467	1,41
660	355,1	1,56	1,08	1,415	1,37
680	362,8	1,43	1,08	1,371	1,34
700	370,2	1,32	1,08	1,333	1,31

			$p=10,0$		
110	1936	—0,64	267,3	0,155	28,73
120	1871	—0,60	246,0	0,171	29,28
130	1813	—0,58	227,7	0,186	29,51
140	1761	—0,58	211,9	0,202	29,56
150	1709	—0,57	196,7	0,219	29,45
160	1654	—0,57	181,5	0,236	29,16
170	1595	—0,56	166,3	0,255	28,70
180	1533	—0,55	151,4	0,276	28,12
190	1471	—0,54	137,1	0,298	27,45
200	1408	—0,52	123,7	0,323	26,72
210	1346	—0,50	111,3	0,349	25,96
220	1286	—0,48	99,85	0,378	25,18
230	1228	—0,45	89,39	0,410	24,38
240	1171	—0,43	79,83	0,444	23,55
250	1115	—0,40	71,07	0,481	22,71
260	1061	—0,37	63,03	0,522	21,83
270	1007	—0,33	55,64	0,567	20,93
280	953,5	—0,29	48,84	0,618	20,00
290	900,5	—0,25	42,59	0,676	19,04
300	847,6	—0,20	36,84	0,748	18,05
310	794,8	—0,14	31,57	0,821	17,03
320	742,0	—0,07	26,77	0,914	15,99
330	689,2	0,02	22,42	1,026	14,92
340	636,6	0,12	18,51	1,164	13,82
350	584,2	0,25	15,04	1,335	12,71

Table II.8 (*Continued*)

T	w	μ	k	$\alpha/\alpha_\jmath$	γ/γ_0
			$p=10,0$		
360	532,4	0,42	12,00	1,548	11,57
370	481,5	0,62	9,37	1,820	10,41
380	431,9	0,89	7,15	2,169	9,25
390	384,4	1,25	5,32	2,621	8,09
400	340,4	1,73	3,88	3,198	6,96
410	302,4	2,34	2,81	3,869	5,91
420	273,9	3,02	2,08	4,461	4,99
430	257,1	3,61	1,65	4,703	4,24
440	250,3	3,99	1,40	4,545	3,66
450	249,7	4,14	1,26	4,176	3,22
460	252,7	4,13	1,19	3,768	2,89
470	257,5	4,02	1,14	3,395	2,64
480	263,2	3,86	1,11	3,079	2,43
490	269,3	3,67	1,10	2,817	2,27
500	275,5	3,48	1,09	2,601	2,14
520	287,6	3,10	1,08	2,272	1,94
540	299,2	2,76	1,07	2,038	1,80
560	310,1	2,47	1,08	1,866	1,68
580	320,3	2,23	1,08	1,735	1,60
600	329,9	2,01	1,08	1,633	1,53
620	338,9	1,83	1,08	1,552	1,47
640	347,5	1,67	1,08	1,485	1,43
660	355,6	1,53	1,08	1,430	1,39
680	363,4	1,40	1,09	1,384	1,35
700	370,8	1,29	1,09	1,345	1,32
			$p=11$		
110	1947	—0,64	245,8	0,155	26,30
120	1878	—0,60	225,5	0,170	26,75
130	1819	—0,58	208,5	0,185	26,92
140	1767	—0,58	194,0	0,201	26,97
150	1715	—0,57	180,2	0,218	26,87
160	1660	—0,57	166,3	0,235	26,61
170	1601	—0,56	152,5	0,254	26,20
180	1540	—0,55	138,9	0,274	25,68
190	1478	—0,54	126,0	0,296	25,07
200	1415	—0,52	113,8	0,320	24,42
210	1354	—0,50	102,4	0,346	23,73
220	1294	—0,48	92,05	0,375	23,03
230	1236	—0,46	82,53	0,405	22,31
240	1180	—0,43	73,82	0,439	21,57
250	1125	—0,40	65,85	0,475	20,82
260	1071	—0,37	58,54	0,515	20,04
270	1018	—0,34	51,82	0,558	19,23
280	965,1	—0,30	45,63	0,607	18,40
290	913,0	—0,26	39,94	0,662	17,54
300	861,2	—0,21	34,71	0,724	16,66
310	809,5	—0,15	29,92	0,797	15,76
320	758,0	—0,09	25,54	0,883	14,83
330	706,7	—0,01	21,58	0,985	13,89
340	655,8	0,08	18,01	1,108	12,93
350	605,4	0,20	14,83	1,257	11,95

Table II.8 (*Continued*)

T	w	μ	k	α/α_0	ν/ν_0

		$p=11$			
360	555,9	0,33	12,04	1,438	10,96
370	507,5	0,51	9,62	1,661	9,96
380	460,8	0,72	7,56	1,935	8,96
390	416,3	0,99	5,84	2,270	7,97
400	374,8	1,34	4,45	2,676	7,01
410	337,9	1,76	3,37	3,143	6,10
420	307,4	2,26	2,57	3,612	5,27
430	285,2	2,75	2,02	3,957	4,56
440	271,8	3,16	1,67	4,068	3,97
450	265,7	3,42	1,46	3,952	3,51
460	264,5	3,53	1,33	3,708	3,14
470	266,4	3,54	1,25	3,425	2,85
480	270,0	3,47	1,20	3,152	2,62
490	274,6	3,35	1,16	2,907	2,44
500	279,7	3,21	1,14	2,695	2,29
520	290,6	2,91	1,12	2,359	2,06
540	301,5	2,62	1,10	2,113	1,89
560	312,0	2,36	1,10	1,929	1,76
580	322,0	2,13	1,10	1,788	1,66
600	331,5	1,93	1,10	1,677	1,58
620	340,5	1,75	1,10	1,589	1,52
640	349,0	1,60	1,10	1,517	1,47
660	357,1	1,47	1,10	1,458	1,42
680	364,9	1,35	1,10	1,408	1,38
700	372,3	1,24	1,10	1,366	1,35

		$p=12$			
110	1957	—0,64	227,8	0,154	24,26
120	1885	—0,60	208,3	0,169	24,62
130	1824	—0,58	192,4	0,184	24,77
140	1772	—0,58	179,0	0,200	24,80
150	1720	—0,57	166,3	0,217	24,71
160	1666	—0,57	153,7	0,234	24,48
170	1607	—0,56	141,0	0,253	24,11
180	1546	—0,55	128,6	0,273	23,63
190	1484	—0,54	116,6	0,294	23,09
200	1422	—0,52	105,4	0,318	22,49
210	1362	—0,50	95,07	0,343	21,87
220	1302	—0,48	85,53	0,371	21,24
230	1245	—0,46	76,79	0,401	20,58
240	1189	—0,44	68,80	0,434	19,92
250	1134	—0,41	61,49	0,469	19,24
260	1081	—0,38	54,78	0,507	18,53
270	1028	—0,35	48,61	0,550	17,81
280	976,4	—0,31	42,94	0,596	17,06
290	925,1	—0,27	37,72	0,648	16,29
300	874,2	—0,22	32,92	0,708	15,50
310	823,6	—0,17	28,52	0,776	14,69
320	773,3	—0,11	24,50	0,855	13,86
330	723,4	—0,04	20,85	0,949	13,02
340	673,9	0,05	17,56	1,059	12,16
350	625,2	0,15	14,63	1,191	11,30

Table II.8 (*Continued*)

T	w	μ	k	α/α_0	γ/γ_0
			$p=12$		
360	577,5	0,27	12,05	1,348	10,42
370	531,2	0,41	9,80	1,536	9,54
380	486,7	0,59	7,87	1,760	8,67
390	444,5	0,81	6,25	2,024	7,80
400	405,1	1,07	4,92	2,331	6,97
410	369,5	1,39	3,84	2,676	6,17
420	338,7	1,75	3,02	3,035	5,44
430	314,1	2,14	2,41	3,351	4,78
440	296,5	2,50	1,98	3,552	4,22
450	285,6	2,79	1,69	3,598	3,75
460	280,2	2,97	1,51	3,507	3,36
470	278,7	3,05	1,39	3,334	3,05
480	279,9	3,06	1,31	3,130	2,80
490	282,6	3,01	1,25	2,925	2,60
500	286,4	2,92	1,21	2,735	2,43
520	295,4	2,70	1,17	2,413	2,17
540	305,2	2,46	1,14	2,166	1,98
560	315,1	2,23	1,13	1,977	1,84
580	324,7	2,02	1,12	1,830	1,73
600	333,9	1,84	1,12	1,714	1,64
620	342,7	1,68	1,11	1,620	1,57
640	351,1	1,53	1,11	1,544	1,51
660	359,2	1,40	1,11	1,481	1,46
680	366,9	1,29	1,11	1,428	1,42
700	374,2	1,18	1,11	1,383	1,38
			$p=13$		
110	1967	—0,64	212,5	0,154	22,53
120	1891	—0,60	193,7	0,169	22,82
130	1830	—0,58	178,8	0,184	22,94
140	1777	—0,58	166,3	0,199	22,97
150	1726	—0,58	154,6	0,216	22,89
160	1671	—0,57	142,9	0,233	22,67
170	1613	—0,56	131,2	0,251	22,34
180	1553	—0,55	119,7	0,271	21,90
190	1491	—0,54	108,7	0,292	21,40
200	1429	—0,52	98,40	0,315	20,86
210	1369	—0,51	88,81	0,340	20,30
220	1310	—0,49	79,99	0,368	19,71
230	1253	—0,46	71,92	0,397	19,12
240	1197	—0,44	64,54	0,429	18,52
250	1143	—0,41	57,78	0,463	17,90
260	1090	—0,38	51,58	0,501	17,26
270	1038	—0,35	45,89	0,541	16,60
280	987,3	—0,32	40,65	0,586	15,92
290	936,8	—0,28	35,82	0,636	15,22
300	886,8	—0,23	31,38	0,692	14,51
310	837,2	—0,19	27,31	0,756	13,78
320	787,9	—0,13	23,59	0,830	13,03
330	739,2	—0,06	20,22	0,916	12,27
340	691,1	0,01	17,17	1,016	11,50
350	643,8	0,10	14,44	1,134	10,72

Table II.8 (*Continued*)

T	ω	μ	k	α/α_0	γ/γ_0
			$p=13$		
360	597,7	0,21	12,03	1,273	9,94
370	553,1	0,34	9,93	1,435	9,16
380	510,3	0,49	8,12	1,623	8,39
390	469,9	0,66	6,58	1,839	7,62
400	432,2	0,87	5,30	2,083	6,88
410	397,8	1,12	4,25	2,352	6,18
420	367,5	1,40	3,41	2,633	5,52
430	341,9	1,70	2,77	2,898	4,92
440	321,9	2,00	2,29	3,108	4,39
450	307,6	2,27	1,95	3,224	3,93
460	298,6	2,47	1,71	3,236	3,55
470	293,9	2,60	1,55	3,161	3,23
480	292,4	2,66	1,43	3,032	2,96
490	293,0	2,66	1,35	2,880	2,74
500	295,1	2,63	1,30	2,723	2,56
520	301,9	2,48	1,23	2,435	2,27
540	310,3	2,29	1,19	2,198	2,07
560	319,3	2,10	1,17	2,010	1,91
580	328,3	1,91	1,15	1,861	1,79
600	337,1	1,74	1,14	1,742	1,69
620	345,7	1,59	1,14	1,645	1,61
640	353,9	1,46	1,13	1,566	1,55
660	361,9	1,34	1,13	1,500	1,49
680	369,4	1,23	1,13	1,445	1,45
700	376,7	1,13	1,12	1,398	1,41
			$p=14$		
110	1976	—0,64	199,2	0,153	21,04
120	1897	—0,60	181,2	0,168	21,27
130	1835	—0,59	167,0	0,183	21,36
140	1782	—0,58	155,4	0,199	21,39
150	1731	—0,58	144,5	0,215	21,32
160	1677	—0,57	133,7	0,232	21,12
170	1619	—0,56	122,8	0,250	20,82
180	1559	—0,55	112,2	0,269	20,42
190	1497	—0,54	102,0	0,290	19,96
200	1436	—0,53	92,35	0,313	19,46
210	1376	—0,51	83,44	0,338	18,94
220	1317	—0,49	75,24	0,364	18,41
230	1260	—0,47	67,73	0,393	17,86
240	1205	—0,44	60,87	0,424	17,31
250	1152	—0,42	54,59	0,458	16,74
260	1099	—0,39	48,83	0,494	16,16
270	1048	—0,36	43,54	0,533	15,56
280	997,9	—0,32	38,66	0,576	14,94
290	948,2	—0,29	34,18	0,624	14,31
300	899,0	—0,25	30,06	0,678	13,65
310	850,2	—0,20	26,27	0,738	12,99
320	802,0	—0,15	22,81	0,807	12,31
330	754,3	—0,09	19,66	0,886	11,62
340	707,3	—0,02	16,81	0,978	10,92
350	661,3	0,07	14,26	1,085	10,22

Table II.8 (*Continued*)

T	w	μ	k	α/α_0	γ/γ_0

$p=14$

360	616,5	0,16	12,00	1,208	9,52
370	573,3	0,27	10,02	1,350	8,81
380	532,1	0,40	8,31	1,512	8,12
390	493,1	0,55	6,85	1,694	7,44
400	456,8	0,72	5,61	1,896	6,78
410	423,5	0,92	4,59	2,114	6,14
420	393,7	1,14	3,76	2,340	5,55
430	368,0	1,38	3,10	2,560	5,00
440	346,8	1,63	2,59	2,752	4,51
450	330,3	1,86	2,21	2,889	4,07
460	318,6	2,06	1,92	2,955	3,70
470	311,0	2,21	1,72	2,949	3,37
480	306,9	2,30	1,58	2,886	3,10
490	305,4	2,34	1,47	2,786	2,87
500	305,7	2,34	1,40	2,668	2,68
520	310,0	2,26	1,30	2,425	2,38
540	316,7	2,12	1,24	2,208	2,15
560	324,6	1,96	1,21	2,028	1,98
580	332,9	1,80	1,19	1,881	1,85
600	341,2	1,65	1,17	1,762	1,74
620	349,4	1,51	1,16	1,664	1,66
640	357,4	1,39	1,15	1,583	1,59
660	365,1	1,27	1,15	1,515	1,53
680	372,5	1,17	1,14	1,458	1,48
700	379,6	1,08	1,14	1,409	1,44

$p=15$

110	1984	—0,64	187,6	0,153	19,74
120	1903	—0,60	170,2	0,168	19,92
130	1839	—0,59	156,8	0,183	19,99
140	1787	—0,58	145,9	0,198	20,02
150	1736	—0,58	135,7	0,214	19,95
160	1682	—0,57	125,6	0,231	19,78
170	1625	—0,56	115,5	0,249	19,50
180	1565	—0,55	105,6	0,268	19,13
190	1504	—0,54	96,06	0,288	18,70
200	1443	—0,53	87,09	0,311	18,24
210	1383	—0,51	78,76	0,335	17,76
220	1325	—0,49	71,11	0,361	17,27
230	1268	—0,47	64,09	0,389	16,77
240	1213	—0,45	57,68	0,420	16,26
250	1160	—0,42	51,81	0,453	15,74
260	1109	—0,39	46,43	0,488	15,21
270	1058	—0,36	41,49	0,526	14,66
280	1008	—0,33	36,94	0,567	14,09
290	959,2	—0,30	32,75	0,613	13,50
300	910,7	—0,26	28,89	0,664	12,91
310	862,8	—0,21	25,35	0,721	12,30
320	815,5	—0,16	22,11	0,786	11,67
330	768,8	—0,11	19,16	0,860	11,04
340	722,8	—0,04	16,49	0,944	10,41
350	677,9	0,03	14,09	1,041	9,77

Table II.8 (*Continued*)

T	ω	μ	k	α/α_0	γ/γ_0

$$\dot{p}=15$$

T	ω	μ	k	α/α_0	γ/γ_0
360	634,3	0,12	11,96	1,152	9,13
370	592,4	0,22	10,09	1,278	8,50
380	552,3	0,33	8,46	1,420	7,87
390	514,5	0,46	7,06	1,577	7,25
400	479,3	0,60	5,88	1,748	6,66
410	447,0	0,77	4,89	1,930	6,08
420	417,9	0,95	4,07	2,118	5,54
430	392,2	1,15	3,40	2,302	5,04
440	370,4	1,35	2,87	2,471	4,58
450	352,7	1,55	2,46	2,607	4,17
460	339,0	1,73	2,14	2,696	3,81
470	329,2	1,87	1,91	2,731	3,49
480	322,9	1,98	1,73	2,716	3,22
490	319,3	2,04	1,60	2,661	2,99
500	317,8	2,07	1,51	2,581	2,79
520	319,4	2,04	1,38	2,389	2,47
540	324,3	1,94	1,30	2,199	2,23
560	330,9	1,82	1,26	2,031	2,05
580	338,1	1,68	1,23	1,890	1,91
600	346,0	1,55	1,21	1,773	1,80
620	353,8	1,43	1,19	1,675	1,70
640	361,4	1,31	1,18	1,594	1,63
660	368,8	1,21	1,17	1,525	1,56
680	376,0	1,11	1,16	1,467	1,51
700	383,0	1,02	1,16	1,418	1,46

$$p=16$$

T	ω	μ	k	α/α_0	γ/γ_0
110	1992	−0,64	177,4	0,152	18,59
120	1908	−0,60	160,5	0,167	18,73
130	1844	−0,59	147,8	0,182	18,79
140	1791	−0,58	137,5	0,197	18,81
150	1740	−0,58	128,0	0,213	18,76
160	1687	−0,57	118,6	0,230	18,60
170	1630	−0,56	109,1	0,247	18,34
180	1571	−0,56	99,82	0,266	18,00
190	1510	−0,54	90,89	0,286	17,61
200	1449	−0,53	82,48	0,308	17,18
210	1390	−0,51	74,66	0,332	16,73
220	1332	−0,49	67,48	0,358	16,28
230	1276	−0,47	60,90	0,386	15,81
240	1221	−0,45	54,88	0,415	15,34
250	1169	−0,42	49,37	0,447	14,86
260	1117	−0,40	44,32	0,482	14,37
270	1067	−0,37	39,68	0,519	13,86
280	1018	−0,34	35,41	0,559	13,34
290	969,8	−0,30	31,48	0,603	12,80
300	922,1	−0,27	27,86	0,651	12,25
310	875,0	−0,22	24,53	0,705	11,69
320	828,5	−0,18	21,49	0,766	11,11
330	782,6	−0,13	18,71	0,835	10,54
340	737,7	−0,07	16,19	0,914	9,95
350	693,8	0,00	13,93	1,003	9,37

Table II.8 (*Continued*)

T	w	μ	k	α/α_0	γ/γ_0
			$p=16$		
360	651,2	0,08	11,92	1,103	8,78
370	610,3	0,17	10,14	1,216	8,20
380	571,3	0,27	8,59	1,342	7,63
390	534,5	0,38	7,25	1,480	7,07
400	500,2	0,51	6,10	1,627	6,53
410	468,7	0,65	5,14	1,783	6,01
420	440,2	0,80	4,33	1,943	5,51
430	414,8	0,96	3,67	2,100	5,05
440	392,7	1,13	3,13	2,247	4,62
450	374,2	1,30	2,70	2,374	4,24
460	359,3	1,45	2,36	2,469	3,89
470	347,8	1,59	2,09	2,526	3,59
480	339,6	1,70	1,89	2,542	3,32
490	334,2	1,78	1,74	2,522	3,08
500	331,1	1,82	1,62	2,475	2,88
520	329,8	1,84	1,46	2,333	2.55
540	332,8	1,78	1,37	2,173	2,31
560	338,1	1,68	1,31	2,022	2,12
580	344,5	1,57	1,27	1,889	1.97
600	351,5	1,46	1,24	1,776	1,84
620	358,7	1,34	1,22	1,681	1,75
640	365,9	1,24	1,21	1,600	1,67
660	373,1	1,14	1,20	1,532	1,60
680	380,0	1,05	1,19	1,473	1,54
700	386,8	0,97	1,18	1,423	1,49
			$p=17$		
110	1999	—0,64	168,3	0,152	17,57
120	1913	—0,60	151,9	0,167	17,68
130	1848	—0,59	139,8	0,181	17,72
140	1795	—0,58	130,1	0,196	17,75
150	1745	—0,58	121,2	0,212	17,70
160	1692	—0,57	112,3	0,229	17,55
170	1635	—0,57	103,5	0,246	17,32
180	1576	—0,56	94,72	0,265	17,00
190	1516	—0,55	86,32	0,284	16,63
200	1456	—0,53	78,40	0,306	16,24
210	1396	—0,51	71,04	0,329	15,82
220	1339	—0,50	64,27	0,355	15,40
230	1283	—0,48	58,07	0,382	14,97
240	1229	—0,45	52,40	0,411	14,53
250	1177	—0,43	47,21	0,442	14,08
260	1126	—0,40	42,45	0,476	13,63
270	1077	—0,37	38,08	0,512	13,16
280	1028	—0,34	34,06	0,551	12,67
290	980,2	—0,31	30,35	0,593	12,18
300	933,2	—0,28	26,94	0,639	11,67
310	886,8	—0,24	23,80	0,691	11,14
320	841,0	—0,19	20,93	0,748	10,62
330	796,0	—0,14	18,31	0,813	10,08
340	751,9	—0,09	15,92	0,886	9,54
350	708,9	—0,02	13,78	0,968	9,00

Table II.8 (*Continued*)

T	w	μ	k	α/α_0	γ/γ_0
			$p=17$		
360	667,3	0,05	11,87	1,060	8,47
370	627,2	0,13	10,17	1,162	7,93
380	589,1	0,22	8,69	1,275	7,41
390	553,2	0,31	7,40	1,397	6,89
400	519,8	0,42	6,29	1,527	6,40
410	489,0	0,54	5,35	1,663	5,92
420	460,9	0,67	4,56	1,‌01	5,46
430	435,8	0,81	3,90	1,938	5,04
440	413,7	0,95	3,36	2,067	4,64
450	394,8	1,10	2,92	2,182	4,28
460	379,0	1,24	2,56	2,276	3,95
470	366,3	1,36	2,28	2,342	3,66
480	356,7	1,47	2,05	2,377	3,39
490	349,7	1,55	1,88	2,381	3,16
500	345,1	1,60	1,74	2,360	2,96
520	341,2	1,65	1,56	2,263	2,63
540	342,2	1,62	1,44	2,133	2,37
560	346,0	1,55	1,37	2,001	2,18
580	351,3	1,46	1,32	1,880	2,02
600	357,6	1,36	1,28	1,773	1,89
620	364,2	1,26	1,26	1,681	1,79
640	371,0	1,17	1,24	1,602	1,70
660	377,8	1,08	1,22	1,534	1,63
680	384,4	1,00	1,21	1,476	1,57
700	391,0	0,92	1,20	1,426	1,52
			$p=18$		
110	2006	−0,64	160,1	0,151	16,66
120	1918	−0,60	144,3	0,166	16,73
130	1852	−0,59	132,7	0,181	16,77
140	1799	−0,58	123,5	0,196	16,80
150	1749	−0,58	115,1	0,211	16,76
160	1697	−0,57	106,8	0,228	16,63
170	1641	−0,57	98,41	0,245	16,40
180	1582	−0,56	90,17	0,263	16,11
190	1522	−0,55	82,24	0,283	15,77
200	1462	−0,53	74,76	0,304	15,40
210	1403	−0,52	67,81	0,327	15,01
220	1346	−0,50	61,41	0,352	14,61
230	1290	−0,48	55,54	0,379	14,21
240	1237	−0,46	50,18	0,407	13,81
250	1185	−0,43	45,28	0,438	13,39
260	1135	−0,41	40,78	0,470	12,97
270	1086	−0,38	36,65	0,505	12,53
280	1038	−0,35	32,85	0,543	12,08
290	990,3	−0,32	29,34	0,583	11,62
300	943,9	−0,28	26,12	0,628	11,14
310	898,2	−0,25	23,15	0,677	10,66
320	853,2	−0,21	20,42	0,731	10,17
330	808,9	−0,16	17,94	0,792	9,67
340	765,6	−0,11	15,68	0,860	9,17
350	723,4	−0,05	13,64	0,936	8,67

Table II.8 (*Continued*)

T	w	μ	k	α/α_0	γ/γ_0

$p=18$

360	682,6	0,02	11,81	1,021	8,18
370	643,4	0,09	10,19	1,114	7,68
380	606,1	0,17	8,77	1,216	7,20
390	570,9	0,26	7,53	1,326	6,73
400	538,1	0,36	6,46	1,442	6,27
410	507,9	0,46	5,54	1,562	5,83
420	480,3	0,57	4,77	1,684	5,40
430	455,5	0,69	4,11	1,804	5,01
440	433,5	0,81	3,57	1,918	4,64
450	414,3	0,94	3,12	2,022	4,30
460	397,9	1,06	2,75	2,111	3,99
470	384,4	1,17	2,45	2,180	3,71
480	373,7	1,27	2,21	2,225	3,45
490	365,5	1,35	2,02	2,245	3,23
500	359,5	1,41	1,87	2,243	3,03
520	353,2	1,47	1,65	2,184	2,69
540	352,2	1,47	1,52	2,084	2,43
560	354,5	1,42	1,43	1,972	2,23
580	358,8	1,35	1,37	1,862	2,07
600	364,1	1,27	1,32	1,763	1,94
620	370,1	1,18	1,29	1,675	1,83
640	376,4	1,10	1,27	1,599	1,74
660	382,8	1,02	1,25	1,533	1,66
680	389,2	0,94	1,24	1,476	1,60
700	395,5	0,87	1,22	1,426	1,54

$p=19$

110	2013	−0,64	152,7	0,151	15,84
120	1922	−0,60	137,3	0,166	15,89
130	1855	−0,59	126,2	0,180	15,92
140	1803	−0,58	117,5	0,195	15,94
150	1753	−0,58	109,6	0,210	15,91
160	1701	−0,57	101,8	0,227	15,79
170	1646	−0,57	93,87	0,244	15,59
180	1587	−0,56	86,09	0,262	15,31
190	1528	−0,55	78,58	0,281	14,99
200	1468	−0,53	71,49	0,302	14,64
210	1409	−0,52	64,90	0,324	14,28
220	1352	−0,50	58,84	0,349	13,91
230	1297	−0,48	53,28	0,375	13,53
240	1244	−0,46	48,19	0,403	13,15
250	1193	−0,44	43,54	0,433	12,77
260	1143	−0,41	39,28	0,465	12,37
270	1094	−0,38	35,36	0,499	11,96
280	1047	−0,36	31,76	0,535	11,54
290	1000	−0,33	28,43	0,575	11,11
300	954,3	−0,29	25,37	0,617	10,67
310	909,3	−0,26	22,55	0,664	10,22
320	864,9	−0,22	19,97	0,716	9,76
330	821,4	−0,17	17,60	0,773	9,30
340	778,8	−0,13	15,45	0,837	8,84
350	737,4	−0,07	13,50	0,907	8,37

Table II.8 (*Continued*)

T	w	μ	k	α/α_0	ν/ν_0
		$p=19$			
360	697,3	—0,01	11,76	0,986	7,91
370	658,8	0,06	10,20	1,071	7,45
380	622,2	0,13	8,83	1,165	7,00
390	587,7	0,21	7,64	1,264	6,56
400	555,5	0,30	6,60	1,369	6,14
410	525,8	0,39	5,71	1,476	5,73
420	498,6	0,49	4,95	1,585	5,34
430	474,1	0,59	4,30	1,692	4,97
440	452,1	0,70	3,76	1,794	4,62
450	432,8	0,80	3,31	1,888	4,30
460	416,1	0,91	2,93	1,971	4,01
470	402,0	1,01	2,62	2,039	3,74
480	390,4	1,10	2,36	2,089	3,50
490	381,2	1,18	2,16	2,119	3,28
500	374,2	1,24	1,99	2,129	3,08
520	365,7	1,31	1,75	2,101	2,75
540	362,7	1,33	1,59	2,028	2,49
560	363,5	1,30	1,49	1,935	2,28
580	366,6	1,25	1,42	1,839	2,11
600	371,2	1,18	1,37	1,748	1,98
620	376,5	1,11	1,33	1,666	1,87
640	382,3	1,03	1,30	1,593	1,77
660	388,3	0,96	1,28	1,529	1,69
680	394,3	0,89	1,26	1,473	1,63
700	400,3	0,82	1,25	1,424	1,57
		$p=20$			
110	2019	—0,64	146,0	0,150	15,09
120	1926	—0,60	131,1	0,165	15,12
130	1859	—0,59	120,4	0,179	15,15
140	1806	—0,58	112,2	0,194	15,17
150	1757	—0,58	104,7	0,210	15,15
160	1705	—0,57	97,25	0,226	15,04
170	1650	—0,57	89,78	0,242	14,85
180	1592	—0,56	82,40	0,260	14,59
190	1533	—0,55	75,28	0,279	14,29
200	1474	—0,54	68,55	0,300	13,97
210	1416	—0,52	62,29	0,322	13,62
220	1359	—0,50	56,52	0,346	13,28
230	1304	—0,48	51,23	0,372	12,92
240	1251	—0,46	46,40	0,399	12,57
250	1200	—0,44	41,97	0,429	12,21
260	1151	—0,41	37,92	0,460	11,83
270	1103	—0,39	34,20	0,493	11,45
280	1056	—0,36	30,77	0,528	11,06
290	1010	—0,33	27,61	0,566	10,66
300	964,5	—0,30	24,69	0,607	10,25
310	920,0	—0,27	22,01	0,652	9,82
320	876,3	—0,23	19,55	0,701	9,40
330	833,5	—0,19	17,29	0,755	8,96
340	791,6	—0,14	15,23	0,815	8,53
350	750,8	—0,09	13,37	0,881	8,10

Table II.8 (*Continued*)

T	w	μ	k	α/α₀	γ/γ₀

$p = 20$

T	w	μ	k	α/α₀	γ/γ₀
360	711,4	−0,04	11,70	0,954	7,67
370	673,6	0,03	10,21	1,033	7,24
380	637,6	0,09	8,89	1,119	6,82
390	603,7	0,17	7,73	1,209	6,41
400	572,0	0,24	6,72	1,304	6,01
410	542,7	0,33	5,85	1,402	5,63
420	515,9	0,41	5,11	1,500	5,27
430	491,6	0,51	4,47	1,596	4,92
440	469,8	0,60	3,93	1,688	4,60
450	450,4	0,69	3,48	1,774	4,30
460	433,5	0,79	3,09	1,851	4,02
470	418,9	0,87	2,78	1,916	3,76
480	406,7	0,96	2,51	1,967	3,53
490	396,7	1,03	2,29	2,003	3,32
500	388,8	1,09	2,11	2,022	3,13
520	378,4	1,17	1,85	2,017	2,80
540	373,7	1,20	1,67	1,967	2,54
560	373,0	1,19	1,55	1,893	2,33
580	374,9	1,15	1,47	1,810	2,15
600	378,6	1,10	1,41	1,728	2,02
620	383,2	1,03	1,37	1,652	1,90
640	388,4	0,97	1,34	1,583	1,80
660	394,0	0,90	1,31	1,522	1,72
680	399,7	0,84	1,29	1,467	1,65
700	405,4	0,77	1,27	1,419	1,59

$p = 21$

T	w	μ	k	α/α₀	γ/γ₀
110	2024	−0,64	139,9	0,150	14,41
120	1929	−0,60	125,3	0,164	14,42
130	1862	−0,59	115,1	0,179	14,45
140	1809	−0,58	107,3	0,194	14,48
150	1761	−0,58	100,2	0,209	14,46
160	1710	−0,57	93,15	0,225	14,36
170	1655	−0,57	86,07	0,241	14,18
180	1598	−0,56	79,06	0,259	13,94
190	1539	−0,55	72,28	0,277	13,66
200	1480	−0,54	65,87	0,298	13,35
210	1422	−0,52	59,91	0,320	13,03
220	1366	−0,50	54,41	0,343	12,70
230	1311	−0,49	49,37	0,369	12,37
240	1259	−0,46	44,77	0,396	12,04
250	1208	−0,44	40,55	0,424	11,70
260	1159	−0,42	36,68	0,455	11,35
270	1111	−0,39	33,13	0,487	10,99
280	1065	−0,37	29,87	0,521	10,62
290	1019	−0,34	26,85	0,558	10,25
300	974,4	−0,31	24,07	0,597	9,86
310	930,5	−0,27	21,51	0,640	9,46
320	887,4	−0,24	19,16	0,687	9,06
330	845,2	−0,20	17,00	0,739	8,66
340	804,0	−0,16	15,03	0,795	8,25
350	763,8	−0,11	13,25	0,857	7,84

Table II.8 (*Continued*)

T	w	μ	k	α/α₀	γ/γ₀

$$p = 21$$

T	w	μ	k	α/α_0	γ/γ_0
360	725,1	—0,06	11,64	0,925	7,44
370	687,8	0,00	10,21	0,998	7,04
380	652,4	0,06	8,93	1,077	6,65
390	619,0	0,13	7,81	1,161	6,27
400	587,8	0,20	6,83	1,248	5,89
410	558,9	0,27	5,98	1,337	5,54
420	532,4	0,35	5,25	1,426	5,20
430	508,3	0,43	4,62	1,513	4,87
440	486,5	0,52	4,09	1,597	4,57
450	467,2	0,60	3,63	1,676	4,28
460	450,1	0,68	3,25	1,747	4,02
470	435,2	0,76	2,92	1,809	3,77
480	422,6	0,83	2,65	1,860	3,55
490	411,9	0,90	2,42	1,898	3,34
500	403,3	0,96	2,23	1,923	3,16
520	391,2	1,04	1,95	1,935	2,84
540	384,9	1,08	1,75	1,904	2,58
560	382,8	1,08	1,62	1,847	2,37
580	383,6	1,06	1,53	1,777	2,19
600	386,3	1,02	1,46	1,705	2,05
620	390,2	0,96	1,41	1,635	1,93
640	394,9	0,90	1,37	1,571	1,83
660	400,0	0,85	1,34	1,512	1,75
680	405,3	0,79	1,32	1,460	1,68
700	410,7	0,73	1,30	1,413	1,62

$$p = 22$$

T	w	μ	k	α/α_0	γ/γ_0
110	2029	—0,64	134,2	0,149	13,79
120	1932	—0,60	120,1	0,164	13,79
130	1864	—0,59	110,3	0,178	13,81
140	1812	—0,58	102,8	0,193	13,84
150	1764	—0,58	96,07	0,208	13,83
160	1714	—0,58	89,41	0,224	13,74
170	1660	—0,57	82,68	0,240	13,57
180	1603	—0,56	76,01	0,257	13,35
190	1544	—0,55	69,55	0,276	13,08
200	1486	—0,54	63,44	0,296	12,79
210	1428	—0,52	57,74	0,317	12,49
220	1372	—0,51	52,49	0,341	12,18
230	1318	—0,49	47,68	0,365	11,87
240	1266	—0,47	43,28	0,392	11,55
250	1215	—0,45	39,25	0,420	11,23
260	1167	—0,42	35,56	0,450	10,90
270	1119	—0,40	32,16	0,481	10,57
280	1073	—0,37	29,04	0,515	10,22
290	1028	—0,34	26,16	0,550	9,87
300	984,1	—0,31	23,50	0,588	9,50
310	940,8	—0,28	21,06	0,629	9,13
320	898,2	—0,25	18,80	0,674	8,75
330	856,6	—0,21	16,74	0,723	8,37
340	815,9	—0,17	14,85	0,776	7,99
350	776,4	—0,13	13,14	0,835	7,61

Table II.8 (*Continued*)

T	w	μ	k	α/α_0	v/v_0
			$p=22$		
360	738,2	—0,08	11,59	0,898	7,23
370	701,5	—0,03	10,20	0,967	6,85
380	666,6	0,03	8,97	1,040	6,49
390	633,7	0,09	7,88	1,117	6,13
400	602,9	0,16	6,93	1,197	5,78
410	574,3	0,23	6,10	1,279	5,44
420	548,0	0,30	5,38	1,361	5,12
430	524,1	0,37	4,76	1,441	4,82
440	502,5	0,44	4,23	1,518	4,53
450	483,2	0,52	3,77	1,590	4,26
460	466,0	0,59	3,39	1,656	4,01
470	450,9	0,66	3,06	1,715	3,77
480	437,9	0,73	2,78	1,764	3,56
490	426,8	0,79	2,55	1,804	3,36
500	417,5	0,85	2,35	1,832	3,18
520	404,0	0,93	2,05	1,856	2,87
540	396,3	0,97	1,84	1,841	2,61
560	392,8	0,99	1,69	1,799	2,40
580	392,5	0,97	1,58	1,742	2,23
600	394,2	0,94	1,51	1,679	2,08
620	397,4	0,89	1,45	1,616	1,96
640	401,5	0,84	1,41	1,556	1,86
660	406,2	0,79	1,38	1,501	1,77
680	411,1	0,74	1,35	1,451	1,70
700	416,2	0,68	1,33	1,406	1,64
			$p=23$		
110	2034	—0,64	129,0	0,148	13,22
120	1935	—0,60	115,2	0,163	13,20
130	1867	—0,59	105,8	0,178	13,22
140	1815	—0,58	98,70	0,192	13,25
150	1768	—0,58	92,31	0,207	13,25
160	1718	—0,58	85,98	0,223	13,17
170	1664	—0,57	79,58	0,239	13,02
180	1608	—0,56	73,22	0,256	12,81
190	1549	—0,55	67,05	0,274	12,55
200	1491	—0,54	61,21	0,294	12,28
210	1434	—0,53	55,76	0,315	11,99
220	1378	—0,51	50,73	0,338	11,70
230	1324	—0,49	46,13	0,362	11,40
240	1273	—0,47	41,91	0,388	11,11
250	1223	—0,45	38,05	0,416	10,80
260	1174	—0,45	34,52	0,445	10,50
270	1127	—0,40	31,27	0,476	10,18
280	1082	—0,38	28,28	0,508	9,86
290	1037	—0,35	25,52	0,543	9,52
300	993,5	—0,32	22,98	0,580	9,18
310	950,7	—0,29	20,63	0,619	8,83
320	908,8	—0,26	18,47	0,662	8,47
330	867,7	—0,22	16,49	0,708	8,11
340	827,6	—0,18	14,67	0,759	7,75
350	788,6	—0,14	13,03	0,814	7,39

Table II.8 (*Continued*)

T	w	μ	k	α/α₀	γ/γ₀

$p=23$

T	w	μ	k	α/α₀	γ/γ₀
360	750,9	—0,10	11,53	0,873	7,03
370	714,8	—0,05	10,19	0,938	6,68
380	680,3	0,00	9,00	1,006	6,33
390	647,8	0,06	7,94	1,078	5,99
400	617,3	0,12	7,01	1,152	5,67
410	589,1	0,18	6,20	1,227	5,35
420	563,1	0,25	5,49	1,303	5,05
430	539,3	0,31	4,88	1,377	4,76
440	517,8	0,38	4,36	1,448	4,49
450	498,4	0,45	3,91	1,515	4,23
460	481,2	0,51	3,52	1,576	3,99
470	465,9	0,58	3,19	1,632	3,77
480	452,6	0,64	2,90	1,680	3,56
490	441,1	0,70	2,66	1,719	3,37
500	431,4	0,75	2,46	1,749	3,20
520	416,8	0,83	2,14	1,781	2,89
540	407,7	0,88	1,92	1,779	2,64
560	403,0	0,90	1,76	1,750	2,43
580	401,5	0,89	1,64	1,704	2,26
600	402,4	0,87	1,56	1,650	2,11
620	404,9	0,83	1,49	1,594	1,99
640	408,4	0,79	1,45	1,539	1,89
660	412,5	0,74	1,41	1,488	1,80
680	417,1	0,69	1,38	1,440	1,72
700	421,9	0,64	1,35	1,397	1,66

$p=24$

T	w	μ	k	α/α₀	γ/γ₀
110	2038	—0,64	124,2	0,148	12,69
120	1937	—0,60	110,8	0,163	12,66
130	1869	—0,59	101,7	0,177	12,68
140	1818	—0,58	94,92	0,192	12,72
150	1771	—0,58	88,85	0,207	12,72
160	1722	—0,58	82,82	0,222	12,65
170	1669	—0,57	76,72	0,238	12,51
180	1612	—0,56	70,65	0,255	12,31
190	1555	—0,55	64,75	0,272	12,07
200	1497	—0,54	59,16	0,292	11,81
210	1440	—0,53	53,94	0,313	11,54
220	1384	—0,51	49,12	0,335	11,26
230	1331	—0,49	44,70	0,359	10,98
240	1279	—0,47	40,66	0,385	10,70
250	1230	—0,45	36,96	0,412	10,41
260	1182	—0,43	33,57	0,441	10,12
270	1135	—0,41	30,45	0,471	9,82
280	1090	—0,38	27,58	0,502	9,52
290	1046	—0,35	24,94	0,536	9,20
300	1003	—0,33	22,50	0,571	8,88
310	960,5	—0,30	20,24	0,609	8,55
320	919,0	—0,27	18,17	0,650	8,21
330	878,5	—0,23	16,26	0,695	7,87
340	838,9	—0,20	14,51	0,743	7,53
350	800,4	—0,16	12,92	0,795	7,19

Table II.8 (*Continued*)

T	w	μ	k	α/α_0	ν/ν_0
			$p=24$		
360	763,3	—0,11	11,48	0,851	6,85
370	727,6	—0,07	10,18	0,911	6,52
380	693,6	—0,02	9,02	0,975	6,19
390	661,4	0,03	7,99	1,042	5,87
400	631,3	0,09	7,09	1,111	5,56
410	603,3	0,15	6,29	1,181	5,26
420	577,5	0,20	5,60	1,251	4,97
430	553,9	0,26	5,00	1,319	4,70
440	532,4	0,33	4,48	1,385	4,44
450	513,1	0,39	4,03	1,448	4,20
460	495,8	0,45	3,64	1,506	3,97
470	480,4	0,50	3,31	1,558	3,76
480	466,9	0,56	3,02	1,604	3,56
490	455,1	0,61	2,78	1,642	3,38
500	445,0	0,66	2,57	1,674	3,21
520	429,4	0,74	2,24	1,712	2,91
540	419,1	0,79	2,00	1,719	2,66
560	413,3	0,81	1,82	1,701	2,46
580	410,8	0,81	1,70	1,665	2,28
600	410,8	0,80	1,61	1,620	2,14
620	412,5	0,77	1,54	1,570	2,01
640	415,4	0,73	1,48	1,521	1,91
660	419,1	0,69	1,44	1,473	1,82
680	423,2	0,65	1,41	1,428	1,74
700	427,7	0,60	1,38	1,387	1,68
			$p=25$		
110	2041	—0,64	119,7	0,147	12,20
120	1939	—0,60	106,6	0,162	12,16
130	1871	—0,59	97,90	0,176	12,18
140	1820	—0,58	91,42	0,191	12,22
150	1774	—0,58	85,65	0,206	12,23
160	1725	—0,58	79,91	0,221	12,17
170	1673	—0,57	74,09	0,237	12,04
180	1617	—0,56	68,28	0,253	11,85
190	1560	—0,55	62,63	0,271	11,62
200	1502	—0,54	57,26	0,290	11,37
210	1446	—0,53	52,25	0,311	11,12
220	1390	—0,51	47,63	0,333	10,85
230	1337	—0,49	43,38	0,356	10,59
240	1286	—0,48	39,50	0,382	10,32
250	1237	—0,45	35,94	0,408	10,05
260	1189	—0,43	32,68	0,436	9,78
270	1143	—0,41	29,69	0,465	9,49
280	1098	—0,38	26,94	0,496	9,20
290	1055	—0,36	24,39	0,529	8,91
300	1012	—0,33	22,05	0,563	8,60
310	970,0	—0,30	19,88	0,600	8,29
320	929,0	—0,27	17,88	0,639	7,97
330	889,0	—0,24	16,04	0,682	7,65
340	849,9	—0,21	14,36	0,727	7,32
350	812,0	—0,17	12,82	0,777	7,00

Table II.8 (*Continued*)

T	w	μ	k	α/α_0	γ/γ_0
			$p=25$		
360	775,3	—0,13	11,43	0,830	6,68
370	740,0	—0,09	10,17	0,886	6,36
380	706,4	—0,04	9,04	0,946	6,05
390	674,6	0,01	8,04	1,009	5,75
400	644,8	0,06	7,15	1,073	5,46
410	617,0	0,11	6,38	1,139	5,17
420	591,4	0,17	5,69	1,204	4,90
430	567,9	0,22	5,10	1,268	4,64
440	546,5	0,28	4,58	1,330	4,40
450	527,2	0,33	4,14	1,388	4,16
460	509,9	0,39	3,75	1,442	3,95
470	494,4	0,44	3,42	1,492	3,74
480	480,7	0,49	3,13	1,535	3,55
490	468,6	0,54	2,88	1,573	3,38
500	458,2	0,58	2,67	1,605	3,22
520	441,7	0,66	2,33	1,646	2,93
540	430,5	0,71	2,07	1,661	2,68
560	423,6	0,74	1,89	1,652	2,48
580	420,1	0,74	1,76	1,626	2,31
600	419,2	0,73	1,66	1,588	2,16
620	420,2	0,71	1,58	1,545	2,04
640	422,5	0,68	1,52	1,501	1,93
660	425,7	0,65	1,48	1,457	1,84
680	429,5	0,61	1,44	1,415	1,76
700	433,6	0,57	1,41	1,376	1,70
			$p=26$		
110	2044	—0,64	115,6	0,147	11,74
120	1941	—0,60	102,8	0,162	11,70
130	1873	—0,59	94,36	0,176	11,72
140	1822	—0,58	88,17	0,190	11,76
150	1777	—0,58	82,68	0,205	11,77
160	1729	—0,58	77,21	0,220	11,72
170	1677	—0,57	71,65	0,236	11,60
180	1622	—0,56	66,09	0,252	11,42
190	1565	—0,56	60,67	0,269	11,21
200	1508	—0,54	55,51	0,288	10,97
210	1451	—0,53	50,70	0,308	10,73
220	1396	—0,51	46,25	0,330	10,48
230	1344	—0,50	42,16	0,354	10,22
240	1293	—0,48	38,42	0,378	9,97
250	1244	—0,46	35,00	0,404	9,72
260	1196	—0,44	31,87	0,432	9,46
270	1151	—0,41	28,99	0,460	9,19
280	1106	—0,39	26,33	0,491	8,91
290	1063	—0,36	23,89	0,522	8,63
300	1021	—0,34	21,63	0,556	8,34
310	979,3	—0,31	19,54	0,591	8,05
320	938,8	—0,28	17,61	0,629	7,74
330	899,3	—0,25	15,84	0,670	7,44
340	860,7	—0,22	14,21	0,713	7,13
350	823,2	—0,18	12,72	0,760	6,82

Table II.8 (*Continued*)

T	w	μ	k	α/α_0	γ/γ_0
			$p=26$		
360	786,9	—0,15	11,38	0,810	6,52
370	752,1	—0,11	10,16	0,863	6,22
380	718,9	—0,06	9,06	0,920	5,92
390	687,4	—0,02	8,08	0,978	5,64
400	657,8	0,03	7,22	1,039	5,36
410	630,3	0,08	6,45	1,100	5,09
420	604,8	0,13	5,78	1,161	4,83
430	581,4	0,18	5,20	1,221	4,58
440	560,1	0,23	4,68	1,279	4,35
450	540,8	0,28	4,24	1,334	4,13
460	523,4	0,33	3,86	1,385	3,92
470	507,8	0,38	3,52	1,432	3,72
480	494,0	0,43	3,23	1,474	3,54
490	481,7	0,47	2,98	1,511	3,37
500	471,0	0,51	2,76	1,542	3,22
520	453,8	0,58	2,41	1,586	2,94
540	441,7	0,64	2,15	1,606	2,70
560	433,9	0,67	1,96	1,605	2,50
580	429,5	0,68	1,81	1,586	2,33
600	427,8	0,67	1,71	1,556	2,18
620	428,1	0,66	1,62	1,519	2,06
640	429,7	0,63	1,56	1,480	1,95
660	432,4	0,60	1,51	1,439	1,86
680	435,8	0,57	1,47	1,401	1,78
700	439,6	0,53	1,44	1,364	1,71
			$p=28$		
110	2050	—0,65	107,9	0,146	10,92
120	1944	—0,61	95,79	0,160	10,86
130	1875	—0,59	87,97	0,175	10,88
140	1826	—0,58	82,33	0,189	10,94
150	1782	—0,58	77,34	0,204	10,96
160	1736	—0,58	72,36	0,218	10,92
170	1685	—0,57	67,27	0,233	10,82
180	1631	—0,57	62,15	0,249	10,66
190	1575	—0,56	57,15	0,266	10,47
200	1518	—0,55	52,37	0,285	10,26
210	1462	—0,53	47,90	0,304	10,03
220	1408	—0,52	43,77	0,325	9,80
230	1356	—0,50	39,97	0,348	9,58
240	1305	—0,48	36,49	0,372	9,35
250	1257	—0,46	33,31	0,397	9,12
260	1210	—0,44	30,40	0,424	8,88
270	1165	—0,42	27,72	0,451	8,64
280	1122	—0,40	25,25	0,480	8,39
290	1079	—0,37	22,97	0,510	8,14
300	1038	—0,35	20,87	0,541	7,88
310	997,4	—0,32	18,92	0,574	7,61
320	957,8	—0,30	17,12	0,610	7,34
330	919,1	—0,27	15,47	0,647	7,06
340	881,4	—0,24	13,94	0,687	6,78
350	844,8	—0,21	12,55	0,729	6,50

Table II.8 (*Continued*)

T	w	μ	k	α/α_0	γ/γ_0
			$p=28$		
360	809,4	—0,17	11,28	0,774	6,23
370	775,3	—0,14	10,12	0,822	5,96
380	742,8	—0,10	9,09	0,873	5,69
390	711,9	—0,06	8,15	0,925	5,43
400	682,8	—0,02	7,32	0,978	5,17
410	655,6	0,03	6,59	1,033	4,93
420	630,4	0,07	5,94	1,087	4,69
430	607,2	0,11	5,36	1,140	4,47
440	586,0	0,16	4,86	1,191	4,25
450	566,7	0,20	4,43	1,240	4,05
460	549,2	0,24	4,04	1,286	3,86
470	533,4	0,29	3,71	1,328	3,68
480	519,3	0,33	3,42	1,367	3,51
490	506,8	0,36	3,16	1,401	3,35
500	495,6	0,40	2,94	1,431	3,21
520	477,3	0,46	2,57	1,477	2,94
540	463,8	0,51	2,30	1,505	2,72
560	454,3	0,54	2,09	1,514	2,52
580	448,3	0,56	1,93	1,509	2,36
600	445,1	0,57	1,80	1,491	2,21
620	443,9	0,56	1,71	1,466	2,09
640	444,5	0,54	1,64	1,435	1,98
660	446,1	0,52	1,58	1,403	1,89
680	448,7	0,49	1,53	1,370	1,81
700	451,7	0,46	1,49	1,338	1,74
			$p=30$		
110	2053	—0,65	101,2	0,144	10,19
120	1945	—0,61	89,62	0,159	10,13
130	1877	—0,59	82,36	0,174	10,15
140	1829	—0,58	77,20	0,188	10,22
150	1787	—0,58	72,67	0,202	10,25
160	1742	—0,58	68,13	0,217	10,23
170	1693	—0,57	63,45	0,231	10,14
180	1639	—0,57	58,72	0,247	10,00
190	1584	—0,56	54,08	0,263	9,83
200	1528	—0,55	49,64	0,281	9,63
210	1473	—0,54	45,47	0,300	9,43
220	1419	—0,52	41,61	0,321	9,22
230	1368	—0,50	38,06	0,343	9,01
240	1318	—0,49	34,81	0,366	8,80
250	1270	—0,47	31,84	0,390	8,59
260	1224	—0,45	29,11	0,416	8,38
270	1180	—0,43	26,61	0,442	8,16
280	1137	—0,40	24,30	0,470	7,94
290	1095	—0,38	22,17	0,498	7,71
300	1054	—0,36	20,20	0,528	7,47
310	1015	—0,33	18,38	0,559	7,23
320	976,0	—0,31	16,69	0,592	6,98
330	938,2	—0,28	15,14	0,627	6,73
340	901,3	—0,26	13,70	0,663	6,47
350	865,4	—0,23	12,39	0,702	6,22

Table II.8 (*Continued*)

T	w	μ	k	α/α_0	γ/γ_0
			$p=30$		
360	830,8	—0,20	11,19	0,743	5,97
370	797,4	—0,16	10,09	0,786	5,72
380	765,5	—0,13	9,10	0,831	5,47
390	735,1	—0,09	8,21	0,878	5,23
400	706,5	—0,06	7,41	0,926	5,00
410	679,6	—0,02	6,70	0,975	4,78
420	654,7	0,02	6,07	1,023	4,56
430	631,6	0,06	5,51	1,071	4,35
440	610,4	0,10	5,02	1,117	4,16
450	591,1	0,13	4,59	1,161	3,97
460	573,5	0,17	4,21	1,203	3,79
470	557,6	0,21	3,87	1,242	3,63
480	543,2	0,24	3,58	1,277	3,47
490	530,4	0,27	3,32	1,309	3,32
500	518,9	0,31	3,10	1,338	3,19
520	499,7	0,36	2,72	1,384	2,94
540	485,1	0,41	2,43	1,415	2,72
560	474,3	0,44	2,21	1,432	2,54
580	466,9	0,46	2,04	1,435	2,38
600	462,3	0,47	1,90	1,427	2,24
620	460,0	0,47	1,80	1,411	2,12
640	459,3	0,46	1,71	1,389	2,01
660	460,0	0,45	1,65	1,364	1,92
680	461,7	0,43	1,59	1,337	1,84
700	464,1	0,40	1,55	1,310	1,77
			$p=32$		
110	2055	—0,65	95,14	0,143	9,54
120	1945	—0,61	84,13	0,158	9,47
130	1878	—0,59	77,36	0,172	9,51
140	1832	—0,58	72,65	0,187	9,58
150	1791	—0,58	68,54	0,201	9,63
160	1748	—0,58	64,39	0,215	9,62
170	1700	—0,57	60,08	0,229	9,55
180	1648	—0,57	55,70	0,245	9,42
190	1593	—0,56	51,38	0,261	9,26
200	1538	—0,55	47,23	0,278	9,08
210	1484	—0,54	43,33	0,296	8,90
220	1431	—0,52	39,71	0,316	8,70
230	1379	—0,51	36,38	0,338	8,51
240	1330	—0,49	33,33	0,360	8,32
250	1283	—0,47	30,54	0,384	8,13
260	1237	—0,45	27,98	0,408	7,94
270	1194	—0,43	25,62	0,434	7,74
280	1151	—0,41	23,46	0,460	7,54
290	1110	—0,39	21,46	0,488	7,33
300	1070	—0,37	19,61	0,516	7,11
310	1032	—0,34	17,89	0,545	6,89
320	993,5	—0,32	16,31	0,576	6,66
330	956,5	—0,30	14,84	0,608	6,43
340	920,3	—0,27	13,49	0,642	6,20
350	885,3	—0,24	12,24	0,678	5,97

Table II.8 (*Continued*)

T	w	μ	k	α/α_0	γ/γ_0

$p = 32$

T	w	μ	k	α/α_0	γ/γ_0
360	851,3	—0,22	11,10	0,715	5,74
370	818,6	—0,19	10,06	0,755	5,51
380	787,2	—0,16	9,11	0,795	5,28
390	757,3	—0,13	8,26	0,838	5,06
400	729,1	—0,09	7,49	0,881	4,85
410	702,5	—0,06	6,80	0,925	4,64
420	677,7	—0,02	6,18	0,969	4,44
430	654,8	0,01	5,64	1,012	4,25
440	633,6	0,04	5,16	1,054	4,07
450	614,2	0,08	4,73	1,094	3,89
460	596,5	0,11	4,35	1,132	3,73
470	580,4	0,14	4,02	1,168	3,57
480	565,9	0,17	3,73	1,201	3,43
490	552,8	0,20	3,47	1,231	3,29
500	541,0	0,23	3,24	1,258	3,16
520	521,2	0,28	2,86	1,303	2,93
540	505,6	0,32	2,56	1,336	2,72
560	493,8	0,35	2,33	1,357	2,54
580	485,3	0,37	2,14	1,366	2,39
600	479,5	0,39	2,00	1,366	2,25
620	476,0	0,39	1,88	1,357	2,13
640	474,3	0,39	1,79	1,343	2,03
660	474,0	0,38	1,72	1,324	1,94
680	474,9	0,37	1,66	1,303	1,86
700	476,5	0,35	1,61	1,280	1,79

$p = 34$

T	w	μ	k	α/α_0	γ/γ_0
110	2056	—0,65	89,67	0,142	8,95
120	1944	—0,61	79,18	0,157	8,89
130	1878	—0,59	72,89	0,171	8,93
140	1834	—0,58	68,59	0,186	9,02
150	1795	—0,58	64,86	0,199	9,08
160	1753	—0,58	61,07	0,213	9,08
170	1707	—0,58	57,09	0,228	9,02
180	1656	—0,57	53,02	0,242	8,91
190	1602	—0,56	48,98	0,258	8,76
200	1548	—0,55	45,10	0,275	8,60
210	1494	—0,54	41,43	0,293	8,42
220	1441	—0,53	38,03	0,312	8,25
230	1391	—0,51	34,89	0,333	8,07
240	1342	—0,49	32,01	0,355	7,90
250	1295	—0,48	29,38	0,378	7,73
260	1250	—0,46	26,97	0,401	7,55
270	1207	—0,44	24,75	0,426	7,37
280	1166	—0,42	22,71	0,451	7,18
290	1125	—0,40	20,82	0,478	6,99
300	1086	—0,38	19,08	0,504	6,79
310	1048	—0,35	17,46	0,532	6,59
320	1011	—0,33	15,96	0,561	6,38
330	974,1	—0,31	14,57	0,591	6,17
340	938,7	—0,29	13,29	0,623	5,96
350	904,3	—0,26	12,11	0,656	5,74

Table II.8 (*Continued*)

T	ω	μ	k	α/α_0	ν/ν_0
			$p=34$		
360	871,0	—0,24	11,02	0,690	5,53
370	838,9	—0,21	10,02	0,726	5,31
380	808,0	—0,18	9,12	0,764	5,11
390	778,6	—0,15	8,30	0,802	4,90
400	750,7	—0,12	7,55	0,841	4,70
410	724,4	—0,09	6,89	0,881	4,51
420	699,8	—0,06	6,29	0,921	4,32
430	676,9	—0,03	5,75	0,960	4,15
440	655,8	0,00	5,28	0,999	3,98
450	636,3	0,03	4,86	1,035	3,81
460	618,5	0,06	4,49	1,071	3,66
470	602,3	0,09	4,16	1,104	3,52
480	587,5	0,11	3,86	1,134	3,38
490	574,2	0,14	3,60	1,162	3,25
500	562,2	0,16	3,37	1,188	3,13
520	541,7	0,21	2,99	1,232	2,91
540	525,4	0,25	2,68	1,265	2,71
560	512,7	0,28	2,44	1.289	2,54
580	503,2	0,30	2,24	1,303	2,39
600	496,4	0,32	2,09	1,308	2,26
620	491,8	0,32	1,96	1,305	2,15
640	489,2	0,32	1,86	1,297	2,04
660	488,0	0,32	1,78	1,284	1,96
680	488,1	0,31	1,72	1,268	1,88
700	489,0	0,30	1,66	1,250	1,81
			$p=36$		
110	2055	—0,65	84,69	0,141	8,43
120	1942	—0,61	74,69	0,156	8,36
130	1877	—0,59	68,85	0,170	8,42
140	1835	—0,58	64,94	0,184	8,51
150	1798	—0,58	61,55	0,198	8,58
160	1758	—0,58	58,09	0,212	8,59
170	1713	—0,58	54,42	0,226	8,55
180	1663	—0,57	50,63	0,240	8,45
190	1611	—0,56	46,84	0.255	8,32
200	1557	—0,55	43,19	0,272	8,17
210	1504	—0,54	39,74	0,289	8,01
220	1452	—0,53	36,52	0,308	7,84
230	1402	—0,51	33,56	0,328	7,68
240	1353	—0,50	30,84	0,349	7,52
250	1307	—0,48	28,35	0,372	7,36
260	1263	—0,46	26,06	0,395	7,20
270	1220	—0,44	23,97	0,419	7,03
280	1179	—0,42	22,03	0,443	6,86
290	1140	—0,40	20,25	0,468	6,69
300	1101	—0,38	18,60	0,494	6,51
310	1063	—0,36	17,07	0,520	6,32
320	1027	—0,34	15,64	0,548	6,13
330	991,2	—0,32	14,33	0,576	5,93
340	956,5	—0,30	13,11	0,605	5,73
350	922,7	—0,27	11,98	0,636	5,54

Table II.8 (*Continued*)

T	w	μ	k	α/α₀	γ/γ₀
				α/α_0	γ/γ_0

$$p = 36$$

360	890,0	—0.25	10,94	0,668	5,34
370	858,4	—0,23	9,99	0,701	5,14
380	828,1	—0,20	9,12	0,735	4,95
390	799,1	—0,18	8,33	0,770	4,76
400	771,5	—0,15	7,61	0,806	4,57
410	745,5	—0,12	6,96	0,842	4,39
420	721,0	—0,09	6,38	0,879	4,22
430	698,2	—0,07	5,86	0,915	4,05
440	677,1	—0,04	5,39	0,950	3,89
450	657,6	—0,01	4,98	0,984	3,74
460	639,6	0,01	4,61	1,016	3,60
470	623,2	0,04	4,28	1,047	3,46
480	608.3	0,06	3,98	1,076	3,33
490	594,7	0,09	3,72	1,103	3,21
500	582,5	0,11	3,49	1,127	3,09
520	561,4	0,15	3,10	1,169	2,88
540	544,4	0,18	2,79	1,203	2,70
560	531,0	0,21	2,54	1,228	2,53
580	520,6	0,24	2,34	1,244	2,39
600	512,9	0,25	2,17	1,253	2,26
620	507,5	0,26	2,04	1,256	2,15
640	503,9	0,27	1,94	1,252	2,05
660	502,0	0,26	1,85	1,244	1,97
680	501,3	0,26	1,78	1,233	1,89
700	501,5	0,25	1,72	1,219	1,82

$$p = 38$$

110	2052	—0,65	80,12	0,139	7,94
120	1939	—0,61	70,60	0,155	7,88
130	1875	—0,59	65,17	0,169	7,95
140	1835	—0,59	61,62	0,183	8,05
150	1801	—0,58	58,57	0,197	8,14
160	1763	—0,58	55,40	0,210	8,16
170	1720	—0,58	52,01	0,224	8,12
180	1671	—0,57	48,47	0,238	8,04
190	1619	—0,57	44,92	0,253	7,92
200	1566	—0,56	41,48	0,269	7,78
210	1514	—0,55	38,21	0,286	7,63
220	1462	—0,53	35,17	0,304	7,48
230	1412	—0,52	32,36	0,324	7,33
240	1365	—0,50	29,78	0,344	7,18
250	1319	—0,48	27,42	0,366	7,03
260	1275	—0,47	25,25	0,388	6,88
270	1233	—0,45	23,26	0,411	6,73
280	1193	—0,43	21,43	0,435	6,58
290	1154	—0,41	19,73	0,459	6,42
300	1116	—0,39	18,16	0,484	6,25
310	1079	—0,37	16,71	0,509	6,08
320	1043	—0,35	15,36	0,535	5,90
330	1008	—0,33	14,10	0,562	5,72
340	973,7	—0,31	12,94	0,589	5,53
350	940,5	—0,29	11,87	0,618	5,35

Table II.8 (*Continued*)

T	w	μ	k	α/α_0	ν/ν_0
			$p = 38$		
360	908,4	—0,27	10,87	0,647	5,16
370	877,3	—0,24	9,96	0,678	4,98
380	847,5	—0,22	9,12	0,709	4,80
390	818,9	—0,20	8,36	0,742	4,62
400	791,6	—0,17	7,66	0,775	4,45
410	765,8	—0,15	7,03	0,808	4,28
420	741,6	—0,12	6,47	0,841	4,12
430	718,8	—0,10	5,95	0,874	3,96
440	697,7	—0,07	5,49	0,907	3,81
450	678,1	—0,05	5,08	0,938	3,67
460	660,0	—0,03	4,72	0,969	3,53
470	643,5	0,00	4,39	0,997	3,40
480	628,4	0,02	4,10	1,024	3,28
490	614,6	0,04	3,84	1,049	3,16
500	602,0	0,06	3,60	1,073	3,05
520	580,4	0,10	3,21	1,113	2,86
540	562,8	0,13	2,89	1,146	2,68
560	548,7	0,16	2,63	1,172	2,52
580	537,6	0,18	2,43	1,191	2,38
600	529,1	0,20	2,26	1,203	2,26
620	522,9	0,21	2,12	1,209	2,15
640	518,6	0,21	2,01	1,209	2,06
660	515,8	0,22	1,91	1,205	1,97
680	514,4	0,21	1,83	1,198	1,90
700	514,0	0,21	1,77	1,188	1,83
			$p = 40$		
110	2049	—0,65	75,91	0,138	7,50
120	1935	—0,61	66,83	0,153	7,45
130	1872	—0,59	61,80	0,168	7,52
140	1835	—0,59	58,60	0,182	7,64
150	1803	—0,58	55,85	0,196	7,73
160	1767	—0,58	52,96	0,209	7,77
170	1726	—0,58	49,82	0,222	7,74
180	1678	—0,57	46,52	0,236	7,67
190	1628	—0,57	43,18	0,250	7,56
200	1575	—0,56	39,93	0,266	7,43
210	1523	—0,55	36,84	0,282	7,29
220	1472	—0,54	33,95	0,300	7,15
230	1423	—0,52	31,28	0,319	7,01
240	1376	—0,51	28,82	0,339	6,87
250	1330	—0,49	26,57	0,360	6,74
260	1287	—0,47	24,51	0,382	6,60
270	1246	—0,45	22,62	0,405	6,46
280	1206	—0,43	20,87	0,428	6,32
290	1167	—0,41	19,26	0,451	6,17
300	1130	—0,40	17,77	0,475	6,01
310	1094	—0,38	16,38	0,499	5,86
320	1058	—0,36	15,10	0,523	5,69
330	1024	—0,34	13,90	0,549	5,52
340	990,3	—0,32	12,79	0,574	5,35
350	957,8	—0,30	11,76	0,601	5,18

Table II.8 (*Continued*)

T	w	μ	k	α/α_0	γ/γ_0
			$p=40$		
360	926,2	—0,28	10,81	0,629	5,01
370	895,7	—0,26	9,93	0,657	4,84
380	866,3	—0,24	9,13	0,686	4,67
390	838,0	—0,22	8,39	0,716	4,50
400	811,1	—0,19	7,71	0,746	4,34
410	785,6	—0,17	7,10	0,777	4,18
420	761,4	—0,15	6,54	0,808	4,02
430	738,8	—0,13	6,04	0,838	3,88
440	717,6	—0,10	5,59	0,868	3,73
450	698,0	—0,08	5,18	0,898	3,60
460	679,8	—0,06	4,82	0,926	3,47
470	663,1	—0,04	4,49	0,953	3,35
480	647,8	—0,02	4,20	0,978	3,23
490	633,8	0,00	3,94	1,002	3,12
500	621,0	0,02	3,71	1,024	3,02
520	598,8	0,05	3,31	1,063	2,83
540	580,7	0,08	2,99	1,096	2,66
560	565,9	0,11	2,73	1,122	2,51
580	554,2	0,13	2,51	1,142	2,38
600	545,0	0,15	2,34	1,156	2,26
620	538,0	0,16	2,19	1,165	2,15
640	533,0	0,17	2,07	1,168	2,06
660	529,5	0,17	1,97	1,168	1,98
680	527,4	0,17	1,89	1,164	1,90
700	526 3	0,17	1,82	1,157	1,84
			$p=45$		
110	2033	—0,65	66,63	0,135	6,53
120	1919	—0,61	58,60	0,150	6,50
130	1862	—0,59	54,48	0,166	6,60
140	1832	—0,59	52,06	0,180	6,75
150	1807	—0,58	50,01	0,193	6,87
160	1777	—0,58	47,74	0,206	6,93
170	1739	—0,58	45,17	0,218	6,93
180	1696	—0,58	42,37	0,231	6,88
190	1647	—0 57	39,48	0,245	6,79
200	1597	—0,56	36,64	0 259	6,69
210	1547	—0,55	33,91	0,275	6,57
220	1497	—0,54	31,35	0,291	6,45
230	1449	—0,53	28,98	0,309	6,34
240	1402	—0,51	26,79	0,328	6,22
250	1358	—0,50	24,78	0,348	6,11
260	1316	—0,48	22,94	0,368	6,00
270	1276	—0,46	21,25	0,389	5,88
280	1237	—0,45	19,69	0,410	5,76
290	1200	—0,43	18,26	0,432	5.64
300	1164	—0,41	16,92	0,454	5,51
310	1129	—0,39	15,68	0,475	5,38
320	1095	—0,38	14,53	0,497	5,25
330	1062	—0,36	13,46	0,520	5,10
340	1030	—0,34	12,46	0,542	4,96
350	998,9	—0,32	11,53	0,565	4,82

Table II.8 (*Continued*)

T	w	$\overset{\cdot}{\mu}$	k	α/α_0	γ/γ_0
			$p=45$		
360	968,7	—0,31	10,67	0,588	4,67
370	939,3	—0,29	9,87	0,612	4,52
380	910,9	—0,27	9,13	0,637	4,38
390	883,7	—0,25	8,44	0,661	4,23
400	857,5	—0,24	7,82	0,687	4,09
410	832,5	—0,22	7,24	0,712	3,95
420	808,8	—0,20	6,72	0,737	3,82
430	786,4	—0,18	6,24	0,763	3,69
440	765,3	—0,16	5,80	0,788	3,56
450	745,6	—0,15	5,41	0,812	3,44
460	727,2	—0,13	5,05	0,836	3,33
470	710,1	—0,11	4,73	0,859	3,22
480	694,3	—0,09	4,44	0,881	3,12
490	679,7	—0,08	4,18	0,902	3,02
500	666,3	—0,06	3,94	0,922	2,93
520	642,9	—0,03	3,54	0,958	2,75
540	623,3	—0,01	3,21	0,989	2,60
560	607,2	0,01	2,93	1,014	2,46
580	593,9	0,03	2,71	1,036	2,34
600	583,2	0,05	2,52	1,053	2,24
620	574,7	0,06	2,36	1,066	2,14
640	568,1	0,07	2,23	1,075	2,05
660	563,0	0,08	2,12	1,081	1,97
680	559,3	0,08	2,03	1,083	1,90
700	556,8	0,08	1,95	1,083	1,84
			$p=50$		
110	2010	—0,66	58,74	0,131	5,72
120	1896	—0,61	51,64	0,147	5,70
130	1846	—0,59	48,35	0,163	5,84
140	1825	—0,59	46,64	0,178	6,03
150	1808	—0,58	45,21	0,191	6,18
160	1785	—0,58	43,48	0,203	6,26
170	1752	—0,58	41,38	0,215	6,28
180	1712	—0,58	39,01	0,227	6,25
190	1666	—0,57	36,50	0,240	6,18
200	1618	—0,57	33,99	0,253	6,09
210	1569	—0,56	31,56	0,267	5,99
220	1521	—0 55	29,26	0,283	5,89
230	1474	—0,53	27,13	0 300	5,79
240	1428	—0,52	25,15	0,318	5,70
250	1385	—0,50	23,34	0,336	5,60
260	1344	—0,49	21,67	0,355	5,51
270	1305	—0,47	20,14	0,375	5,41
280	1267	—0,46	18,74	0,395	5,31
290	1231	—0,44	17,44	0,415	5,21
300	1196	—0,42	16,23	0,435	5,11
310	1163	—0,41	15,11	0,455	5,00
320	1130	—0,39	14,07	0,475	4,88
330	1098	—0,37	13,09	0,495	4,76
340	1068	—0,36	12,18	0,515	4,64
350	1038	—0,34	11,34	0,535	4,52

Table II.8 (*Continued*)

T	w	μ	k	α/α_0	γ/γ_0
			$p=50$		
360	1009	−0,33	10,55	0,556	4,39
370	980,3	−0,31	9,81	0,576	4,26
380	953,0	−0,30	9,13	0,597	4,13
390	926,6	−0,28	8,49	0,618	4,01
400	901,2	−0,27	7,91	0,639	3,88
410	876,8	−0,25	7,37	0,660	3,76
420	853,5	−0,24	6,87	0,682	3,64
430	831,4	−0,22	6,41	0,703	3,53
440	810,5	−0,21	5,99	0,724	3,42
450	790,7	−0,19	5,61	0,745	3,31
460	772,2	−0,18	5,26	0,765	3,21
470	754,8	−0,17	4,95	0,785	3,11
480	738,6	−0,15	4,66	0,804	3,01
490	723,6	−0,14	4,40	0,823	2,92
500	709,6	−0,13	4,16	0,841	2,84
520	684,9	−0,10	3,74	0,873	2,68
540	664,1	−0,08	3,40	0,902	2,54
560	646,5	−0,06	3,12	0,927	2,42
580	631,9	−0,04	2,89	0,948	2,30
600	619,8	−0,03	2,69	0,966	2,20
620	609,9	−0,02	2,52	0,981	2,11
640	601,9	−0,01	2,38	0,994	2,03
660	595,5	0,00	2,26	1,003	1,96
680	590,5	0,01	2,15	1,010	1,89
700	586,7	0,01	2,07	1,014	1,84
			$p=55$		
110	—	−0,66	—	0,127	5,01
120	—	−0,61	—	0,144	5,02
130	—	−0,59	—	0,161	5,20
140	1814	−0,59	42,04	0,176	5,42
150	1807	−0,58	41,17	0,189	5,60
160	1791	−0,58	39,93	0,201	5,71
170	1764	−0,58	38,25	0,212	5,75
180	1727	−0,58	36,23	0,223	5,73
190	1685	−0,57	34,05	0,235	5,68
200	1639	−0,57	31,81	0,247	5,61
210	1591	−0,56	29,63	0,261	5,52
220	1544	−0,55	27,55	0,276	5,43
230	1498	−0,54	25,61	0,291	5,35
240	1453	−0,53	23,81	0,308	5,26
250	1411	−0,51	22,15	0,326	5,18
260	1371	−0,50	20,63	0,344	5,10
270	1332	−0,48	19,23	0,363	5,02
280	1296	−0,46	17,94	0,381	4,94
290	1261	−0,45	16,76	0,400	4,86
300	1227	−0,43	15,66	0,419	4,77
310	1195	−0,42	14,63	0,438	4,68
320	1163	−0,40	13,68	0,456	4,58
330	1133	−0,39	12,79	0,474	4,48
340	1103	−0,37	11,95	0,492	4,37
350	1074	−0,36	11,17	0,510	4,27

Table II.8 (*Continued*)

T	w	μ	k	α/α_0	γ/γ_0

$$p=55$$

T	w	μ	k	α/α_0	γ/γ_0
360	1046	—0,35	10,45	0,528	4,16
370	1019	—0,33	9,77	0,546	4,05
380	992,8	—0,32	9,13	0,564	3,93
390	967,3	—0,31	8,54	0,582	3,82
400	942,6	—0,29	7,99	0,600	3,71
410	918,9	—0,28	7,48	0,618	3,60
420	896,1	—0,27	7,01	0,637	3,50
430	874,4	—0,26	6,57	0,655	3,39
440	853,7	—0,24	6,17	0,673	3,29
450	834,0	—0,23	5,80	0,691	3,19
460	815,4	—0,22	5,46	0,708	3,10
470	797,8	—0,21	5,14	0,725	3,01
480	781,4	—0,20	4,86	0,742	2,92
490	766,0	—0,18	4,60	0,758	2,84
500	751,6	—0,17	4,36	0,774	2,76
520	725,8	—0,15	3,94	0,803	2,62
540	703,6	—0,13	3,59	0,830	2,49
560	684,8	—0,12	3,30	0,854	2,37
580	668,8	—0,10	3,05	0,875	2,26
600	655,4	—0,09	2,84	0,893	2,17
620	644,2	—0,08	2,67	0,909	2,08
640	634,9	—0,07	2,52	0,923	2,01
660	627,3	—0,06	2,39	0,934	1,94
680	621,0	—0,05	2,28	0,944	1,88
700	616,0	—0,04	2,18	0,951	1,82

$$p=60$$

T	w	μ	k	α/α_0	γ/γ_0
110	—	—0,66	—	0,123	4,38
120	—	—0,62	—	0,141	4,42
130	—	—0,59	—	0,159	4,65
140	—	—0,59	—	0,174	4,91
150	1804	—0,58	37,71	0,187	5,12
160	1796	—0,58	36,92	0,199	5,25
170	1774	—0,58	35,61	0,209	5,30
180	1742	—0,58	33,91	0,220	5,30
190	1702	—0,58	31,99	0,231	5,26
200	1659	—0,57	30,00	0,242	5,20
210	1613	—0,56	28,02	0,255	5,13
220	1567	—0,55	26,13	0,269	5,05
230	1522	—0,54	24,34	0,283	4,97
240	1478	—0,53	22,69	0,299	4,90
250	1436	—0,52	21,16	0,316	4,83
260	1397	—0,50	19,75	0,333	4,76
270	1359	—0,49	18,47	0,351	4,70
280	1324	—0,47	17,28	0,369	4,63
290	1289	—0,46	16,19	0,387	4,56
300	1257	—0,44	15,17	0,405	4,49
310	1225	—0,43	14,23	0,422	4,41
320	1195	—0,41	13,35	0,439	4,33
330	1166	—0,40	12,53	0,456	4,24
340	1137	—0,39	11,76	0,473	4,15
350	1110	—0,37	11,04	0,489	4,06

Table II.8 (*Continued*)

T	w	μ	k	α/α_0	γ/γ_0
			$p=60$		
360	1083	—0,36	10,36	0,505	3,96
370	1056	—0,35	9,73	0,521	3,86
380	1031	—0,34	9,14	0,537	3,76
390	1006	—0,33	8,58	0,552	3,66
400	982,3	—0,32	8,06	0,568	3,56
410	959,2	—0,30	7,58	0,584	3,47
420	937,0	—0,29	7,13	0,599	3,37
430	915,6	—0,28	6,71	0,615	3,28
440	895,2	—0,27	6,33	0,630	3,18
450	875,7	—0,26	5,97	0,646	3,10
460	857,2	—0,25	5,64	0,661	3,01
470	839,6	—0,24	5,33	0,676	2,93
480	823,0	—0,23	5,05	0,690	2,85
490	807,3	—0,22	4,79	0,705	2,77
500	792,6	—0,21	4,55	0,719	2,70
520	765,8	—0,19	4,13	0,745	2,56
540	742,6	—0,18	3,77	0,769	2,44
560	722,5	—0,16	3,47	0,792	2,33
580	705,3	—0,15	3,21	0,812	2,23
600	690,6	—0,14	3,00	0,830	2,14
620	678,1	—0,12	2,81	0,846	2,06
640	667,5	—0,11	2,65	0,861	1,98
660	658,7	—0,10	2,51	0,873	1,92
680	651,2	—0,10	2,39	0,884	1,86
700	645,1	—0,09	2,29	0,894	1,81
			$p=65$		
110	—	—0,67	—	0,118	3,80
120	—	—0,62	—	0,138	3,88
130	—	—0,60	—	0,158	4,16
140	—	—0,59	—	0,173	4,46
150	—	—0,58	—	0,186	4,70
160	1799	—0,58	34,33	0,197	4,85
170	1784	—0,58	33,35	0,207	4,93
180	1757	—0,58	31,93	0,217	4,94
190	1720	—0,58	30,25	0,227	4,91
200	1678	—0,57	28,46	0,237	4,86
210	1634	—0,57	26,67	0,249	4,79
220	1589	—0,56	24,92	0,262	4,73
230	1545	—0,55	23,28	0,276	4,66
240	1502	—0,54	21,74	0,291	4,59
250	1461	—0,52	20,32	0,307	4,53
260	1422	—0,51	19,02	0,324	4,48
270	1386	—0,49	17,82	0,341	4,42
280	1351	—0,48	16,72	0,358	4,36
290	1317	—0,46	15,70	0,375	4,31
300	1286	—0,45	14,76	0,392	4,24
310	1255	—0,43	13,89	0,408	4,18
320	1226	—0,42	13,07	0,424	4,11
330	1197	—0,41	12,31	0,440	4,03
340	1170	—0,40	11,59	0,455	3,96
350	1143	—0,38	10,92	0,470	3,88

Table II.8 (*Continued*)

T	w	μ	k	α/α_0	γ/γ_0

$p=65$

360	1117	—0,37	10,29	0,485	3,79
370	1092	—0,36	9,70	0,499	3,70
380	1067	—0,35	9,14	0,513	3,62
390	1044	—0,34	8,62	0,527	3,53
400	1020	—0,33	8,13	0,541	3,44
410	998,0	—0,32	7,68	0,555	3,35
420	976,3	—0,31	7,25	0,568	3,26
430	955,4	—0,30	6,85	0,582	3,18
440	935,4	—0,29	6,47	0,595	3,09
450	916,1	—0,29	6,13	0,609	3,01
460	897,7	—0,28	5,81	0,622	2,93
470	880,2	—0,27	5,51	0,635	2,85
480	863,5	—0,26	5,23	0,647	2,78
490	847,7	—0,25	4,97	0,660	2,71
500	832,8	—0,24	4,73	0,672	2,64
520	805,4	—0,23	4,31	0,696	2,51
540	781,2	—0,21	3,95	0,718	2,39
560	760,0	—0,20	3,64	0,739	2,29
580	741,6	—0,19	3,38	0,758	2,19
600	725,6	—0,17	3,15	0,775	2,11
620	711,9	—0,16	2,95	0,791	2,03
640	700,1	—0,15	2,79	0,806	1,96
660	690,0	—0,15	2,64	0,819	1,90
680	681,4	—0,14	2,51	0,831	1,84
700	674,1	—0,13	2,40	0,842	1,79

$p=70$

110	—	—0,67	—	0,113	3,27
120	—	—0,62	—	0,135	3,39
130	—	—0,60	—	0,156	3,72
140	—	—0,59	—	0,173	4,07
150	—	—0,58	—	0,185	4,35
160	1802	—0,58	32,07	0,196	4,52
170	1794	—0,58	31,40	0,205	4,61
180	1771	—0,58	30,24	0,214	4,63
190	1738	—0,58	28,77	0,223	4,61
200	1698	—0,57	27,16	0,233	4,57
210	1656	—0,57	25,51	0,244	4,51
220	1612	—0,56	23,90	0,256	4,45
230	1568	—0,55	22,37	0,269	4,39
240	1526	—0,54	20,94	0,284	4,33
250	1486	—0,53	19,61	0,299	4,28
260	1448	—0,51	18,39	0,315	4,23
270	1411	—0,50	17,27	0,331	4,18
280	1377	—0,49	16,24	0,347	4,13
290	1344	—0,47	15,29	0,364	4,09
300	1313	—0,46	14,41	0,380	4,03
310	1284	—0,44	13,59	0,396	3,98
320	1255	—0,43	12,83	0,411	3,92
330	1228	—0,42	12,12	0,426	3,86
340	1201	—0,40	11,45	0,440	3,79
350	1176	—0,39	10,82	0,454	3,72

Table II.8 (*Continued*)

T	w	μ	k	α/α_0	γ/γ_0

$p=70$

360	1151	—0,38	10,23	0,468	3,64
370	1126	—0,37	9,68	0,481	3,57
380	1103	—0,36	9,16	0,493	3,49
390	1080	—0,35	8,66	0,506	3,41
400	1057	—0,35	8,20	0,518	3,33
410	1035	—0,34	7,77	0,530	3,25
420	1014	—0,33	7,36	0,542	3,17
430	993,9	—0,32	6,97	0,554	3,09
440	974,3	—0,31	6,62	0,566	3,01
450	955,4	—0,30	6,28	0,577	2,94
460	937,2	—0,30	5,97	0,589	2,86
470	919,8	—0,29	5,67	0,600	2,79
480	903,2	—0,28	5,40	0,611	2,72
490	887,4	—0,27	5,15	0,622	2,65
500	872,3	—0,27	4,91	0,633	2,59
520	844,4	—0,25	4,49	0,654	2,47
540	819,5	—0,24	4,12	0,674	2,36
560	797,4	—0,23	3,81	0,693	2,26
580	778,0	—0,22	3,54	0,710	2,16
600	760,9	—0,21	3,30	0,727	2,08
620	746,0	—0,20	3,10	0,743	2,01
640	733,0	—0,19	2,92	0,757	1,94
660	721,7	—0,18	2,77	0,770	1,88
680	712,0	—0,17	2,63	0,783	1,82
700	703,5	—0,16	2,51	0,794	1,77

$p=75$

110	—	—0,68	—	0,107	2,77
120	—	—0,63	—	0,132	2,93
130	—	—0,60	—	0,156	3,33
140	—	—0,58	—	0,173	3,73
150	—	—0,58	—	0,185	4,04
160	1804	—0,58	30,09	0,195	4,23
170	1803	—0,58	29,71	0,203	4,33
180	1785	—0,58	28,78	0,211	4,37
190	1755	—0,58	27,49	0,220	4,35
200	1718	—0,57	26,04	0,229	4,32
210	1677	—0,57	24,52	0,239	4,27
220	1634	—0,56	23,03	0,251	4,21
230	1592	—0,56	21,60	0,263	4,16
240	1550	—0,54	20,25	0,277	4,11
250	1510	—0,53	19,00	0,291	4,06
260	1473	—0,52	17,85	0,306	4,02
270	1437	—0,51	16,79	0,322	3,97
280	1403	—0,49	15,82	0,338	3,93
290	1371	—0,48	14,93	0,354	3,89
300	1341	—0,46	14,10	0,369	3,85
310	1312	—0,45	13,33	0,385	3,81
320	1284	—0,44	12,62	0,399	3,76
330	1258	—0,42	11,95	0,413	3,70
340	1232	—0,41	11,33	0,427	3,64
350	1207	—0,40	10,74	0,440	3,58

Table II.8 (*Continued*)

T	w	μ	k	α/α_0	ν/ν_0
			$p=75$		
360	1183	—0,39	10,19	0,452	3,52
370	1159	—0,38	9,66	0,464	3,45
380	1137	—0,37	9,17	0,476	3,38
390	1114	—0,36	8,70	0,487	3,31
400	1093	—0,36	8,26	0,498	3,23
410	1072	—0,35	7,85	0,509	3,16
420	1051	—0,34	7,46	0,520	3,09
430	1031	—0,33	7,09	0,530	3,02
440	1012	—0,33	6,75	0,541	2,94
450	993,5	—0,32	6,42	0,551	2,87
460	975,7	—0,31	6,12	0,561	2,81
470	958,5	—0,31	5,84	0,571	2,74
480	942,0	—0,30	5,57	0,581	2,67
490	926,2	—0,29	5,32	0,590	2,61
500	911,1	—0,29	5,09	0,600	2,55
520	883,0	—0,28	4,66	0,618	2,43
540	857,6	—0,26	4,30	0,636	2,33
560	834,8	—0,25	3,98	0,653	2,23
580	814,5	—0,24	3,70	0,669	2,14
600	796,4	—0,23	3,46	0,685	2,06
620	780,5	—0,22	3,24	0,700	1,99
640	766,4	—0,22	3,06	0,714	1,92
660	754,0	—0,21	2,90	0,727	1,86
680	743,1	—0,20	2,75	0,739	1,81
700	733,5	—0,19	2,62	0,751	1,76
			$p=80$		
110	—	—0,69	—	0,100	2,28
120	—	—0,63	—	0,129	2,48
130	—	—0,60	—	0,156	2,96
140	—	—0,58	—	0,174	3,42
150	—	—0,58	—	0,186	3,77
160	—	—0,58	—	0,195	3,98
170	1813	—0,58	28,23	0,202	4,09
180	1800	—0,58	27,50	0,209	4,14
190	1773	—0,58	26,39	0,217	4,13
200	1738	—0,58	25,07	0,226	4,10
210	1699	—0,57	23,67	0,235	4,06
220	1657	—0,57	22,28	0,246	4,00
230	1615	—0,56	20,93	0,257	3,95
240	1574	—0,55	19,66	0,270	3,91
250	1535	—0,54	18,48	0,284	3,87
260	1497	—0,52	17,39	0,299	3,83
270	1462	—0,51	16,39	0,314	3,79
280	1428	—0,50	15,46	0,329	3,76
290	1397	—0,48	14,62	0,344	3,73
300	1367	—0,47	13,84	0,360	3,69
310	1339	—0,45	13,11	0,374	3,65
320	1312	—0,44	12,44	0,389	3,61
330	1286	—0,43	11,81	0,402	3,56
340	1262	—0,42	11,22	0,415	3,51
350	1238	—0,41	10,67	0,427	3,46

Table II.8 (*Continued*)

T	w	μ	k	α/α₀	? ?:

$p = 80$

T	w	μ	k	α/α₀	
360	1214	—0,40	10.14	0,439	3,40
370	1192	—0,39	9,65	0,450	3.34
380	1169	—0,38	9,18	0,461	3,28
390	1148	—0,37	8,74	0,471	3,22
400	1127	—0,37	8,32	0,481	3,15
410	1107	—0,36	7,93	0,491	3,08
420	1087	—0,35	7,56	0,501	3,02
430	1067	—0,34	7,21	0,510	2,95
440	1049	—0,34	6,87	0,519	2,88
450	1031	—0,33	6,56	0,528	2,82
460	1013	—0,33	6,27	0,537	2,75
470	996,2	—0,32	5,99	0,546	2,69
480	980,0	—0,31	5,73	0,554	2,63
490	964,3	—0,31	5,48	0,563	2,57
500	949,3	—0,30	5,25	0,571	2,51
520	921,1	—0,29	4,83	0,587	2,40
540	895,4	—0,28	4,47	0,603	2,30
560	872,1	—0,27	4,14	0,619	2,21
580	851,1	—0,26	3,86	0,634	2,12
600	832,2	—0,26	3,61	0,648	2,04
620	815,3	—0,25	3,39	0,662	1.97
640	800,2	—0,24	3,20	0,675	1.90
660	786,8	—0,23	3,03	0,687	1,85
680	774,8	—0,22	2,88	0,700	1,79
700	764,2	—0,22	2,74	0,711	1,74

$p = 85$

T	w	μ	k	α/α₀	
130	—	—0,60	—	0,158	2,62
140	—	—0,58	—	0,177	3,15
150	—	—0,58	—	0,187	3,53
160	—	—0,58	—	0,195	3,76
170	1822	—0,58	26,93	0,201	3,89
180	1815	—0,58	26,40	0,208	3,94
190	1791	—0,58	25,43	0,215	3,94
200	1759	—0,58	24,24	0,222	3.91
210	1720	—0,57	22,94	0,231	3.87
220	1680	—0,57	21,63	0,241	3,82
230	1639	—0,56	20,36	0,252	3,78
240	1598	—0,55	19,16	0,264	3,73
250	1559	—0,54	18,03	0,277	3,70
260	1522	—0,53	16,99	0,291	3,66
270	1487	—0,52	16,03	0,306	3,63
280	1454	—0,50	15,15	0,321	3,60
290	1423	—0,49	14,35	0,336	3.58
300	1393	—0,47	13,61	0,351	3,55
310	1366	—0,46	12,92	0,365	3.51
320	1340	—0,45	12,28	0,379	3,48
330	1314	—0,43	11,69	0,392	3.44
340	1290	—0,42	11,13	0,404	3.10
350	1267	—0,41	10,60	0,416	3.35

Table II.8 (*Continued*)

T	w	η	k	α/α_0	γ/γ_0
			$p=85$		
360	1245	—0,40	10,11	0,427	3,30
370	1223	—0,40	9,64	0,438	3,25
380	1201	—0,39	9,20	0,448	3,19
390	1181	—0,38	8,78	0,457	3,14
400	1160	—0,37	8,38	0,467	3,08
410	1141	—0,37	8,01	0,475	3,02
420	1121	—0,36	7,65	0,484	2,96
430	1103	—0,35	7,31	0,492	2,89
440	1084	—0,35	6,99	0,501	2,83
450	1067	—0,34	6,69	0,509	2,77
460	1050	—0,34	6,40	0,516	2,71
470	1033	—0,33	6,14	0,524	2,65
480	1017	—0,33	5,88	0,532	2,59
490	1002	—0,32	5,64	0,539	2,54
500	986,8	—0,32	5,42	0,546	2,48
520	958,7	—0,31	5,00	0,561	2,38
540	932,9	—0,30	4,64	0,575	2,28
560	909,3	—0,29	4,31	0,589	2,19
580	887,7	—0,28	4,02	0,602	2,10
600	868,2	—0,27	3,77	0,615	2,03
620	850,6	—0,27	3,55	0,628	1,96
640	834,6	—0,26	3,35	0,640	1,89
660	820,2	—0,25	3,17	0,652	1,84
680	807,3	—0,24	3,01	0,664	1,78
700	795,7	—0,24	2,87	0,675	1,73
			$p=90$		
130	—	—0,59	—	0,164	2,29
140	—	—0,58	—	0,181	2,91
150	—	—0,57	—	0,190	3,32
160	—	—0,57	—	0,196	3,58
170	1832	—0,57	25,79	0,201	3,71
180	1830	—0,58	25,43	0,207	3,77
190	1810	—0,58	24,60	0,213	3,77
200	1779	—0,58	23,51	0,220	3,75
210	1743	—0,57	22,31	0,227	3,71
220	1703	—0,57	21,07	0,237	3,67
230	1662	—0,56	19,87	0,247	3,62
240	1622	—0,55	18,72	0,258	3,58
250	1583	—0,54	17,64	0,271	3,55
260	1546	—0,53	16,64	0,284	3,51
270	1511	—0,52	15,73	0,299	3,49
280	1479	—0,51	14,89	0,313	3,46
290	1448	—0,49	14,11	0,328	3,44
300	1419	—0,48	13,40	0,342	3,42
310	1392	—0,46	12,75	0,356	3,39
320	1366	—0,45	12,14	0,370	3,36
330	1342	—0,44	11,58	0,382	3,33
340	1318	—0,43	11,05	0,395	3,30
350	1296	—0,42	10,55	0,406	3,26

Table II.8 (*Continued*)

T	w	μ	k	α/α_0	$\gamma\,\gamma_0$
			$p=90$		
360	1274	—0,41	10,08	0,417	3,21
370	1253	—0,40	9,64	0,427	3,17
380	1232	—0,39	9,22	0,436	3,12
390	1212	—0,39	8,82	0,445	3,07
400	1193	—0,38	8,44	0,454	3,01
410	1174	—0,37	8,08	0,462	2,96
420	1155	—0,37	7,74	0,470	2,90
430	1137	—0,36	7,41	0,477	2,84
440	1119	—0,36	7,10	0,484	2,79
450	1102	—0,35	6,81	0,492	2,73
460	1085	—0,35	6,54	0,499	2,67
470	1069	—0,34	6,27	0,505	2,62
480	1053	—0,34	6,03	0,512	2,56
490	1038	—0,33	5,79	0,519	2.51
500	1024	—0,33	5,57	0,525	2,46
520	995,7	—0,32	5,16	0,538	2,36
540	969,9	—0,31	4,80	0,551	2,26
560	946,1	—0,30	4,48	0,563	2,17
580	924,3	—0,30	4,19	0,575	2,09
600	904,3	—0,29	3,93	0,587	2,02
620	886,0	—0,28	3,70	0,598	1,95
640	869,4	—0,28	3,49	0,609	1,89
660	854,2	—0,27	3,31	0,621	1,83
680	840,4	—0,26	3,15	0,632	1,78
700	827,9	—0,26	3,00	0.642	1,73
			$p=95$		
140	—	—0,57	—	0,188	2,68
150	—	—0,57	—	0,193	3,14
160	—	—0,57	—	0,197	3,41
170	—	—0,57	—	0,201	3,56
180	1846	—0,57	24,58	0,206	3,62
190	1829	—0,58	23,87	0,211	3,63
200	1801	—0,58	22,88	0,217	3,61
210	1765	—0,57	21,76	0,224	3,57
220	1726	—0,57	20,59	0,233	3,53
230	1686	—0,56	19,44	0,242	3,48
240	1646	—0,56	18,34	0,253	3,45
250	1608	—0,55	17,30	0,265	3,41
260	1571	—0,54	16,34	0,278	3,38
270	1536	—0,52	15,46	0.292	3,36
280	1503	—0,51	14,65	0,306	3,34
290	1473	—0,50	13,91	0,320	3,32
300	1444	—0,48	13,23	0,334	3.30
310	1418	—0,47	12,60	0,348	3,28
320	1392	—0,46	12,02	0,361	3,26
330	1369	—0,44	11,48	0,374	3,23
340	1346	—0,43	10,97	0,386	3,20
350	1324	—0,42	10,50	0,397	3,17

Table II.8 (*Continued*)

T	w	μ	k	α/α_0	γ/γ_0
			$p=95$		
360	1303	—0,41	10,05	0,407	3,13
370	1282	—0,40	9,63	0,417	3,09
380	1262	—0,40	9,23	0,426	3,05
390	1243	—0,39	8,85	0,434	3,00
400	1224	—0,38	8,49	0,442	2,95
410	1206	—0,38	8,15	0,450	2,90
420	1188	—0,37	7,82	0,457	2,85
430	1170	—0,37	7,51	0,464	2,80
440	1153	—0,36	7,21	0,470	2,74
450	1136	—0,36	6,93	0,477	2,69
460	1120	—0,35	6,66	0,483	2,64
470	1104	—0,35	6,41	0,489	2,59
480	1089	—0,34	6,17	0,495	2,54
490	1074	—0,34	5,94	0,501	2,48
500	1060	—0,34	5,72	0,507	2,43
520	1032	—0,33	5,32	0,518	2,34
540	1006	—0,32	4,96	0,529	2,25
560	982,6	—0,32	4,64	0,540	2,16
580	960,6	—0,31	4,35	0,551	2,08
600	940,3	—0,30	4,09	0,561	2,01
620	921,6	—0,30	3,85	0,572	1,94
640	904,4	—0,29	3,64	0,582	1,88
660	888,6	—0,28	3,46	0,592	1,82
680	874,1	—0,28	3,28	0,603	1,77
700	860,8	—0,27	3,13	0,613	1,72
			$p=100$		
140	—	—0,56	—	0,201	2,49
150	—	—0,56	—	0,199	2,99
160	—	—0,56	—	0,200	3,27
170	—	—0,57	—	0,202	3,42
180	1862	—0,57	23,84	0,205	3,49
190	1849	—0,57	23,25	0,209	3,50
200	1822	—0,58	22,34	0,215	3,48
210	1788	—0,57	21,28	0,221	3,45
220	1750	—0,57	20,17	0,229	3,40
230	1711	—0,57	19,07	0,238	3,36
240	1671	—0,56	18,01	0,248	3,32
250	1632	—0,55	17,01	0,259	3,29
260	1595	—0,54	16,08	0,272	3,27
270	1561	—0,53	15,23	0,285	3,25
280	1528	—0,51	14,45	0,299	3,23
290	1498	—0,50	13,73	0,313	3,21
300	1469	—0,49	13,07	0,327	3,20
310	1443	—0,47	12,47	0,340	3,18
320	1418	—0,46	11,91	0,353	3,16
330	1395	—0,45	11,39	0,365	3,14
340	1372	—0,44	10,91	0,377	3,12
350	1351	—0,43	10,46	0,388	3,09

Table II.8 (*Continued*)

T	w	μ	k	α/α_0	γ/γ_0
			$p=100$		
360	1331	—0,42	10,03	0,398	3,06
370	1311	—0,41	9,63	0,408	3,02
380	1292	—0,40	9,25	0,416	2,98
390	1273	—0,39	8,88	0,424	2,94
400	1255	—0,39	8,54	0,432	2,90
410	1237	—0,38	8,21	0,439	2,85
420	1220	—0,38	7,90	0,446	2,81
430	1203	—0,37	7,60	0,452	2,76
440	1186	—0,37	7,31	0,458	2,71
450	1170	—0,36	7,04	0,464	2,66
460	1154	—0,36	6,78	0,469	2,61
470	1139	—0,36	6,53	0,475	2,56
480	1124	—0,35	6,30	0,480	2,51
490	1109	—0,35	6,08	0,485	2,46
500	1095	—0,34	5,86	0,491	2,42
520	1068	—0,34	5,47	0,501	2,32
540	1042	—0,33	5,11	0,510	2,24
560	1019	—0,33	4,79	0,520	2,16
580	996,6	—0,32	4,51	0,530	2,08
600	976,1	—0,31	4,24	0,539	2,01
620	957,1	—0,31	4,01	0,548	1,94
640	939,5	—0,30	3,80	0,558	1,88
660	923,2	—0,30	3,60	0,567	1,82
680	908,1	—0,29	3,43	0,577	1,77
700	894,1	—0,29	3,27	0,586	1,72

PART
THREE

REFERENCES

1. Anisimov, M. A., Beketov, V. G., Voronov, Ya. P. et al. Experimental investigation of the isochoric specific heat of propane in the single-phase and two-phase regions. In: Thermophysical Properties of Substances and Materials, No. 16, Izd. Standartov, Moscow, pp. 48–59.
2. Barkan, E. S. Review and refining of the experimental data on the second virial coefficient of propane and propylene, Zh. Fiz. Khimii, 1983, Vol. 57, No. 6, pp. 1351–1355.
3. Vasserman, A. A. and Kreizerova, A. Ya. Optimization of the number of coefficients in the equation of state. Teplofiz. Vys. Temp., 1978, Vol. 16, No. 6, pp. 1185–1188.
4. Golovskoi, E. A., Zagoruchenko, V. A., and Tsymarny, V. A. Measurements of propane density at temperatures 88.24–272.99 K and pressures up to 609.97 bar, Deposited at the Institute VNIIEGasprom, 1978, No. 45 M, 03.10.
5. Golovskoi, E. A., Mitskevich, E. P., and Tsymarny, V. A. Measurements of ethane density at temperatures 90.24–270.21 K and pressures up to 604.09 bar, Deposited at the Institute VNIIEGasprom, 1978, No. 39M, 25.07.
6. Guigo, E. I., Ershova, N. S., and Margolin, M. F. Investigation of the thermal properties of propane, Kholod. Tekhnika, 1978, No. 11, pp. 29–30.
7. Ershova, N. S. and Kletskii, A. V. Thermodynamic properties of propane. In:

Thermophysical Properties of Substances and Materials (Collection of Physical Constants and Properties of Substances), GSSD, Izd. Standartov, Moscow, 1985, No. 22, pp. 75–86.

8. Shmakov, N. G., Gorbunova, V. G., and Chernova, G. N. Propane: Isochoric specific heat in the two-phase region for the temperature range 90–350 K, GSSD, Izd. Standartov, Moscow, 1983, pp. 38–82.

9. Sychev, V. V., Vasserman, A. A., Kozlov, A. D., Spiridonov, G. A., and Tsymarny, V. A. Thermodynamic properties of nitrogen, Hemisphere, Washington, 1987.

10. Sychev, V. V., Vasserman, A. A., Kozlov, A. D., Spiridonov, G. A., and Tsymarny, V. A. Thermodynamic properties of helium, Hemisphere, Washington, 1987.

11. Sychev, V. V., Vasserman, A. A., Zagoruchenko, V. A., Kozlov, A. D., Spiridonov, G. A., and Tsymarny, V. A. Thermodynamic properties of methane, Hemisphere, Washington, 1987.

12. Anderson, R. P. Gasoline from natural gas. I. Method of removal. J. Ind. Eng. Chem. 1920, Vol. 12, No. 6, pp. 547–549.

13. Babb, S. E., Jr. and Robertson, S. L. Isotherms of ethylene and propane to 10,000 bar. J. Chem. Phys. 1970, Vol. 53, No. 3, pp. 1097–1099.

14. Beattie, J. A., Kay, W. C., and Kaminsky, J. The compressibility of and an equation of state for gaseous propane. J. Amer. Chem. Soc. 1937, Vol. 59, No. 9, pp. 1589–1590.

15. Beattie, J. A., Poffenberger, N., and Hadlock, C. The critical constants of propane. J. Chem. Phys. 1935, Vol. 3, pp. 96–97.

16. Beeck, O. The exchange of energy between organic molecules and solid surfaces. Part 1. Accommodation coefficients and specific heats of hydrocarbon molecules. J. Chem. Phys. 1936, Vol. 4, pp. 680–689.

17. Bottomley, G. A., Massie, D. S., and Whytlaw, R. A comparison of the compressibilities of some gases with that of nitrogen at pressures below one atmosphere. Proc. Roy. Soc., London, 1950, Vol. A200, pp. 201–208.

18. Brever, J. Determination of mixed virial coefficients. Midwest Res. Inst., Kansas City. December, 1967, No. 64110.

19. Burgoyne, J. H. Two-phase equilibrium in binary and ternary systems. IV. The thermodynamic properties of propane. Proc. Roy. Soc. 1940, Vol. A176, No. 965, pp. 280–295.

20. Burrell, G. A. and Jones, G. W. Pressure-volume deviation of methane, propane, and carbon dioxide at elevated pressures. Bur. Mines., Rept. Investig. 1921, Ser. 2276.

21. Burrell, G. A. and Robertson, I. W. The vapor pressures of propane, propylene, and normal butane at low temperatures. J. Amer. Chem. Soc. 1915, Vol. 37, pp. 2188–2193.

22. Carney, B. R. Density of liquefied petroleum gas hydrocarbons, their mixtures, and three natural gasolines. Petro. Ref. 1942, Vol. 21, No. 9, p. 274.

23. Carruth, G. F. Determination of vapor pressures of n-paraffins and extension of a corresponding states correlation to low reduced temperatures. Thesis. Dept. Chem. Eng., Rice Univ., Houston, TX, 1970.

24. Carruth, G. F. and Kobayashi, R. Vapor pressure of normal paraffins ethane

through n-decane from their triple points to about 10 mm Hg. J. Chem. Eng. Data. 1973, Vol. 18, No. 2, pp. 115–126.

25. Chao, J., Wilhoit, R. c., and Zwolinski, B. J. Ideal gas thermodynamic properties of ethane and propane. J. Phys. Chem. Ref. Data. 1973, Vol. 2, No. 2, pp. 427–437.

26. Cherney, B. J., Marchman, H., and York, R., Jr. Equipment for compressibility measurements. Ind. Eng. Chem. 1949, Vol. 41, No. 11, pp. 2653–2658.

27. Chui, C. and Canfield, F. B. Liquid density and excess properties of argon-krypton and krypton-xenon binary liquid mixtures and density of ethane. Trans. Faraday Soc. 1971, Vol. 67, No. 10, pp. 2933–2940.

28. Clegg, H. P. and Rowlinson, J. S. The physical properties of some fluorine compounds and their solutions. 2. The system sulphur hexafluoride + propane. Trans. Faraday Soc. 1955, Vol. 51, No. 10, pp. 1333–1340.

29. Cutler, A. J. and Morrison, J. A. Excess thermodynamic functions for liquid mixtures of methane + propane. Trans. Faraday Soc. 1965, Vol. 61, Part 3, No. 507, pp. 429–437.

30. Dailey, B. P. and Felsing, W. A. The heat capacities at higher temperatures of ethane and propane. J. Amer. Chem. Soc. 1943, Vol. 65, No. 1, pp. 42–44.

31. Dana, L. J., Jenkins, A. C., Burdick, J. N., and Timm, R. C. Thermodynamic properties of butane, isobutane and propane. Refrig. Eng. 1926, Vol. 12, No. 12, pp. 387–405.

32. Das, T. R. and Eubank, P. T. Thermodynamic properties of propane. 1. Vapor-liquid coexistence curve. Adv. Cryog. Eng. 1973, Vol. 18, pp. 208–219.

33. Dawson P. P., Jr. and McKetta, J. J. Z_s for propane and methylacetylene. Petro. Ref. 1960, Vol. 39, No. 4, pp. 151–154.

34. Delaplace, R. Tension de vapeur des carbures satures et non satures aux basses temperatures. Compt. Rend. 1937, Vol. 204, No. 493, pp. 1940–1941.

35. Deshner, W. W. and Brown, G. G. P-V-T relations for propane. Ind. Eng. Chem. 1940, Vol. 32, No. 6, pp. 836–840.

36. Dittmar, P., Schulz, F., and Strese, G. Druck, Dichte, Temperaturwerte für Propan und Propylen. Chem. Ing. Techn. 1962, Vol. 34, No. 6, pp. 437–441.

37. Ely, J. F. and Kobayashi, R. Isochoric pressure-volume-temperature measurements for compressed liquid propane. J. Chem. Eng. Data. 1978, Vol. 23, No. 3, pp. 221–223.

39. Ernst, G. Zur universellen Darstellung des Realantels der spezifischen Wärme von Gasen. Chem. Ing. Techn. 1969, No. 41, pp. 544–551.

40. Ernst, G. and Busser, J. Ideal and real gas state heat calacitves c_p of C_3H_8, i-C_4H_{10}, C_2F_5Cl, CH_2ClCF_3, $CF_3ClCFCl_2$, and CHF_2Cl. J. Chem. Thermodynamics. 1970, Vol. 2, pp. 787–791.

41. Francis, A. W. and Robbins, G. W. The vapor pressures of propane and propylene. J. Amer. Chem. Soc. 1933, Vol. 55, No. 11, pp. 4339–4342.

42. Gilliland, E. R. and Scheeline, H. W. High-pressure vapor-liquid equilibrium for the systems propylene-isobutane and propanehydrogen sulfide. Ind. Eng. Chem. 1940, Vol. 32, No. 1, pp. 48–54.

43. Goodwin, R. D. Specific heats of saturated and compressed liquid propane. J. Res. NBS. 1978, Vol. 83, No. 5, pp. 449–458.

44. Goodwin, R. D. and Haynes, W. M. Thermophysical properties of propane from

85 to 700 K at pressures to 70 MPa. U.S. Dept. of Commer. NBS. 1982. Monogram. No. 170.

45. Gunn, R. D. The volumetric properties of nonpolar mixtures. M.S. Thesis. Univ. of Calif., Berkeley. 1958.

46. Hahn, R., Schäfer, K., and Schramm, B. Messungen zweiten Virialkoeffizienten im Temperaturbereich von 200–300 K. Ber. Bunsenges. Phys. Chem. 1974, No. 78, No. 3, pp. 287–289.

47. Haynes, W. M. Measurements of densities and dielectric constants of liquid propane from 90 to 300 K at pressures to 35 MPa. J. Chem. Thermodynamics. 1983, Vol. 15, No. 3, pp. 419–424.

48. Haynes, W. M. and Hiza, M. J. Measurements of the orthobaric liquid densities of methane, ethane, propane, isobutane, and normal butane. J. Chem. Thermodynamics. 1977, Vol. 9, No. 2, pp. 179–187.

49. Helgeson, N. L. and Sage, B. H. Latent heat of vaporization of propane. J. Chem. Eng. Data. 1967, Vol. 12, No. 1, pp. 47–49.

50. Heuse, W. Molvolumen von Kohlenwasserstoffen und einigen anderen Verbindungen bei tiefem Temperatur. Z. Phys. Chem. 1930, Vol. 147, pp. 266–274.

51. Hicks-Bruun, M. M. and Bruun, J. H. The freezing point and boiling point of propane. J. Amer. Chem. Soc. 1936, Vol. 58, pp. 810–812.

52. Hirschfelder, J. O., McClure, F. T., and Weeks, I. F. Second virial coefficients and the forces between complex molecules. J. Chem. Phys. 1942, Vol. 10, No. 4, pp. 201–211.

53. Huang, E. T. S., Swift, G. W., and Kurata, F. Viscosities of methane and propane at low temperatures and high pressures. AIChE Journal. 1966, Vol. 12, No. 5, pp. 932–936.

54. Huff, Y. A. and Reed, T. M. Second virial coefficients of mixtures of nonpolar molecules from correlations on pure components. J. Chem. Eng. Data. 1963, Vol. 8, No. 3, pp. 306, 311.

55. Jensen, R. H. and Kurata, F. Density of liquefied natural gas. J. Petrol. Technol. June, 1969, pp. 683–691.

56. Jensen, F. W. and Lightfoot, J. H. Compressibility of butane-pentane mixtures below one atmosphere. Ind. Eng. Chem. 1938, Vol. 30, pp. 312–314.

57. Kahre, L. C. Liquid density of light hydrocarbon mixtures. J. Chem. Eng. Data. 1973, Vol. 18, No. 3, pp. 267–270.

58. Kahre, L. C. and Livingston, R. J. More accuracy in liquid propane compressibility. Petrol. Refin. 1964, Vol. 43, No. 4, pp. 119–121.

59. Kapallo, W., Lund, N., and Schäfer, K. Zwischenmolekulare Kräfte zwischen gleichen und ungleichen Molekulen aus Virialkoeffizienten. Z. Phys. Chem. Neue Folge. 1963, Vol. 37, pp. 196–209.

60. Kay, W. B. Vapor-liquid equilibrium relations of binary systems. The propane-n-alkane systems. N-butane and n-pentane. J. Chem. Eng. Data. 1970, Vol. 15, No. 1, pp. 46–52.

61. Kemp. J. D. and Egan, C. J. Hindered rotation of the methyl groups in propane. The heat capacity, vapor pressure, heats of fusion, and vaporization of propane. Entropy and density of the gas. J. Amer. Chem. Soc. 1938, Vol. 60, No. 7, pp. 1521–1525.

63. Kistiakowski, G. B., Lecher, J. R., and Ransom, W. W. The low temperature

gaseous heat capacities of certain C_3 hydrocarbons. J. Chem. Phys. 1940, Vol. 8, pp. 970–977.

64. Kistiakowski, G. B. and Rice, W. W. Gaseous heat capacities. J. Chem. Phys. 1940, Vol. 8, pp. 610–618.

65. Klosek, J. and McKinley, C. Densities of liquefied natural gas and of low molecular weight hydrocarbons. Pap. 22, sess. 5. 1st Int. Conf. on LNG, Chicago, Apr. 1968.

66. Kratzke, H. Thermodynamic quantities for propane. 1. The vapor pressure of liquid propane. J. Chem. Thermodyn. 1980, Vol. 12, pp. 305–309.

67. Kratzke, H. and Muller, S. Thermodynamic quantities for propane. 3. The thermodynamic behavior of saturated and compressed propane. J. Chem. Thermodyn. 1984, Vol. 14, No. 12, pp. 1157–1174.

68. Kretschmer, C. B. and Wiebe, R. The solubility of propane and the butanes in ethanol. J. Amer. Chem. Soc. 1951, Vol. 73, pp. 3778–3781.

69. Kudchadker, A. P., Alani, G. H., and Zwolinsky, B. J. The critical constants of organic substances. Chem. Rev. 1968, Vol. 68, No. 6, pp. 659–735.

70. Lacam, A. Etude experimentale de la propagation des ultrasons dans les fluides. En function de la pression (1200 atmospheres) et de la temperature (200 °C). J. Rech. CNRS. 1956, No. 34, pp. 25–56.

71. Lebeau, P. Sur quelques properietes physiques du propane. Compt. Rend. Acad. Sci. Paris, 1905, Vol. 201, pp. 1454–1456.

72. Maass, O. and Wright, C. H. Some physical properties of hydrocarbons containing two and three carbon atoms. J. Amer. Chem. Soc. 1921, Vol. 43, pp. 1098–1111.

73. Massie, D. S. and Whytlaw-Gray, R. The normal density of propane and its expansion coefficients between 0° and 20°. J. Chem. Soc. 1949, Vol. 7, pp. 2874–2877.

74. McClure, C. R. Measurement of the densities of liquefied hydrocarbons from 92 to 173 K. Cryogenics. 1976, Vol. 16, No. 5, pp. 289–295.

75. McGlashan, M. L. and Potter, D. J. An apparatus for the measurement of the second virial coefficients of vapors. The second virial coefficients of some n-alkanes and of some mixtures of n-alkanes. Proc. Roy. Soc. London. 1962, Vol. A267, pp. 478–500.

76. Meyers, C. H. An equation for the isotherms of pure substances at their critical temperatures. J. Res. NBS. 1942, Vol. 29, pp. 157–176.

77. Moussa, A. H. H. The physical properties of highly purified samples of propane and n-hexane. J. Chem. Thermodyn. 1977, Vol. 9, pp. 1063–1065.

78. National Gasoline Assoc. of Amer. Techn. Committee. Densities of liquefied petroleum gases. Ind. Eng. Chem. 1942, Vol. 34, No. 10, pp. 1240–1243.

79. Orrit, J. and Laupretre, J. M. Density of liquefied natural gas components. Adv. Cryog. Eng. 1978, Vol. 23, pp. 573–579.

80. Orrit, J. and Olives, J. F. Density of liquefied natural gas and its components. 4th Int. Conf. LNG, Proc. Algeria, 1974.

81. Pavese, F. Some thermodynamic properties of ethane between its double solid-to-solid transition and its triple-point temperature. J. Chem. Thermodyn. 1978, Vol. 10, pp. 369–379.

82. Pavese, F. and Besley, L. M. Triple-point temperature of propane: measurements

on two solid-to-liquid transition and one solid-to-solid transition. J. Chem. Thermodyn. 1981, Vol. 13, pp. 1095–1104.

83. Rao Seshadri, M. G. S. Temperature variation of ultrasonic velocity & related thermodynamic parameters in liquid propane and n-butane. Indian J. Pure Appl. Phys. 1971, Vol. 9, pp. 169–170.

84. Reamer, H. H., Sage, B. H., and Lacey, W. N. Phase equilibria in hydrocarbon systems. Volumetric behavior of propane. Ind. Eng. Chem. 1949, Vol. 41, No. 3, pp. 482–484.

85. Reeves, L. E., Scott, G. J., and Babb, S. E., Jr. Melting curves of pressure-transmitting fluids. J. Chem. Phys. 1964, Vol. 40, No. 12, pp. 3662–3666.

86. Rodosevich, J. B. and Miller, R. C. Experimental liquid mixture densities for testing and improving correlations for liquefied natural gas. AIChE Journal. 1973, Vol. 19, No. 4, pp. 729–735.

87. Sage, B. H., Evans, B. H., and Lacey, W. N. Phase equilibria in hydrocarbon systems. Latent heat of vaporization of propane and n-pentane. Ind. Eng. Chem. 1939, Vol. 31, No. 6, pp. 763–767.

88. Sage, B. H. and Lacey, W. N. Phase equilibria in hydrocarbon systems. IX. Specific heats of n-butane and propane. Ind. Eng. Chem. 1935, Vol. 27, No. 12, pp. 1484–1488.

89. Sage, B. H., Schaafsma, J. G., and Lacey, W. N. Phase equilibria in hydrocarbon systems. V. Pressure-volume-temperature relations and thermal properties of propane. Ind. Eng. Chem. 1934, Vol. 26, No. 11, pp. 1218–1224.

90. Sage, B. H., Webster, D. C., and Lacey, W. N. Phase equilibria in hydrocarbon systems. XX. Heat capacity isobaric of gaseous propane, n-butane, isobutane, and n-pentane. Ind. Eng. Chem. 1937, Vol. 29, No. 11, pp. 1309–1314.

91. Schäfer, K. Schramm, B., and Navarro, J. S. U. Bestimmung von zwischenmolekularen Potentialen aus Messwerten des zweiten Virialkoeffizienten und Berechnung der Wärmeleitfähigkeiten von Gasmischungen. Z. Phys. Chem. Neue Folge. 1974, Vol. 93, pp. 203–216.

92. Seshadri, D. N. and Viswanath, D. S. Thermodynamic properties of propane. Proc. 5th Bien. Int. CODATA Conf., Boulder, Colo., 1976, Oxford, e.a., 1977, pp. 413–423.

93. Shana'a, M. Y. and Canfield, F. B. Liquid density and excess volume of liquid hydrocarbon mixtures at 165 °C. Trans. Faraday Soc. 1968, Vol. 64, No. 549, pp. 2281–2286.

94. Sliwinski, P. die Lorentz-Lorenz Funktion von dampfförmigen und flüssigen Äthan, Propan, und Butan. Z. Phys. Chem., Neue Folge. 1969, Vol. 63, pp. 263–270.

95. Sliwinski, P. Die Clausius-Mossotti-Funktion fuer die gesaettigten Daempfe und Fluessigkeiten des Aethans und Propans. Z. Phys. Chem., Frankfurt. 1969, Vol. 68, pp. 91–98.

96. Staveley, L. A. K. and Tupman, W. I. Entropies of vaporization and internal order in liquids. J. Chem. Soc. 1950, Vol. 54, pp. 1950–1953.

97. Straty, G. C. and Palavra, A. M. F. Automated high-temperature pvT apparatus with data for propane. J. Res. NBS. 1984, Vol. 89, No. 5, pp. 375–383.

98. Strein, V. K., Lichtenchalter, R. N., Schramm, B., and Schäfer, K. Messwerte des zweiten Virialkoeffizienten einiger gesaettiger Kohlenwasserstoffe von 300–500 K. Ber. Bunsenges Phys. Chem. 1971, Vol. 75, No. 12, pp. 1308–1313.

99. Teichmann, J. Pressure-density-temperature measurements of liquid propane and benzene. Ph.D. Diss. Ruhr Univ., Bochum, 1978.
100. Thermodynamic functions of gases. Ed. F. Din. Vol. 2, Butterworth's Sci. Publ. London, 1956.
101. Thomas, R. H. P. and Harrison, R. H. Pressure-volume-temperature relations of propane. J. Chem. Eng. Data. 1982, Vol. 27, No. 1, pp. 1–12.
102. Tickner, A. W. and Lossing, F. P. The measurement of low vapor pressures by means of a mass spectrometer. J. Phys. Colloid. Chem. 1951, Vol. 55, pp. 733–740.
103. Timmermans, J. Physico-chemical constants of pure organic compounds. Intersci. New York, 1950.
104. Tomlinson, J. R. Liquid densities of ethane, propane, and ethanepropane mixtures. Techn. Publ. TP-1. Natural Gas Processor Assoc. Tulsa, Okla., 1971.
105. Vet, A. P. van der. Density, compressibility, expansion of light hydrocarbons and of light hydrocarbon blends. Congress Mond. Petrol. Paris. 1937, Vol. 2, p. 515.
106. Warowny, W., Wielopolski, P., and Stecki, J. Compressibility factors and virial coefficients for propane, propene, and their mixtures by the Burnett method. Physica. 1978, Vol. 91A, pp. 73–87.
107. Yesavage, V. F., Katz, D. L., and Powers, J. E. Thermal properties of propane. J. Chem. Eng. Data. 1969, Vol. 14, No. 2, pp. 197–204.
108. Younglove, B. A. Velocity of sound in liquid propane. J. Res. NBS. 1981, Vol. 86, No. 2, pp. 165–170.
109. Seeman, V. F. and Urban, M. Die Dichte des fluessigen Propans. Erdoel und Kohle-Erdgas-Petrochem., 1963, Vol. 16, No. 2, pp. 117–123.

INDEX